INFANT MORTALITY: A CONTINUING SOCIAL PROBLEM

Sir George Newman
(1870-1948)

Author of
Infant Morality: a Social Problem
1906
[Methuen & Co., London]

Infant Mortality: A Continuing Social Problem

A volume to mark the centenary of the 1906 publication of *Infant Mortality: a Social Problem* by George Newman

Edited by

EILIDH GARRETT
University of Cambridge, UK

CHRIS GALLEY
Barnsley College, UK

NICOLA SHELTON
University College London, UK

ROBERT WOODS
University of Liverpool, UK

LONDON AND NEW YORK

First published 2006 by Ashgate Publishing

Reissued 2018 by Routledge
2 Park Square, Milton Park, Abingdon, Oxon, OX14 4RN
605 Third Avenue, New York, NY 10017

First issued in paperback 2021

Routledge is an imprint of the Taylor & Francis Group, an informa business

A Library of Congress record exists under LC control number: 2006031603

ISBN 13: 978-0-815-38973-6 (hbk)
ISBN 13: 978-1-351-15564-9 (ebk)
ISBN 13: 978-1-138-35817-1 (pbk)

DOI: 10.4324/9781351155649

Contents

List of Figures *vii*
List of Tables *xi*
List of Contributors *xiii*

INTRODUCTION

1 Infant Mortality: A Social Problem?
Eilidh Garrett, Chris Galley, Nicola Shelton and Robert Woods 3

PART I

2 George Newman – A Life in Public Health
Chris Galley 17

3 Newman's *Infant Mortality* as an Agenda for Research
Robert Woods 33

PART II

4 Place and Status as Determinants of Infant Mortality in England c. 1550-1837
Richard Smith and Jim Oeppen 53

5 A Double Penalty? Infant Mortality in the Lincolnshire Fens, 1870-1900
Sam Sneddon 79

6 Infant Mortality in Northamptonshire: A Vaccination Register Study
Tricia James 99

7 Urban-rural Differences in Infant Mortality: A View from the Death Registers of Skye and Kilmarnock
Eilidh Garrett 119

8 Diarrhoea: The Central Issue?
Eric Hall and Michael Drake 149

9 Infant Mortality, a Spatial Problem: Notting Dale Special Area in George Newman's London
Graham Mooney and Andrea Tanner 169

10 Health Visitors and 'Enlightened Motherhood'
Alice Reid 191

PART III

11 Infant Mortality and Social Progress in Britain, 1905-2005
Danny Dorling 213

12 The Health of Infants at the Beginning of the Twenty-first Century
Yvonne Kelly 229

13 Conclusion: The Social Dimension of Infant Well-being
Nicola Shelton 249

References *261*
Index *287*

List of Figures

Figure 1.1 Infant mortality rates in France, England and Wales and Scotland, 1800-2006 6
Figure 1.2 Mortality rates amongst infants, those in early childhood, and children under the age of 10: England 1580-2000 7
Figure 4.1 English life expectancy at age 25: sexes combined, 1600-1900 59
Figure 4.2 Biometric analysis of infant mortality: 26 English parishes, 1580-1837 60
Figure 4.3 The relative movements of endogenous and exogenous infant mortality: 26 English parishes, 1580-1837 60
Figure 4.4 A comparison of endogenous and exogenous infant mortality in the last years of reconstitution data with data from the Registrar General's returns for the 1840s 62
Figure 4.5 Female life expectancy at age 25: cohort data plotted at mean age of death for British Peerage and English parish populations and England and Wales 71
Figure 4.6 Male life-expectancy at age 25: cohort data plotted at mean age of death for British Peerage, MPs and Scottish Ministers, English parish populations and England and Wales 71
Figure 4.7 Maternal mortality among English parishioners, British Peers and England and Wales 73
Figure 4.8 Relative risk of mortality: maternal over paternal among English parishioners and England and Wales 73
Figure 4.9 The relative risk of maternal mortality for British Peers 74
Figure 4.10 Maternal mortality: English data and international comparisons 74
Figure 5.1 Average infant mortality rate in the three Fenland counties, per decade 1850s-1890s, compared with the national average infant mortality rate 84
Figure 5.2 Infant mortality rates for the registration districts within the three Fenland counties, mapped according to four nested means, 1850s-1890s 87
Figure 5.3 Boundary maps showing both registration districts and sub-districts of the three Fenland counties, superimposed with a geological survey map (Waller, 1994) to indicate which particular districts lay within the Fens 88
Figure 5.4 The mean quinquennial infant mortality rate for the registration sub-districts of Lincolnshire, 1870-1899 89
Figure 5.5 Quinquennial infant mortality rates for registration sub-districts in Lincolnshire, 1870-1899, mapped using four nested means 90
Figure 5.6 The location of the registration sub-districts in Lincolnshire, indicating whether urban fen, rural fen, urban non-fen or rural

non-fen 92
Figure 6.1 Annual infant mortality rates in the Higham Ferrers sub-district, 1871-1890 106
Figure 6.2 Annual cohort infant mortality rate in the Higham Ferrers sub-district, 1880-1889 108
Figure 6.3 The 3-year moving average IMR for Rushden, other urban areas and rural areas: Higham Ferrers sub-district, 1880-1889 108
Figure 6.4 Comparison of infant mortality rates, neonatal mortality rates and post-neonatal mortality rates for the towns and villages of the Higham Ferrers sub-district, 1880-1889 109
Figure 6.5 Comparisons of IMR, P-NMR and NMR for infants of shoemakers by place of birth, 1880-1889 113
Figure 6.6 Comparisons of IMR, P-NMR and NMR of infants of agricultural labourers by place of birth, 1880-1889 113
Figure 6.7 Comparisons of IMR, P-NMR and NMR of infants of men working in occupations other than shoemaking and agriculture by place of birth, 1880-1889 114
Figure 7.1 The location of Skye and Kilmarnock 122
Figure 7.2 Annual IMR and 3 year moving average IMR: Kilmarnock and Skye, 1861-1900 125
Figure 7.3 The annual percentage of infant deaths and all deaths which were medically certified: Skye and Kilmarnock, 1861-1890 127
Figure 7.4 The cumulative percentage of infant deaths registered by number of days elapsed after death: Skye 1860s and 1890s and Kilmarnock, 1860s-1890s 135
Figure 7.5 The seasonality of infant mortality: Kilmarnock and Skye, by decade 1860s-1890s 140
Figure 7.6 Cumulative IMR in the first four weeks and following 11 months of life: Kilmarnock and Skye, 1860s-1890s 141
Figure 7.7 Infant mortality rates in the first four weeks of life: Skye, its parishes and Kilmarnock, 1860s-1890s 145
Figure 7.8 Concentration of second and third week deaths, by 'birth clusters': Skye 1861-1900, showing IMR(7<21 days) per 1,000 live births 145
Figure 8.1 Five year moving average of the IMR and net of diarrhoea/enteritis deaths: Ipswich, 1876-1930 152
Figure 8.2 Monthly deaths from diarrhoea as a percentage of annual deaths from the disease in the Eastern and Western sub-registration districts of Ipswich, 1876-1890 154
Figure 8.3 Monthly deaths from diarrhoea as a percentage of annual deaths from the disease in the Eastern and Western sub-registration districts of Ipswich, 1891-1910 155
Figure 8.4 Monthly deaths from diarrhoea as a percentage of annual deaths from the disease in the Eastern and Western sub-registration districts of Ipswich, 1911-1930 155
Figure 8.5 The IMR resulting from diarrhoea/enteritis, by class:

	Ipswich, 1876-1930	163
Figure 10.1	The administrative units of early twentieth century Derbyshire	195
Figure 10.2	A and B: Example pages from the Derbyshire health visitor ledgers	196
Figure 10.3	The mean age of infant at first visit by a health visitor, by Derbyshire administrative unit, 1917-1922	202
Figure 10.4	Hazard ratios for the risk of being visited: Derbyshire, 1917-1922	204
Figure 10.5	Hazard ratios for the risk of being artificially fed: Derbyshire, 1917-1922	206
Figure 10.6	Hazard ratios for the risk of post neonatal mortality: Derbyshire, 1917-1922	208
Figure 11.1	1905: The meaning of poverty	215
Figure 11.2	Infant mortality and affluence in the UK: six views of 160 years (1841-2001)	219
Figure 11.3	(a) and (b) Infant mortality for the largest cities of England, 2001. (c) A key to the location of the major cities in England – by population space	222
Figure 11.4	(a) The relationship between IMR around 1931 and SMR under age 65 by 1990-1992. (b) the same relationship for 1921 and 1981-1985	223
Figure 11.5	(a-d) Infant mortality in England and Wales, 1881-1951	224
Figure 11.5	(e-h) Infant mortality in England and Wales, 1982-2000	225
Figure 11.6	Infant mortality 1921-2002 in England and Wales by age of death (a) IMRs per 1,000 births per year. (b) deaths by age as a proportion of all infant deaths per year (c) deaths by age per 1,000 births per year (d) deaths by age per 1,000 births per year	227
Figure 12.1	Breastfeeding initiation and continuation rates by mother's occupation	239
Figure 12.2	Exclusive breastfeeding rates at 1 and 4 months by mother's occupation	240
Figure 12.3	Breastfeeding initiation and continuation rates by mother's ethnic group	241
Figure 12.4	Exclusive breastfeeding rates at 1 and 4 months by mother's ethnic group	241
Figure 13.1	Infant survival percentages and the relative risk of dying in infancy: England and Wales, 1850-2006	253

List of Tables

Table 2.1	The major publications of George Newman	19
Table 3.1	Preventive methods for the reduction of infant mortality as discussed by George Newman in *Infant Mortality: a Social Problem*	43
Table 4.1	Illegitimacy ratios, legitimate and overall infant mortality rates: 26 English parishes, 1580-1837	56
Table 4.2	Infant and early childhood mortality (1000q_x)in 26 English parishes, 1580-1837	57
Table 4.3	Infant mortality in the 26 parishes in 1675-1749 and in the registration districts in which they were located in the 1840s	63
Table 4.4	Age-specific mortality for specific age groups 0-9 years, plus endogenous and exogenous infant mortality among London Quakers, 1650-1849; and values indexed against 1650-1674	66
Table 4.5	Maternal mortality, early infant mortality and endogenous infant mortality: 26 English parishes, 1580-1837	69
Table 5.1	Key for Figure 5.6, indicating the identifying number and name of each sub-district and whether rural fen, rural non-fen, urban fen and urban non-fen	93
Table 5.2	Average quinquennial infant mortality rates per 1,000 live births for urban non-fen, urban fen, rural non-fen and rural fen registration sub-districts of Lincolnshire	94
Table 6.1	The decline of smallpox in England and Wales, by decade, 1851-1900	103
Table 6.2	A comparison of annual deaths recorded in the Vaccination Birth Registers and Infant Death Registers for the Higham Ferrers sub-district, 1880-1890	105
Table 7.1	The population returned at each census, 1861-1901, for the Isle of Skye and town of Kilmarnock, and the number of births and infant deaths occurring in the intervening decades in each community, along with IMRs	123
Table 7.2	The distribution of male heads of household by occupational sector: Kilmarnock and Skye, 1881	124
Table 7.3	The percentage distribution of selected causes of death amongst neonates, post-neonates and all infants, where causes of death are medically certified and uncertified: Kilmarnock, 1861-1900	128
Table 7.4	The percentage distribution of selected causes of death amongst neonates, post-neonates and all infants, where causes of death are medically certified and uncertified: Skye, 1861-1900	131

Table 7.5	Infant mortality rates, by selected cause of death, showing percentage of the deaths from each cause which were medically certified: Skye and Kilmarnock, 1860s-1890s	134
Table 7.6	The percentage of selected causes of death certified by doctors certifying over 50 infant deaths in Kilmarnock, 1891-1900	138
Table 8.1	The relationship between temperature (degrees centigrade) and diarrhoea/enteritis deaths in the 11 hottest years in Ipswich between 1876 and 1930	153
Table 8.2	Recorded births in Ipswich for selected social classes, 1876-1930: number born to each class and the percentage of all births in Ipswich	157
Table 8.3	Changes in the IMR by social class in 5 year periods: Ipswich, 1876-1930, indexed against the IMR per 1,000 live births for 1876-1880	158
Table 8.4	Number of infant deaths and IMRs per 1,000 live births by cause of death in Classes I & II: Ipswich, 1876-1930	160
Table 8.5	Infant deaths and IMRs by cause of death in Class V, Ipswich, 1876-1930	161
Table 8.6	Number of infant deaths and IMRs per 1,000 live births in Class IX by cause of death: Ipswich, 1876-1930	162
Table 8.7	Percentage of deaths from diarrhoea/enteritis by property value band in years of high and low numbers of such deaths: Ipswich 1890-1911	164
Table 9.1	Overcrowding and crude death rate (CDR per 1,000 population) in Notting Dale Special Area, 1896	173
Table 9.2	Infant mortality rates (IMR per 1,000 live births) in London and Kensington, 1896-1906	177
Table 9.3	Legitimate births surviving to one year and infant mortality rates for all causes of death and prematurity, by employment status of mothers: Kensington, 1910-1914	180
Table 9.4	Legitimate births surviving to one year and infant mortality rates, by tenement size: Kensington, 1912-1914	182
Table 9.5	Mode of feeding at onset of illness for infant deaths between two days and six months: Kensington, 1910-1911	182
Table 9.6	Charitable provision in Notting Dale before 1914	184
Table 10.1	Mean age of first visit for different groups of infants born in Derbyshire, 1917-1922	200
Table 12.1	Birthweight of MCS cohort members by mother's occupation and ethnic group	235
Table 12.2	Gestational age in days and proportion of preterm births in the MCS by mother's occupation and ethnic group	237
Table 12.3	Visit to GP, health centre, health visitor or phone call to NHS Direct and hospital admission during the first nine months of life among MCS cohort members by mother's occupation (NS-SEC) and ethnic group	243

Table 12.4 Socio-demographic characteristics of the mothers of MCS cohort members 245
Table 13.1 Early-age mortality differentials by social class: England and Wales, early 1990s 256

List of Contributors

Danny Dorling is Professor of Human Geography at the University of Sheffield and visiting Professor in the Department of Social Medicine, University of Bristol. He lists his research interests as 'trying to understand and map the changing social, political and medical geographies of Britain and further afield, concentrating on social and spatial inequalities to life chances and how these may be narrowed'. He has authored several important works such as *The Social Atlas of Britain, The Widening Gap: Health Inequalities and Policy in Britain* and *Health, Place and Society* and most recently *People and Places: A 2001 Census Atlas of the UK.*

Michael Drake, Emeritus Professor at the Open University, has spent much of his career encouraging people to adopt a more scholarly approach to family and community history. Formerly on the editorial board of *Local Population Studies* and co-editor of *Family and Community History*, he has recently, with Dr. Peter Razzell, directed 'The decline of infant mortality in England and Wales 1871-1948: a medical conundrum', a project that received support from The Open University and The Wellcome Trust and involved 20 Open University research students.

Chris Galley is Lecturer in Geography at Barnsley College. He is the author of *The Demography of Early Modern Towns: York in the Sixteenth and Seventeenth Centuries* (Liverpool, 1998) and he has published articles in *Economic History Review, Population Studies, Population and Social History of Medicine.*

Eilidh Garrett works part time as a Senior Research Associate at the Cambridge Group for the History of Population and Social Structure at Cambridge University. Her main research interest is the historical demography of nineteenth century Britain. A co-author of *Changing Family Size in England and Wales: Place, Class and Demography, 1891-1911* (Cambridge, 2001) she has most recently been working with Alice Reid and Ros Davies on an ESRC funded project *Determining the Demography of Victorian Scotland through Record Linkage.*

Eric Hall worked for many years as a civil servant at both HM Customs and Excise and the Ministry of Agriculture. On retiring, he undertook a BA degree through the Open University. He followed this with postgraduate study and was recently awarded his PhD for a thesis entitled 'Aspects of infant mortality in Ipswich, 1871-1930'.

Yvonne Kelly is a Lecturer in Epidemiology and Public Health. Her research interests include health inequalities in children and young people, and the influence of early life factors. She is currently working on the ESRC Millennium Cohort Study, a prospective study of infants, aged 9 months, and their families, the principal aim of which is to chart the life trajectories (in terms of health, development and socio-economic factors) of children born in the year 2000.

Tricia James is one of a team of mature postgraduate students, co-ordinated by Professor Michael Drake at the Open University, researching infant mortality using smallpox vaccination registers. She is a Senior Lecturer in Healthcare Studies at University College Northampton, and has recently successfully completed her doctoral thesis on infant mortality in shoe-making communities in Northamptonshire in the nineteenth century.

Graham Mooney has research interests spanning public health in the 19th and 20th centuries, historical epidemiology and historical demography as well as disease surveillance and risk. He has written extensively on the historical geography of health and medicine, with a particular interest in London. Graham recently moved to take up an appointment at the Institute of the History of Medicine, Johns Hopkins University, Baltimore.

Jim Oeppen holds the title of Research Scholar at the Max Planck Institute for Demographic Research. He is also a Senior Research Associate with the Cambridge Group for the History of Population and Social Structure. He has specialized in the development and application of analytic methods in demography, particularly with regard to mortality, contributing to many major books and articles.

Alice Reid is a Senior Research Associate at the Cambridge Group for the History of Population and Social Structure at Cambridge University. She has worked extensively on the subject on infant and child health in nineteenth and twentieth century Britain, with articles in Medical History, Population Studies, International Journal of Population Geography, and Historical Research. She has contributed to several edited volumes, and co-authored *Changing Family Size in England and Wales: Place, Class and Demography, 1891-1911* (Cambridge, 2001).

Nicola Shelton is a Research Fellow in the Department of Epidemiology and Public Health at University College London. She is the co-author (with Robert Woods) of *An Atlas of Victorian Mortality* (Liverpool, 1997) and has published in several journals including *Area*, *International Journal of Population Geography*, *Population Studies* and *Historical Methods.*

Richard Smith is Professor of Historical Demography and Geography in the University of Cambridge and Vice-Master of Downing College. Since 1994 he has been Director of the Cambridge Group for the History of Population and Social Structure. His research focuses on the history of marriage and inheritance practices both in medieval and early modern England and Europe, the economic consequences of demographic change and more recently on the demographic correlates of historic welfare systems He is currently writing a history of mortality in England from the thirteenth to the nineteenth century.

Sam Sneddon first became interested in demographic history and historical epidemiology during her time working at the Cambridge Population Group as a Research Assistant. She pursued her interest at postgraduate level both in Oxford and at Queen Mary College, University of London, where she obtained a doctorate for her thesis entitled 'Infant and Early Childhood Mortality in the Fens, 1850-1900'. Although she is currently working on the history of polio at Nottingham University, she continues her research on nineteenth century infant and childhood mortality whenever possible.

Andrea Tanner has research interests which concentrate on poverty and child health in London during the 19th and early 20th centuries. She is currently working on creating a database of Victorian in-patients at Great Ormond Street Hospital, and is honorary archivist at Fortnum and Mason.

Robert Woods is John Rankin Professor of Geography at the University of Liverpool. He is the author of *The Demography of Victorian England and Wales* (Cambridge, 2000) and of papers in a wide range of international journals including *Population and Development Review*, *Journal of Economic History, Past and Present.* He is currently an editor of the journal *Population Studies*. In 2003 he was elected a Fellow of the British Academy.

INTRODUCTION

Chapter 1

Infant Mortality: A Social Problem?

Eilidh Garrett, Chris Galley, Nicola Shelton and Robert Woods

Sir George Newman (1870-1948) was a pioneer in public and child health during the first half of the twentieth century. After training as a medical doctor he became lecturer in public health at St Bartholomew's Hospital, London. He subsequently held a succession of public offices, becoming Medical Officer of Health (MOH) for Bedfordshire in 1897 and then for the London Borough of Finsbury in 1900. In 1907 he was appointed Chief Medical Officer to the Board of Education and twelve years later he took up the post of Chief Medical Officer to the newly created Ministry of Health. The latter two positions were key public appointments, which meant Newman was well placed to influence the shaping of British public health policy in the aftermath of the Great War. Throughout his career Newman sought to promote better public health and in 1939 he summarised his views in *The Building of a Nation's Health*. He is best remembered, however, for his 1906 publication *Infant Mortality: a Social Problem*. Based on his experiences as a local MOH, this pioneering work awakened wider general interest in thc long-established problem of high infant mortality. Newman used the book to develop a framework whereby the nature and scale of the problem could be fully understood and, more importantly, he proposed a series of policies aimed towards reducing infant mortality rates (IMRs, infant deaths per 1,000 live births) throughout the country. Although its publication was, as a far as can be judged, greeted with no great fanfare *Infant Mortality: a Social Problem* has emerged as a landmark volume. Even at the beginning of the twenty-first century Newman's text, research methods, analyses and conclusions still have much to teach both students and researchers, and the present volume has been conceived as a means of commemorating the centenary of its publication and bringing it to the attention of a modern audience.

While working as MOH for Finsbury, George Newman had become aware of a public health enigma – over the course of the previous fifty years IMRs in England and Wales had remained stubbornly high despite the fact that other sections of the population had experienced clear gains in health. As he himself put it:

> whilst during the last half-century, a time of marvellous growth of science and of preventive medicine, human life has been saved and prolonged, and death made more remote for the general population, infants still die every year much as they did in former times. Indeed, in many places it appears that they die in greater numbers, and more readily than in the past (Newman, 1906: 2).

National statistics, compiled by the Registrar General during the nineteenth century, bear witness to Newman's concerns, indicating that the average IMR for the 1890s decade was virtually identical to that of the 1850s, whereas mortality rates among those aged between 5 and 24 had fallen by between 40 and 50 per cent, and among those aged 1 to 4 and those aged 25 to 34 by 30 to 40 per cent.[1] Nor was Newman alone in his observations. The opening years of the twentieth century saw the reasons for the poor survival chances of the nation's youngest citizens become a focus of national debate, particularly after a connection was made with the poor levels of health and physical fitness among the lower social classes revealed by the number of men who had to be rejected by those recruiting troops to fight in the Boer War (Dwork, 1987: 6-21). In 1903 the Interdepartmental Committee on Physical Deterioration was set up to enquire into the reasons for the poor state of the nation's health. Among its many recommendations was a call for more research on the underlying causes of infant mortality (British Parliamentary Papers (BPP), 1904: xxxii).

As MOH, Newman had made a number of attempts to reduce the IMR in Finsbury, and in 1906 he published *Infant Mortality: a Social Problem*, the first book-length study to consider this issue. In 356 pages he surveyed national and local patterns of infant mortality; examined the fatal diseases of infancy, with special emphasis being given to epidemic diarrhoea; discussed how social factors, such as the occupation of women, domestic conditions and infant management, affected an infant's chances of survival; and, finally, he showed how preventive measures relating to the mother, her child and the environment could bring about infant mortality decline. Newman did not offer a radically new approach to tackling high infant mortality, instead he provided an extensive survey of what others had written on this subject and using his considerable experience as someone whose daily work entailed a constant fight to improve infant health, he proposed a series of simple practical measures which he believed would address this issue. As the book's title proclaims, Newman believed that high infant mortality was essentially a *social* rather than a *medical* problem and consequently his recommendations for improving infant health were firmly rooted in the social sphere, with the mother being placed as the single most important influence on an infant's survival chances: 'The problem of infant mortality is not one of sanitation alone, or housing, or indeed poverty as such, *but is mainly a question of motherhood*' (Newman, 1906: 257). Such a conclusion remains controversial since, by implication, it held the mothers themselves to blame for high IMRs. Nevertheless, Newman provided his contemporaries with a framework in which to develop policies aimed at reducing infant mortality and many of his ideas about infant health have proved to be enduring.

Over the course of the twentieth century most of Newman's proposals for improving infant health have been implemented to some extent at least, although there is still no consensus as to how the various social and medical advances that occurred in this period affected the long-term decline in infant mortality shown in Figure 1.1. Newman's importance, however, extends far beyond whether or not his proposals brought about improvements in infant survival chances. In particular, his stress on the social causes of infant mortality and his comprehensive discussion of the various direct and indirect factors responsible for causing infant deaths resonates

1 Age specific mortality rates for decades calculated from figures collated by Robert Woods and Nicola Shelton and deposited in the UK Data Archive: ESRC study 3552, *Causes of death in England and Wales, 1851-60 to 1891-1900.*

both with social scientists seeking to understand secular changes in infant mortality and researchers analysing health inequalities in the twenty-first century. The centenary of the publication of *Infant Mortality* therefore allows us the opportunity to celebrate Newman's achievement and his continuing influence on those who are researching the various aspects of infant health. This volume does not seek to provide a comprehensive reassessment of Newman's work in bringing about infant mortality decline, although such a study is long overdue; rather the eleven specially commissioned chapters, together with this introduction and a conclusion, show how his ideas concerning the causes of infant mortality and the means by which infant deaths could be prevented have influenced and remain relevant to those considering infant mortality in a wide variety of contexts. Each contribution demonstrates that while we may now know far more about certain aspects of infant health than we did 100 years ago, the core of Newman's thesis remains unchallenged and *Infant Mortality* is a key text that can still be read with profit.

The wealth of studies carried out on infant mortality testifies to its continuing fascination for the modern researcher. Infants are the most vulnerable members of society, they rely exclusively on others for their survival and hence the rate at which they perish is often taken to be a critical measure of that society's well-being. According to Newman (1906: 1). 'A low rate, other things being equal, indicates a healthy community, a high rate the reverse' and, with the IMR being simple to calculate, it is frequently used as a surrogate for a wide variety of social, environmental and medical indicators. During the nineteenth century, as the science of vital statistics developed, information about births and deaths became readily available and it gradually began to emerge that, with the exception of the very old, infants were subjected to the greatest risk of dying. Moreover, once the considerable spatial variations in infant mortality were revealed, it was obvious that many infant deaths were preventable and an increasing number of concerned individuals began to investigate the reasons why so many infants did not survive their first year. Despite these efforts the IMR remained high and by the early twentieth century, with the birth rate declining and notions of national deterioration to the fore, infant mortality developed into an issue of national importance (see Chapter 2). As Newman was writing *Infant Mortality* the IMR in England and Wales was still over 150, which represented a considerable drain on national resources. In 1905 'there was a loss to the nation of 120,000 dead infants in England and Wales alone, a figure which is exactly one quarter of all the deaths in England and Wales in that year' (Newman, 1906: 2). However, unbeknown to Newman, the IMR had already begun a decline that would continue throughout the twentieth century (Figure 1.1). Newman's book was therefore written at an interesting and significant turning point in demographic history – IMRs were still horrendously high, but they were just beginning to be brought under control. Thus, *Infant Mortality*'s significance is threefold: it provides historical demographers with an important and comprehensive analysis of high infant mortality during the late nineteenth and early twentieth centuries; it provides a backcloth, or benchmark against which subsequent events can be seen to unfold; and, additionally, the final chapters can be used as a model to assist the evaluation of other explanations of infant mortality decline.

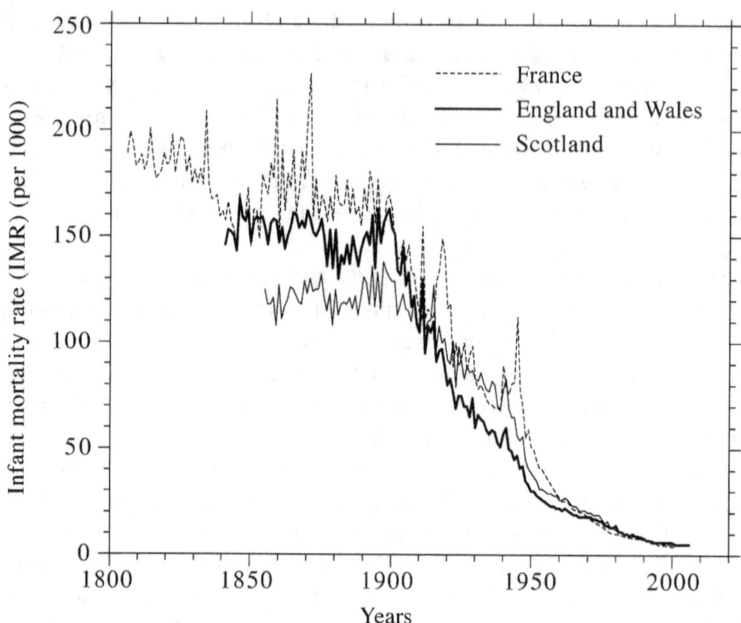

Figure 1.1 Infant mortality rates in France, England and Wales and Scotland, 1800-2006

Figure 1.1 shows nineteenth- and twentieth-century IMRs for England and Wales, together with Scottish and French rates for comparison. In retrospect the twentieth century decline is clear in all three series, but viewed from the perspective of 1906, while the IMR in England and Wales had passed its peak and started to decline, the rise in the 1890s masks this subsequent downturn, making any trend difficult to determine. Throughout the nineteenth century IMRs remained stable over the long term, although there were substantial annual variations, especially in France. Even though all three countries followed a similar path, national differences are apparent. IMRs in Scotland started at a lower level than in England and Wales, increased slightly towards 1900 and then declined more slowly. By contrast, France experienced higher levels for most of the period, far greater annual variations and rates increased significantly as a consequence of both World Wars. After 1950 all three series rapidly converged and it would be hard to escape the conclusion that the secular decline in infant mortality in Britain was part of a wider European phenomenon (Schofield *et al.,* 1991; Corsini and Viazzo, 1997; Bideau *et al.*, 1997).

Figure 1.2 provides both a longer-term perspective and a comparison between infant and childhood mortality rates. It plots decadal rates for England and Wales from the 1580s, thereby removing any annual fluctuations. Until the end of the nineteenth century IMRs remained relatively stable. There was an increase from around 150 to 200 per 1,000 live births between the mid-seventeenth and early-eighteenth centuries followed by a slow, but sustained, decrease until the early-nineteenth century when rates returned to levels in the region of 150. IMRs remained virtually stationary for much of the nineteenth century and Figure 1.2 further highlights the dramatic turning point in the series that occurred around 1900, just as Newman was beginning to become interested in the problem. For much of the period the infant and childhood mortality series follow the same trends, which may suggest that similar influences were affecting both infants and young children. The series begin to diverge significantly during the nineteenth century, although there is speculation that the early childhood (1-4 years) mortality series may not capture the national trend in the early nineteenth century (Wrigley *et al.*, 1997: 258-260). It is also clear that childhood mortality (0-9 years) began to decrease from the 1870s, which further emphasises the increase in infant mortality at the end of the nineteenth century.

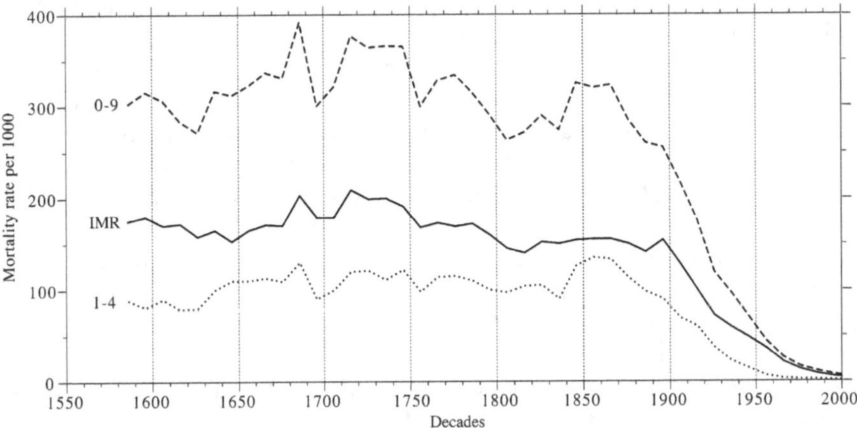

Figure 1.2 Mortality rates amongst infants, those in early childhood, and children under the age of 10: England, 1580-2000

Figure 1.2 was constructed using two sets of series joined together in the 1830s. The post-1837 rates are based on national data collected by the General Register Office (Higgs, 2004), while before 1837 it is necessary to employ some form of ecclesiastical register in order to provide estimates of mortality. The early rates are Wrigley *et al.*'s (1997: 215) attempt to provide national population estimates using family reconstitution, a technique that involves the linking together of vital events, on a representative sample of 26 parish registers (see Chapter 4). The production of this series required heroic efforts – family reconstitution is a time-consuming process that is very wasteful of data since a high degree of completeness and quality in the source material is needed to produce worthwhile results. Even within this set of registers, which were selected because of their apparently excellent quality, not all were deemed sufficiently reliable to yield accurate rates for the entire period. While the amount of effort involved in undertaking family reconstitution ensures that an alternative series will not be forthcoming in the foreseeable future, internal consistency within Wrigley *et al.*'s sample suggests that the broad overall trend revealed in Figure 1.2 is sufficiently robust to be reflective of overall national patterns.

Due to the nature of pre-civil registration sources in England and Wales much research on this period has focused on establishing levels and trends. Thus, while the national IMR was not high, especially in comparison with regions in Europe, where rates could exceed 250 (Corsini and Viazzo, 1997: 1), there were still considerable local variations. Within Wrigley *et al.*'s (1997: 270-271) parish sample, between 1675 and 1749 rates varied from 92 per 1,000 live births in the Devon parish of Bridford to 311 in March in the Cambridgeshire Fens. It is still not entirely clear what factors were responsible for these wide variations, but the low rates in Bridfordare thought to be reflective of a low disease burden in a remote rural settlement, while the high rates in March were due to the unhealthy conditions prevalent throughout fenland areas (West, 1974; see Chapter 5). There were also substantial urban-rural differences in IMRs with large towns, such as York, and even smaller ones, such as Gainsborough, suffering rates in the region of 250 (Galley, 1998: 99). Much still needs to be done to establish the extent of geographical variation in infant mortality across the country and there are whole counties, such as Cornwall, about which little is known (Galley and Shelton, 2001). The possibility of filling gaps in our knowledge is severely restricted by the survival of sources of sufficient quality to allow reliable mortality estimates to be produced. As this is often the result of chance, it means that virtually nothing is known about the emerging industrial towns of the eighteenth century, precisely those places that made Britain unique in this period and where, due to the deteriorating environment, IMRs may not have followed national trends. Here levels of in-migration and church non-attendance were sufficiently high to ensure that the ecclesiastical registration process broke down and estimates of mortality are difficult to make. Moreover, according to Wrigley *et al.* (1997: 73-118) there was a general deterioration in registration during the eighteenth century and this makes the phenomenon of eighteenth-century infant mortality decline, revealed by Figure 1.2, difficult to examine since one of the main indicators of poor registration is a low IMR. Clearly, much remains to be resolved about patterns of infant mortality in the pre-registration period and a careful examination of the wide range of existing sources would undoubtedly prove rewarding.

By comparison with what is known about levels and trends, very little can be discovered about why variation and change occurred. Causes of death were not consistently recorded in parish registers and those given for infants tend to be difficult to interpret. Consequently, it is necessary to adopt alternative approaches, such as examining burial seasonality, in order to explain prevailing patterns (Landers, 1993: 203-241). Likewise, calculating IMRs within special groups, such as multiple births or illegitimates, can be revealing and there is also much to be learnt about the relationship between rates of stillbirth, neonatal mortality and post-neonatal mortality, as indeed there is between infant and early childhood mortality rates (Woods, 2005). An examination of the distribution of infant deaths, by breaking up the IMR into its neonatal and post-neonatal or endogenous and exogenous components, can also be productive (Galley and Shelton, 2001; see Chapter 6). There is also the possibility that considering rates for unconventional age groups may aid the understanding of demographic change, since, for example, deaths associated with weaning may have been transferred from one age group to another if the length of breastfeeding altered (Wrigley *et al.,* 1997: 343-347). Data about many aspects of infant health are sadly lacking, however, and inferences often have to be made from qualitative sources, such as journals, or by recourse to comparative studies. In spite of these seemingly insurmountable obstacles, there is still much that can be usefully discovered about infant mortality in the pre-registration period.

After 1837, and the inception of the civil registration system, a much wider range of data become available to researchers in England and Wales. However, this is normally in the form of pre-digested, published abstracts and reports as access to material from individual civil registration certificates is not usually available in a form conducive to demographic research. The reports do allow geographical variation in infant mortality and associated variables to be studied in some detail at the county, registration district and even sub-registration district level, as Chapter 5 illustrates. The chapters by James, Hall and Drake, and Reid (Chapters 6, 8 and 10) indicate, however, that in a few cases alternative sources, such as the vaccination registers and health visitors' records have been identified, which allow individual infants and their families to be followed. The chapter by Garrett (Chapter 7) demonstrates that research using civil register material is possible in Scotland. MOH reports also provide windows onto local conditions and the resulting levels of infant mortality, as shown in the chapter by Mooney and Tanner (Chapter 9).

Figures 1.1 and 1.2 both demonstrate quite clearly that there has been a very substantial decline in IMR during the twentieth century. It is not a difficult matter to list the causes. Fertility has declined. Ante- and post-natal health schemes have targeted pregnant women, mothers with babies and the infants themselves. The standard of living and the general quality of the health environment have improved affecting the quality of both diet and housing. Successful medical interventions are now possible in the form of, for example, antibiotics, blood transfusion, mass vaccination against the childhood infections, ultrasound scanning, and even foetal surgery. Each of these advances has had a positive impact that in sequence, individually and in combination have helped to push the IMR down. Of course, there are new problems. Several medically advanced Western countries currently have a late-foetal mortality rate that is higher than their IMR and even when IMR appears to be at an irreducibly low level, with the few infant deaths caused by the effects of premature births, by accidents, and sudden infant death syndrome (SIDS), there are still significant differences between the health 'haves' and 'have-nots' in each society (see Chapters 11 and 12).

In 1906 Newman divided his study into eleven chapters which focused on, among other topics: the factors affecting IMRs; infant and childhood diseases; ante-natal influences; female employment; domestic and social conditions; and preventive measures involving the mother, the child and the environment. While this structure could not be mirrored exactly in the current volume, contributors do address the same topics. Beyond this introduction, the volume is made up of three parts. The first, comprising Chapters 2 and 3, considers Newman, his work and his influence on those who came after him in the study of infant mortality. The second consists of seven chapters, which report recent research addressing factors considered by Newman to have had an impact on infant mortality. The final part consists of three chapters, the first two reporting on infant mortality experience in the twentieth century, to act as a reminder of the phenomenal gains in survival chances which have been won for infants during the hundred years since *Infant Mortality: a Social Problem* was published. The final chapter reviews some of the principal changes that occurred in social attitudes, legislation, public health and medical practice as the IMR fell. It also suggests that a complete explanation of how and why infant mortality remained high until the end of the nineteenth century remains elusive, along with the factor, or combination of factors, which finally set it on its downward path. The simple pattern of change shown in Figures 1.1 and 1.2 has often misled researchers into believing that a significant change or changes must have occurred at or around 1900 when rates began to decline. However, as Newman showed, the influences on infant mortality were complex and interconnected and their unravelling has remained tantalisingly difficult to achieve. Moreover, with virtually every aspect of living conditions showing significant improvements during the course of the twentieth century, it has remained difficult to determine the precise weight that must be given to each factor in the decline; the apparent simplicity of the overall trend sometimes masks rather than illuminates the pattern of change. However, all contributors have found it instructive to consult Newman to see how he viewed the problems they have been considering.

Galley, in Chapter 2, discusses the development of infant mortality as a problem of national importance at the end of the nineteenth century. He then outlines the history of Newman's work and his contribution to the debate on the social dimensions of infant mortality, explaining how and why Newman came to write *Infant Mortality*. A life-long Quaker, Newman had strong ideological views on society and welfare, and he developed his interest in public health through medical training. His work as MOH for several localities, as well as other voluntary positions he held, exposed him to society's worst problems and conditions and led to his strong belief in social change as the operative condition for health improvement. He had begun to write on infant mortality and feeding in Finsbury by 1904. The publication of the Physical Deterioration Report in that year focused the nation's interest on infant mortality, and the publication of *Infant Mortality* in 1906 was thus very timely, although it took time for many of Newman's recommendations to be implemented.

In Chapter 3, Woods considers the role of *Infant Mortality* as an agenda for research, both at the time of its writing and as a benchmark for work in the twentieth century and beyond. He argues that Newman's study was especially influential in establishing infant and child health as a social problem, as well as a medical one. In seeking to explain persistently high levels of infant mortality, Newman focused on conditions within the towns and cities and on the timing of death within the first year of life; both issues which continue to be pursued by contemporary analysts. Newman then moved on to deal with epidemiological matters, especially the incidence of diarrhoea. Woods discusses how Newman's approach focused attention on the role of the mother and the home environment, which proved controversial. Newman, and his contemporaries, gave special prominence to the dangers of poor infant feeding thus rendering mothers centrally responsible for infant care. Newman's reorientation of concern away from disease, and onto the care of individual infants marked a departure in medical science and allowed mothers to be made the focus of blame for high rates of infant mortality. However, in channelling policy towards the welfare of mothers he paved the way for many of the maternal and infant health and welfare policies developed during the twentieth century and contributing to the, by historic standards, impressively low rates of infant mortality found in Britain today.

Smith and Oeppen, in Chapter 4, take a deliberately long-term perspective. They illustrate and discuss at length the various background studies that have made the construction of a time-series for infant mortality, such as the one in Figure 1.2, possible. They outline the problems of using ecclesiastical parish registers, of employing family reconstitution techniques, and of charting variations over time and between different localities. For example, they emphasise the need to distinguish between endogenous and exogenous mortality within total infant mortality because the two components moved in different patterns, especially during the eighteenth century. They show that urban-rural mortality differentials were especially important and that they have a long history – well before the industrial revolution of the nineteenth century. They also suggest that foetal mortality may have varied in past centuries and that late-foetal mortality (stillbirths) affected both fertility and infant deaths. Their discussion of maternal mortality highlights the need to appreciate the extent to which the life chances of women were affected by the risks of pregnancy, and how the social elite was not always able to secure a health advantage in the past.

In Chapter 5, Sneddon brings to light some of the deficiencies to be found in the analyses of late-nineteenth century observers of infant health. In 1906 Newman had shown that population density was not the only cause of high infant mortality since London did not have the highest IMR nationally, but these deviations were not explored. Possibly, if high non-urban IMRs had been highlighted Newman's thesis concerning the baleful influence of towns on infant survival would have been compromised, so they were ignored. However, it is more likely that, as MOHs were primarily attracted to large population centres (Shelton, 2000), their reports failed to provide sufficient focus on infant mortality in rural areas to bring this to national attention. The mapping of infant mortality at registration district level by Woods and Shelton (1997: 47-64) makes it very clear that there were areas in England and Wales that had much higher levels of infant mortality than their population density would predict. Sneddon uses data from the essentially rural and agricultural areas of the Fens in the mid-nineteenth century to show that a 'fenland penalty' existed with rural levels of infant mortality that were more comparable with towns such as Leeds, Liverpool or Manchester.

Chapter 6, by James, uses some especially revealing sources to unravel the effects of parental employment and residential location on infant mortality during the mid-nineteenth century. She demonstrates that while there was, in general, an urban penalty and a rural advantage in terms of early-age mortality, employment in certain occupations could have a deleterious effect on the health of one's children even in the countryside. Her use of nominal record linkage with vaccination registers and census enumerators' books provides an especially detailed picture of health and mortality in Northamptonshire, one that the author of *Infant Mortality* would particularly have relished.

Newman argued that the distribution of infant mortality in England and Scotland suggested 'that there may be, broadly, a line of cleavage between urban and rural conditions', the 'study of which may throw some light on to the causes of infant mortality'. In Chapter 7, Garrett examines the civil registers of births, marriages and deaths for the seven registration districts covering the Isle of Skye, Inverness-shire, and the registration district of Kilmarnock, during the later nineteenth century. She attempts a more detailed comparison of infant mortality in urban and rural surroundings than even Newman could construct from late nineteenth-century reports. However, this attempt is partially thwarted when it is realized that a large proportion of deaths, and particularly infant deaths, did not have their cause medically certified. Garrett suggests that even the official sources do not reveal a full picture of the risks run by infants in different settings, and argues that a great deal of work remains to be done at the local level before the nature of differences and the dimensions of inequalities can be fully understood.

In Chapter 8, Hall and Drake focus on the causes and effects of infant diarrhoea, a disease to which Newman devoted much space in his own account. They demonstrate that diarrhoea had an especially important place among the causes of infant death; that its impact was markedly seasonal; that there was also an environmental dimension; but that social class appears to have played an additional, significant role. Their research also stresses the need to consider micro-variations, to deal with the experiences of individual families in detail and to move away from the aggregate approach encouraged by the Registrar General's vital statistics. They show exactly what needed to be done in the field of public health for infant mortality to secure a downward path in the twentieth century. Infant diarrhoea had to be conquered.

Mooney and Tanner, in Chapter 9, consider the Notting Dale Special Area in Kensington, London. They describe how a network of five streets was identified as a blemish on the reputation of the larger district in which it was situated. Interventions directed at the Special Area were conceived as spatially focused remedies for socio-pathological evils. Overcrowding, especially in common lodging houses, was reduced; street paving and road asphalting were improved and access to clean water and provision of lavatories were both increased. Few policies were directly aimed at infants, although the Special Area experienced extremely high IMRs, as high as one in two births in 1899. Commentators emphasized the relationship between poor health and low morality, remarking that people were having too many children and noting that a better class of mother was required. The implication being that parental neglect was to blame for the extremely high IMRs, although some observers acknowledged that mothers were not deliberately neglectful, but rather that their ignorance led to neglect. In their chapter Mooney and Tanner illustrate one of the few schemes that were aimed at infants: the provision of crèches, as suggested by Newman. Some critics argued that such facilities encouraged mothers to work, whereas others acknowledged that female employment in the Special Area's laundries and in cleaning was crucial both to the women themselves and to their employers; the upper classes of Kensington.

In Chapter 10, the last in Part II, Reid discusses the implementation of one of Newman's recommendations: the instruction of mothers by domiciliary health visitors. The increased regulation of the registration of births, as recommended by

Newman, was instrumental in the health visitors knowing where and when to visit. Reid argues, from her work on Derbyshire in the early twentieth century, that it is difficult to establish the impact the schemes had in reducing infant mortality since health care provision varied greatly and those living in the worst conditions were generally visited more often. Despite this, certain groups in Derbyshire were found to have lower than expected post-neonatal mortality; including those infants who were both artificially fed and received an early visit.

In Chapter 11, which opens Part III, Dorling discusses how patterns of infant mortality changed in the twentieth century, highlighting the fact that the largest decline was not seen until 1951. He argues that development, a reduction in poverty and access to affordable health care for all brought about declines in the IMR for all social groups. Key factors in this decline were the introduction of infant welfare schemes and, in the post-1945 period, the National Health Service. Despite very low levels of infant death in contemporary England and Wales, inequalities still remain and these Dorling attributes to poverty.

Kelly, in the penultimate chapter, explores results from the 1946, 1958 and 1970 British Birth Cohort studies and the 1990s Avon Longitudinal Study of Pregnancy and Childbirth. These have shown how social and health inequality throughout the lifecourse may be influenced by environment early in life. Using a sample of over 18,000 babies born during a 12-month period spanning the years 2000-01 and recruited into a Millennium Cohort Study, Kelly shows how the demographic make-up of the United Kingdom population has changed dramatically since the 1970 cohort were born. Infant mortality is no longer a significant problem in the new millennium, she argues, although social inequalities in infant health remain and new inequalities, such as that based on ethnicity, can now be identified.

In our final chapter, Shelton asks how the issue of inequality, which Newman raised, may be pursued in the twenty-first century. In the hundred years since 1906 IMRs in the United Kingdom in all social groups have fallen dramatically and levels are now among some of the lowest in the world. Does inequality still remain? One of the challenges in answering this question is how inequality across time, space and between social groups should be measured. The challenge in measuring inequality across time is two-fold. For example, it is difficult to argue that a society in which one section of the population has an IMR of 200 while another section suffers only 100 infant deaths per 1,000 births experiences half the inequality compared with a society where the maximum IMR is, say, 4 and the minimum 1. Regardless of the level of inequality, the differences between mortality rates in the best and worst areas, or – and these may not always be interchangeable – the richest and poorest groups, highlight the fact that potentially avoidable infant deaths still occur as a result of social inequity a hundred years after Newman identified infant mortality as a social problem. In the remainder of her chapter Shelton discusses Newman's legacy in terms of how contemporary policy relates to his agenda for prevention. She considers how contemporary debate on the social aspects of infant mortality and wider infant health inequalities differ in their focus and policy implications. How, Shelton asks, is blaming mothers for the poor survival of their offspring supposed to help? Does apportioning them the blame form a first step towards their education?

If so, why is the necessary knowledge so partial and so divided by social class or, at least, why is this perceived to be the case?

This volume, in marking the centenary of Newman's *Infant Mortality: a Social Problem*, wishes to celebrate his achievement and lasting influence, but also to call attention to the fact that there still remains a great deal to be learned about infant mortality and its relation to society in Britain, in both historic and contemporary settings. The charts show how infant mortality declined, but not why it declined; they demonstrate social inequalities without fully explaining why those inequalities exist. Our hope, as editors, is that the chapters in this volume will excite and inspire further research on a topic that remains, as Newman described it a century ago, 'a most intricate problem' (Newman, 1906: vi).

PART I

Chapter 2

George Newman – A Life in Public Health

Chris Galley

Introduction – Newman's Career

> During the last five years my work in Finsbury has necessitated a careful study of the problem of infant mortality. This book is part of the outcome. It is an attempt to state in a plain way the chief facts concerning a question which is not without national importance.
>
> … It will be understood that the book is of an elementary character only, and it is hoped that it may serve as an introduction to the intricate problem with which it is concerned (Newman, 1906: v-vi).

These modest sentences appear in the preface to *Infant Mortality: a Social Question.* They summarise Newman's reasons for writing his book and they also give some flavour of his personality, hard work and dedication to improving public health. By surveying a wide range of evidence Newman identified the extent of the problem of high infant mortality and then following careful analysis he concluded that it was essentially 'a social problem' that could be tackled via a hierarchical set of factors relating to the mother, her child and the environment (Newman, 1906: chapters IX-XI). Newman was convinced that public health could best be improved, not via some instant miracle cure, but by the careful and diligent application of the latest scientific principles. He also believed that one of the main roles of the public health official was to disseminate these ideas in a straightforward manner to the widest public possible. Such principles permeate most of Newman's work and by placing *Infant Mortality* within the wider context of his other publications a proper appreciation of its value, originality and influence can be made. This chapter will therefore examine Newman's distinguished career in public health administration, his working methods and motivation; it will also outline how high infant mortality came to be seen as a problem of national importance, Newman's role within the early child welfare movement and finally both the immediate and longer-term impact of the book's publication.

George Newman (1870-1948) was born in Leominster into a prominent Quaker family.[1] He was the fourth child out of three boys and three girls and his parents devoted much of their time and efforts to Quaker matters. Whilst at school, first in

1 This and the following paragraphs are based on entries in *The Dictionary of National Biography* (Sturdy, 2004; Glover, 1959) and Hammer (1995), which examines Newman's

Gloucestershire and then in York, he felt neglected by his parents and resented their Quaker work. This emotional void was filled to some extent by his mother's younger sister who, 'nurtured his early Quaker faith and instilled in him an unshakeable belief in the "duty of service"' (Hammer, 1995: 4). Newman's initial ambition had been to do missionary work in India; however, following the death of his closest sister he decided to become a 'medical missionary' instead. He failed his entrance exams to study medicine at Edinburgh University, but after six months of further study he took them again and was accepted. Whilst at Edinburgh, alongside his studies he held religious meetings, did dispensary work in slum areas and led a temperance crusade. During his fourth year he attended a course in public health and, this together with his work in the slums, had a profound influence on his future career. He entered for the Bachelor of Medicine in 1892 and shortly after graduating left for London.

Newman began work as assistant physician at the London Medical Mission, although he resigned after only four months. He then became warden of Chalfont House, a Quaker institution in Bloomsbury and he continued to undertake voluntary work, running the Frideswade Girls Club where he was to meet his future wife. During his first five years in London Newman held a number of posts associated with public health, such as working part-time for the Strand Board on a special enquiry into working class living conditions. He entered for the Cambridge Diploma in Public Health at Kings College and passed the exam in 1895. In the same year he also graduated as Doctor of Medicine with first class honours from Edinburgh University.[2] By 1896 Newman had decided where his future lay. In March of that year he applied to become Medical Officer of Health (MOH) for the London Borough of Clerkenwell, finishing fourth on the list of candidates, and in January 1897 he was appointed part-time MOH to the Holborn Board of Guardians. He also continued to pursue his academic interests, becoming lecturer in public health at St Bartholomew's Hospital in 1896 and in 1897 part-time demonstrator of pathology and bacteriology at Kings College. In 1898 Newman found time to marry Adelaide Thorp, an accomplished artist, who was 'renowned to be "the most popular girl in the Society of Friends in London"' (Hammer, 1995: 58). Although it produced no children, his marriage brought Newman happiness, stability and respectability which all proved beneficial to his medical career.[3]

Until 1900 Newman had undertaken a range of often part-time employment, but from this date his career began to take off. In March he became part-time MOH for Bedfordshire, in April temporary MOH for Clerkenwell and in June MOH for the newly-created London Borough of Finsbury. This last position allowed Newman to relinquish most of his other responsibilities and concentrate on public health. Much later there followed two important public appointments: Chief Medical Officer to the Board of Education (1907) and Chief Medical Officer to the newly

---

career to 1921. Newman's career would appear to be ripe for the type of intellectual biography undertaken on Newman's contemporary, Arthur Newsholme (see Eyler, 1997).

2 In order to qualify for this degree he had spent one year studying leprosy in England and India.

3 His marriage also allowed him to become independent of his father and after marrying Newman never returned to Leominster.

created Ministry of Health (1919). These positions, which were held concurrently, were key public appointments that helped him to shape post-war public health policy. Throughout his career Newman was a strong advocate of state medicine. He was responsible for many innovations including the creation of a new medical service, the introduction of medical inspection for school children administered by local education authorities and the establishment of the Medical Research Council. Newman also pursued reforms in medical education which brought about the transfer of clinical training from elite private practitioners to university based academic teachers and he helped secure the creation of the then London School of Tropical Medicine (Newman, 1918, 1923; Bynum, 1995). Whilst working for the Board of Education Newman developed close working relationships with Sir Robert Morant, the permanent secretary to the Board, and Christopher Addison, the Minister for Reconstruction, and it was largely through their influence that he was appointed Chief Medical Officer, alongside both Morant and Addison, when the Ministry of Health was established. However, shortly after taking office, in March 1920 Morant died suddenly and when Addison was forced to resign in the spring of 1921 after he had failed to get a Bill through Parliament, Newman's influence was undermined and subsequently most of his energies were devoted to summarising the state of the nation's health in a series of substantial annual reports for the Ministry of Health and Board of Education. Throughout his long career Newman always sought to promote better public health, largely through social and preventive medicine, and according to Glover (1959: 625), 'no man of his generation did more in this country for public health, medical education and the child'.

Table 2.1 The major publications of George Newman

(1895) *On the History of the Decline and Final Extinction of Leprosy as an Endemic Disease in the British Islands* (New Sydenham Society, London)
(1899) *Bacteria* (John Murray, London)
(1903) (with Harold Swithinbank) *Bacteriology of Milk* (John Murray, London)
(1905) (with Arthur Whitelegge) *Hygiene and Public Health* (Cassell and Co., London)
(1906) *Infant Mortality: a Social Problem (Methuen, London)*
(1907) *The Health of the State* (Headley Brothers, London)
(1927) *Interpreters of Nature* (Faber & Gwyer, London)
(1928) *Citizenship and the Survival of Civilization* (Yale University Press, New Haven)
(1931) *Health and Social Evolution* (Allen & Unwin, London)
(1932) *The Rise of Preventive Medicine* (Oxford University Press, London)
(1939) *The Building of a Nation's Health* (Macmillan, London)
(1941) *English Social Services* (Collins, London)
(1946) *Quaker Profiles* (Bannisdale Press, London)

An examination of Newman's major publications, laid out in Table 2.1, reveals his breadth of interests. These may be categorised under four headings: public health, medicine, history of medicine and Quakerism, although elements of all four can usually be found in virtually everything he wrote. His first book, a study in the historical epidemiology of leprosy, was a revised version of his medical degree thesis and gained him a Fellowship of the Royal Historical Society. Little was said about the nature of leprosy itself and Newman concluded that the extinction of this disease was due, 'to a general and extensive social improvement in the life of the people, to a complete change in the poor and insufficient diet ... and to agricultural advancement, improved sanitation, and land drainage' (Newman, 1895: 109). While such conclusions may be challenged today, it is interesting to note that from the beginning of his career Newman was acknowledging the importance of social change in the fight against disease. His next three books, two of which were co-authored, dealt with the latest scientific advances, although they were not just aimed at an academic audience. *Bacteria* was, 'an attempt ... to set forth a popular scientific statement of our present knowledge of bacteria' (Newman, 1899: vii). As with many of his publications it began by placing the subject within its historical context, in this case starting in the sixteenth century. In *Bacteriology of Milk* Newman drew on his experiences as MOH and was responsible for the six chapters that dealt with pathogenic organisms in milk and the control of the milk supply (Swithinbank and Newman, 1903: 210-374, 452-543; Newman, 1903). His next book, written with Arthur Whitelegge, *Hygiene and Public Health*, was a revised manual for students of medicine, in particular MOHs and school medical officers. It dealt with a wide range of issues such as theories of disease transmission, the prevention of epidemic disease, public health legislation and the duties of the MOH.

Newman's belief in the value of education probably found its greatest outlet in *The Health of the State* (1907b). This was aimed at laymen and was intended to be, 'an elementary handbook dealing in plain language'. It neatly summarized Newman's over-riding philosophy, 'Progress seems, therefore, to depend upon a wide dissemination of some of the more important findings in this department of science (public health)' (Newman, 1907b: v-vi). Newman argued that, 'The people perish for lack of knowledge' (Newman, 1907b: 177) and he went on to outline his idea of 'health conscience',

> Fifty years of improvement of environment are now behind us; the future calls for a corresponding reform of personal life ... Much requires to be done in England in the direction of educational life in public health. Especially important is the training of girls in domestic hygiene, food values, and infant management; the personal guidance and teaching by well-qualified health visitors in the homes of the people; the awakening of a well-informed public opinion as to the inestimable value to the State of physical well-being; and the creation of a "health conscience" (Newman, 1907b: 194).

Infant Mortality, published in 1906, should therefore be viewed as one of a succession of publications dealing with the promotion of better public health. As with Newman's other work it was written in a plain no nonsense style; the background to the problem was discussed first, the latest scientific understanding was then given and finally a solution was proposed.

Table 2.1 appears to show a hiatus of twenty years between 1907 and 1927 in Newman's publishing activities but, while no *books* were forthcoming, this was far from the case. During this period Newman continued to give talks and he published both official and unofficial papers. However, his major efforts, from 1907 until his retirement in 1935, were devoted towards the 26 annual reports he produced as Medical Officer to the Board of Education and the 15 he produced for the Ministry of Health. These reports were perhaps Newman's greatest achievements and whilst an analysis of these documents is well beyond the scope of this chapter, cumulatively they provide a unique insight into many aspects of public health in Britain during the first half of the twentieth century. For instance, the 1924 Ministry of Health annual report, entitled 'On the State of the Public Health', begins by providing an overall assessment of the state of public health in England; this is then followed by a large section on infectious disease with individual chapters being devoted to 'Tuberculosis', 'Venereal Disease' and 'those of an exotic origin' (Newman, 1925). Subsequent chapters deal with maternity and child welfare, the insurance and poor law medical services, food hygiene, local sanitation and a summary of important recent research. Newman ends the report with a series of conclusions, although these are little more than his personal recommendations for improving the public's health: 'If it be asked: Where can one find the outline of a national health policy? The answer is not remote. It may be found every year in the contents table of this report' (Newman, 1925).

Newman's later work essentially falls into two categories; edited collections of previously published papers or publications that allowed Newman to expound his medical philosophy that the best way of improving a nation's health was through preventive medicine (Newman, 1928; 1931; 1932; 1941). *Interpreters of Nature* (1927) brings together a diverse collection of essays on medical history ranging over topics from the Paduan School of Medicine to biographical studies of Thomas Sydenham, Louis Pasteur and John Keats' career as apothecary and poet. Newman was always keen to place contemporary developments firmly within their historical context; 'No medical man can afford to be ignorant of the history of his calling or of his own special branch of medicine' (Newman, 1939: 99). His historical writings are however imbued with a Whiggish view of history as the cumulative work of successive generations of doctors is seen to bring disease gradually under control (Newman, 1931; 1932). In some cases, as in the following example about Joseph Lister, there is a tendency towards hagiography:

> His diligent persistence and patience, his caution and integrity, his thoroughness and accuracy, his experimental acumen, his faithfulness to truth were his overpowering characteristics. It was like a steady flame of truth burning in the temple of faith. There is no outcrop of unexpected genius as there is the dawn of truth in a prepared mind. In … Lister there is a steady flow of purposive dedication to the duty to which (he) was called (Newman, 1930: 67-68).

In the same piece Newman argues that Lister's career was a fine example of, 'Quakerism at work' (Newman, 1930: 68). Newman was active in the Quaker movement throughout his life and for forty years he anonymously edited the *Friends' Quarterly Examiner*. His Quaker beliefs had a profound influence on both his attitude to work and on many of his conclusions concerning how best to improve public health. Writing about the role of Quakers in the medical profession Newman concluded:

> On the whole, I think it must be said that the sense of service, sympathy, benevolence, solid sobriety, humanism, and sweet reasonableness of Quaker doctors of the past *has been attractive and advantageous to the patient.* Further, on the whole, the inductive training, habitual diligence, thoroughness, sincerity in seeking truth, disinterestness, independence of judgement and persistence of Quaker doctors in the past has proved of *substantial scientific value in research and investigation* (Newman, 1930: 69).

Newman's promotion of social medicine could therefore be interpreted as an attempt to translate his Quaker beliefs into practical medicine. Newman believed that the central Quaker principle necessitated a 'fearless, comprehensive life-long education' and this could best be achieved through hard work, dedication and self-improvement (Newman, 1930: 58). According to Hammer (1995: 78):

> As far as Newman was concerned, true education involved the cultivation of character – the 'drawing out' of the 'Inward Light' which Quakers believed was present in all human beings, for it was character which would enable the triumph over individual circumstances and the recognition of individual duty.

These core tenets provided a powerful stimulus for Newman's career and they influenced many of the conclusions he reached concerning how best to improve public health.

After Newman retired a suggestion was made that he should publish a selection from his annual reports. He decided against this and instead wrote *The Building of a Nation's Health* (1939) which provides an overview of his contribution to public health and serves as his personal account of the history of public health in Britain. Above all Newman will be best remembered for his advocacy of preventive medicine:

> We know two certainties about disease: first, that it is not something arbitrary, capricious, occult or accidental, but an *effect* of definite causes and conditions; and secondly, that these causes and conditions are in large and increasing measure controllable by man' (Newman, 1920: 4; Bynum, 1995).

However, during his tenure at the Ministry of Health his espousal of preventive, social medicine often seemed at odds with a medical profession keen to adopt more interventionist methods and his influence began to wane shortly after he became Chief Medical Officer to the Ministry of Health. Summing up Newman's career Bynum (1995: 45) argues that, 'Obiturists were barely able to disguise the disappointments of his later career, when he was remote from staff and rather dreamily optimistic in his beloved annual reports'.[4]

---

4 Newman was a flexible and canny political operator who realised that in order to succeed it was necessary to court favour with the political establishment. However, his relationships with some of his colleagues could be complex. This can be illustrated by his dealings with Arthur Newsholme, the other great figure in the field of infant welfare during

While many of Newman's writings may now appear 'old fashioned' and 'out of date', the publication of *Infant Mortality* in 1906 must still rank as a major achievement and its emphasis on the social causes of infant mortality seems especially apposite 100 years on. It was written to address a specific issue and the conclusions it reached were influenced by Newman's working methods and his personal convictions, which in turn were influenced by his religious beliefs. Newman's work as MOH for Finsbury had seen him directly address the problem of high infant mortality at the local level, but the timing of the book's publication was also related to the development of infant mortality as an issue of national importance at the beginning of the twentieth century, when increasing numbers of individuals began to realise that high infant mortality rates (IMRs) were both unacceptable and to a large extent controllable (Wright, 1988).

Newman's Reasons for Writing *Infant Mortality*

When Newman was appointed MOH for Finsbury in 1900 infant health was still a relatively neglected area of research. While a number of notable studies had been carried out, there was still no consensus as to what measures were likely to bring about infant mortality decline with most individuals assuming that this would occur as a natural consequence of sanitary improvement (Ashby, 1884; Jones, 1894). To understand how this situation had arisen it is necessary to examine the work of the General Register Office (GRO), the organisation responsible for disseminating information concerning demographic trends following the introduction of civil registration in 1837. From this date a systematic analysis of national and local IMRs had been possible, but the early annual reports of the Registrar General are notable for their absence of interest in infants.[5] William Farr, the long-serving first Statistical Superintendent to the GRO, was responsible for much of the content of the annual reports between 1837 and 1878, but he was initially concerned with identifying preventable deaths and developing the means by which sanitary measures could be

---

the early twentieth century. Newman quoted Newsholme's work in *Infant Mortality* and in 1907 called him his 'great friend' when, along with the Webbs and Morant, he recommended Newsholme for the post of MOH to the Local Government Board (LGB) (Searle, 1971: 244; Eyler, 1997: 218-219). Newsholme vigorously supported Newman over the medical inspection of school children; however, they came into conflict when in 1914 Newman proposed that that the Board of Education, rather than the LGB, should become responsible for the welfare of infants from four or six weeks. Newman and Newsholme remained in conflict and when the LGB was transformed into the Ministry of Health, Newsholme was forced to retire. Newsholme never forgave Newman for his role in this episode and in his post-retirement publications, whenever the subject of British public health was discussed, Newman's name is conspicuous by its absence (Eyler, 1997: 316-337).

5 As early as the second and third annual reports the proportions dying under one year to both 1,000 births and deaths had been given, but the term infant mortality was not employed and this information was used to check the quality of the registration system rather than as a means of assessing infant health (HMSO, 1840: 10; 1841: 12). Note: the year quoted in the reference refers to the year of the report rather than the year of publication. Each report took about two years to prepare and publish.

promoted (Higgs, 2004; Szreter, 1991a: 403). To facilitate these aims Farr employed the crude death rate (number of deaths per 1,000 population) to measure levels of mortality throughout the country. While this statistic was unsatisfactory to a certain extent, it quickly became adopted as the key demographic indicator and was used to determine the extent of sanitary improvement throughout the Victorian period (Farr, 1859, 1864; Mooney, 1997; Szreter, 1991b).[6]

Alternatives to the crude death rate were discussed and Edwin Chadwick (1860: 580) in an address to the National Association for the Promotion of Social Science stated, 'I have always held [an infant death rate] to be the single best test of the sanitary condition of a population' (also see Newmach, 1861; Tripe, 1873). However, it becomes apparent in a subsequent article published in the same volume that it is not the IMR (infant deaths divided by infant births in a time period *x* 1,000) that is being referred to, but the infant death rate (infant deaths divided by the number of infants alive in a time period *x* 1,000) (Gairdner, 1860). Similar problems affect the work of John Simon, the first MOH for London and later medical officer to the Privy Council, who maintained an interest in infant mortality throughout his career (Simon, 1897; Lambert, 1963). Simon commissioned three major investigations that focussed on issues relating to infants: two by Edward Greenhow – one on the prevalence of diarrhoea especially in infants (1860), the second considering excessive infant mortality in manufacturing places (1862) – and a third by Henry Hunter, examining high infant mortality in rural East Anglia (1864); yet each report employed a different definition of infant mortality.[7] Thus, with no consensus as to what was being measured it was difficult to determine the factors responsible for high infant mortality or formulate appropriate policies to combat them.

Infant health only began to achieve a higher profile once Farr became interested in the problem and by the late 1860s he was acknowledging that many infant deaths must be preventable:

> to discover the means whereby the heavy annual tribute of infant life can be reduced from 203,129 – the number who died in England in 1868 – to the number that would have died at the healthy district rate, viz., 122,524, and thus affect a saving of 80,605 or more lives, is one of the great sanitary *desiderata* of the age (HMSO, 1868: xxiv-xxv).

---

6 Farr also used life expectancy, but this measure required too many calculations to make it easily adaptable for widespread use (Farr, 1859; 1865).

7 The proportion of diarrhoea deaths under the age of one to total diarrhoea deaths was used as the main means of comparison in Greenhow's first report. In his second report, Greenhow calculated the proportion of deaths under 1 to 1,000 persons, but since he could not estimate the number of infants under one, he used the average number of births instead which meant that he had inadvertently calculated the IMR. However, Hunter's later report used infant deaths per 100,000 persons living. These reports are published together with conclusions by Simon in *Second, Fourth and Sixth Reports of the Medical Officer of the Privy Council,* (British Parliamentary Papers (1860) XXIX; (1862) XXII; (1864) XXVIII); also see Lambert, 1963: 320-321, 336-337. For other criticisms of these reports see Greenwood, 1948: 93.

In Farr's 1870 annual report for the very first time a definition of infant mortality was published, 'real infant mortality may be satisfactorily measured by its proportion to births registered' (HMSO, 1870: xi). The national IMR was given together with rates in the previous year and the highest and lowest in 17 large English towns. Thus, the extent of the variations in IMRs was published in an accessible form and a simple means was available to make local comparisons possible. The next year's report contained Farr's first serious attempt at understanding the causes of high infant mortality. It provided a detailed discussion of infant deaths and included, as a supplement, a report that had been requested by Farr from the Obstetrical Society concerning the extent and causes of infant mortality together with suggestions as to how such deaths could be reduced (HMSO, 1871: 225-229; this was also published as Anon., 1870). The culmination of Farr's interest in this topic appeared in the 1875 report. The term 'infantile mortality' was formally defined and annual rates for the period 1870-75 were given alongside a substantial discussion of the causes and influences on infant mortality (these included age at death, illegitimacy, women's work and the special circumstances in individual large towns) (HMSO, 1875: xl-li; Armstrong, 1986). According to Farr:

> the agencies which destroy infant life are many, and they vary in different localities. Some of the principal causes are improper and insufficient food, bad management, use of opiates, neglect, early marriages, and debility of mothers ... it must also be borne in mind that a high death rate is in a great measure also due to bad sanitary arrangements (HMSO, 1875: xli).

This sophisticated analysis represents the first major assessment of the problems caused by high infant mortality at the national level.[8] Unfortunately, Farr's awakening interest in the subject of infant mortality coincided with the end of his career and his final three annual reports, those for 1876-1878, contain little about this subject. Farr's achievements in the field of infant health were considerable, but his suggestions concerning the prevention of infant mortality fell largely on deaf ears and it would take a further twenty-five years before a concerted effort was made to reduce national IMRs.

Although the GRO had taken the lead in disseminating information about national health problems it was the MOHs who were responsible for formulating policies to deal with them at the local level. These officials had been appointed in each sanitary authority following the implementation of the 1872 Public Health Act and Farr had an optimistic view of the value of their work:

8 It is probably no coincidence that Farr's decision to provide a more detailed analysis of infant mortality occurred following the implementation of the 1874 Registration Act which aimed to improve birth registration by transferring its responsibility from the Registrar to the parents, occupier of the house where the infant was born or those having responsibility for the child. For problems before 1874 (see Humphrey's comment in Jones, 1894: 101; Farr, 1865).

> It is gratifying, however, to know that the causes which lead to the regrettable high rate of mortality among children are now receiving full attention from the medical officers of health (HMSO, 1873: xv).

Unfortunately, by 1873 only about a quarter of MOHs were employed full time and writing later John Brownlee (1917: 223) suggested that most of these officials were 'men of narrow outlook' (Waller, 1983: 273, 283). Thus, with MOHs also being responsible for a wide range of issues relating to public health few appear to have devoted much of their efforts towards reducing IMRs. Many MOH reports did record the IMR, but almost exclusively MOHs tended to measure their success in improving the sanitary environment by recourse to the crude death rate which was in general decline throughout the late Victorian period (Galley, 2004). The GRO also provided the MOHs with little leadership after Farr's retirement. When Brydges Henniker, an Etonian Guardsman with personal limitations, became Registrar General (1878-1898) the GRO lost direction and its status declined significantly (Higgs, 2004: 90-128; Szreter, 1991b: 454-462; 1996: 85-107). The format of the annual reports changed; they now contained less discussion of general issues relating to public health and much of the evangelical support for sanitary reform, typical of Farr's reports, is absent from these later ones. Issues relating to infants were again neglected and typically during the 1880s the annual reports provided only the barest information about infant mortality: the national IMR, some comparison with previous years and rates for various counties (for example see HMSO, 1884: x-xi, lvii).[9] Consequently, most MOHs appear to have ignored the subject of infant health and their reports show little indication that infant mortality was considered to be an important issue or that it could easily be addressed by direct action.

During the 1890s a number of enlightened individuals such as Alfred Hill, Birmingham's MOH, began to realise that in spite of widespread sanitary improvement the IMR had failed to decline; indeed in some large cities it had even risen (Hill, 1893; Galley, 2004: 30-35). Studies were undertaken to determine why this was the case, articles about infants began to appear regularly in journals such as *Public Health* and many sensible suggestions were made regarding appropriate preventive measures (Anon., 1894-95; Anderson, 1894-95; Porter, 1894-95; Thomas, 1898-99). Unfortunately, no real consensus emerged concerning the precise set of measures that needed to be adopted in order to combat high IMRs; Dr Kenwood (1896-97: 192) had even concluded, 'in preventable disease affecting the first year or two of life there is practically little scope for legislative machinery and municipal control'. By the turn of the century the main causes of high IMRs had been debated; however, the discussion was dominated by urban MOHs and attention became increasingly focussed on the causes of infantile diarrhoea, the one major infant disease that was universally considered to be preventable (Newsholme, 1899, 1902; Hope, 1899; Gordon, 1902; Richards, 1903).

It was therefore against a background of increasing awareness and understanding of issues relating to infants that in 1900 Newman was appointed MOH for Finsbury,

9 For a rare exception, see William Ogle's analysis of urban and rural IMRs (HMSO, 1891: xiv-xv; Woods, 2000: 258-259).

a London borough experiencing some of the highest rates of mortality. However, there is little in Newman's early reports as MOH to suggest that he was especially interested in infant health. For instance, Newman's annual report for 1902 runs to an impressive 229 pages making it one of London's bulkiest, yet only four pages were given over to infant mortality (1902: 17-20).[10] Newman simply reported the extent of infant mortality throughout the borough and gave a breakdown by cause, although he did highlight that the IMR could be used as an important demographic indicator, '[the IMR] is one of the most reliable tests of the health of a community and of the sanitary condition of a district' (1902: 17).[11] Newman's next two annual reports were virtually identical in format, as was the case with many other series of MOH reports, and similar amounts of space were devoted to infants (1903: 17-20; 1904: 18-20). The 1904 report also contained an appendix which discussed the relationship between feeding and infant mortality (1904: 82-86). In November of the same year an infants' milk depot was established in Finsbury under the auspices of the Social Workers' Association and Newman published a detailed account of the work of this institution (Newman, 1905). He argued that, 'antenatal conditions and lack of maternal care combined with bad artificial feeding are among the chief' causes of infant mortality and a 'great requirement then is the breastfeeding of infants' (Newman, 1905: 7). For those mothers resident in Finsbury who were unable, unwilling or for whom it was impractical to breastfeed the depot supplied clean, modified milk paid for in advance at the market rate. Infants were introduced to the depot by medical practitioners, hospitals, nurses, sanitary inspectors and birth registrars. Only about 30 babies per month were dealt with throughout 1905 and the depot made little impact on the IMR in the borough, although it did appear to focus Newman's mind on infant health.[12] It is clear from Newman's 1905 annual report that by now he was devoting more of his energies towards infant mortality, although the sections dealing with this topic were not much larger than in previous years (1905: 21-28; 52-59). Newman (1905: 16, 23) reproduced a copy of a leaflet about infant feeding that was distributed to all mothers and he revealed that an enquiry had been made into infant deaths in that year (out of a total of 429 infants who had died, 374 had been visited by lady inspectors and details obtained about their circumstances). More importantly perhaps, at least as far as Newman's thinking was concerned, towards the end of the section on infant mortality he was able to conclude, 'the problem of infant mortality was mainly one of motherhood' (1905: 27). Newman's MOH reports chart his developing interest in infant mortality and by the time that the 1905 report was published, early in 1906, Newman must have working on his book. The main arguments that run through *Infant*

10 Another six dealt with diarrhoea and were mainly concerned with infants (1902: 73-79). Thus, at most only 4.4 per cent of the report discussed infants, although 24 per cent of deaths in Finsbury were infants (558 out of 2,283).

11 This quote had appeared verbatim in Newman's previous report (1901: 18). Similar phrases appear in the Registrar General's annual reports from the same date, (HMSO, 1901: lxviii; 1902: lxviii). For an earlier use, see Newsholme (1891: 79).

12 Newman (1905: 23) reported that of the 129 infants admitted to the scheme 62 withdrew before being discharged aged one. The depot eventually closed in 1909, being replaced by two health visitors - it had proved expensive to run (Ashby, 1922: 136).

Mortality are evident in Newman's MOH reports; however, there is little in these documents to suggest that he was about to embark on such a major publication.[13]

While Newman was working in Finsbury high infant mortality suddenly developed into an issue of national importance. This was partly a consequence of the political fallout following the discovery that up to 40 per cent of recruits to the British Army during the Boer War (1899-1902) were physically unfit for service (Harris, 2004: 155-157). In 1903 the Government set up a Committee with wide terms of reference to determine whether the nation's health was deteriorating. Part of their investigations focussed on infant health and the resulting Physical Deterioration Report published in 1904 noted that, 'infantile mortality in this country has not decreased materially during the last twenty-five years, notwithstanding that the general death rate has fallen considerably' and it concluded, 'the means by which infant mortality can be averted present a social problem of the first order' (Parliamentary Papers Vol. 1, 1904: 44; also see Brunton, 1905; Gilbert, 1965; Davin, 1978; Lewis, 1980). The Report's publication stimulated considerable interest in the subject, both at the GRO where, from 1905, increasing amounts of space in the annual reports were given over to infants and young children, and elsewhere (HMSO, 1905: xlii-xlvii, cxviii-cxxxiii).[14] A number of important studies also appeared at around this date (Moore 1904; Hill, 1905; Newsholme, 1905; Nash, 1905) and in 1906 the first in a series of national conferences on infant mortality was organised (McCleary, 1933: 105-110; 151-168). The Local Government Board also began to take greater interest in the problem and from 1905 it required MOHs to provide returns of all infant deaths in their districts on special forms (Eyler, 1997: 298).[15] This resulted in some MOHs compiling special reports on infant mortality (for example, see Newman, 1907a) which was also printed verbatim in Newman's 1906 annual MOH report (HMSO, 1906: 35-77). Consequently, by the mid 1900s infant mortality was a public health issue of the highest order; George McCleary (1933: 112) even suggested that it had become 'fashionable' and had developed 'news value'.

The Physical Deterioration Report also appears to have made an important impression on Newman. Writing to his sister in September 1904 he said of the Report, 'It is good and bad – in some parts weak & inadequate. It also strikes me as

13 Comparison with other series of MOH reports, such as those for Birmingham and Sheffield, reveal that Newman's are in no way exceptional (Galley, 2004).

14 A more sophisticated analysis was provided in 1906 (HMSO, 1906: xxxvii-xliv; cxv-cxxxii). Some measure of overall decline was first identified and most of the key features of high IMRs discussed:

> since the close of the century, however, the subject of the waste of infant life, formally treated with apathy, has received close and increasing attention from all classes of the community, and to this awakening may fairly be ascribed some portion of the decline in the rate of infantile mortality that has taken place during the past few years (HMSO, 1906: xxxvii).

15 The appointment of John Burns as President of the LGB in December 1905 was also significant. He presided over the first National Conference on Infant Mortality, did much to secure the Notification of Births Act (1907), appointed Arthur Newsholme as medical officer to the Board in 1908 and commissioned and published Newsholme's series of special reports into infant and childhood mortality (Newsholme, 1910; 1913; 1914; 1917-18).

a little bit amateurish. But it is the most *human* blue book I've ever seen' (quoted in Hammer, 1995: 82). In October 1904 Newman wrote an editorial about the report for the *Friends' Quarterly Examiner*. Here he was critical of the quality of some of the medical evidence which, 'is very unprofitable reading, and abounds in superficial opinions and obvious fallacies', while,

> of the fifty three separate social proposals recommended to the Government – a Government which has proved itself marvellously unsympathetic to social legislation – there are very few which can be considered really practical at the present time (Newman, 1904: 442, 443).

In spite of these criticisms Newman applauded the Report's publication, calling it a 'Domesday Book of human life in England at the beginning of the twentieth century' (Newman, 1904: 443). He also welcomed the report's acknowledgement that the, 'strength of the State depends, broadly upon the health of the people' (Newman, 1904: 450). Although Newman does not explicitly state that *Infant Mortality* was written in response to the Physical Deterioration Report, the timing of the book's publication must surely be no coincidence.[16] Here presented itself an opportunity for Newman to marshal his various talents and medical expertise and to expound his medical philosophy of prevention via a substantial case study. *Infant Mortality* allowed Newman to correct what he thought were a large number of errors in the report and it also allowed him to transfer his extensive local knowledge to the national stage. That he was to do so in such detail was also characteristic of this hardworking man. Newman's solution to the problem, which essentially lay with educating the mothers, can be also viewed as an extension of his Quaker beliefs that the measures needed to reduce IMRs; good maternal caring practices, lay within each individual mother and that the role of the public health official was to 'draw' these out from her.

In conclusion Newman's decision to write *Infant Mortality* can be related to three sets of interlocking factors: (1) his duties as MOH; (2) the social construction of infant mortality as an issue of national importance that needed to be addressed; and (3) his Quakerism. *Infant Mortality* represents a major achievement in the field of preventive public health and its publication both boosted Newman's career – he became Chief Medical Officer to the Board of Education in the following year – and did much to seal his place as the leading authority on infant welfare (Hammer, 1995: 93).

Conclusion – The Impact of the Book's Publication

With impeccable timing *Infant Mortality* was published just days before the first national conference on infant mortality was held in London in June 1906. It was the first book length publication to deal with this subject and was successful in drawing the attention of the medical establishment to this important and complicated problem.

16 Newman makes constant reference to the report throughout *Infant Mortality* (Newman, 1906: 15-16; 52-53; 55; 75; 86-87; 119; 125; 186; 219-220; 316).

Reviews were favourable (see for example, *Lancet*, 28 July 1906: 230-232; *Journal of the Royal Statistical Society*, 1906 (69): 610-611), but it is difficult to judge the precise impact that the book made on publication. Standards of referencing were different during the first decade of the twentieth century and it is impossible today to determine exactly who read *Infant Mortality* and who was influenced by it. Certainly none of Newman's fellow London MOHs provided a reference to it in their annual reports for 1907. However, writing in 1933 George McCleary, MOH for Battersea and a leading figure in the infant welfare movement, called the publication of *Infant Mortality* an 'outstanding event' which,

> gave in a compact but very readable form an account of practically everything of importance that was known about infant mortality at that time. It was, however, very much more than a compilation of facts. It discussed with a fine appreciation of the issues involved the various conditions making for the destruction of infant life, and the main principles of prevention. The book still ranks high among many contributions made by its distinguished author to the literature of preventive medicine (McCleary, 1933: 113).[17]

Newman's book did not herald a change in the methods adopted by the infant welfare movement; for instance, there was no sudden increase in health visiting or widespread adoption of measures that targeted mothers following its publication. Instead, the importance of *Infant Mortality* lies with the large amount of disparate material it collected and made readily available for the first time, its synthesis of available knowledge and Newman's provision of a clear set of guidelines that could be easily adopted. During the 1900s there was considerable debate as to what needed to be done to tackle infant mortality and while IMRs declined steadily throughout the twentieth century it has consistently remained difficult to disentangle the various factors responsible for this phenomenon (Woods, 2000: 247-309).[18] After 1906 there was a growing realisation that if IMRs were to be controlled then efforts needed to focus on mothers, but it was not until much later that many of Newman's recommendations were implemented.

Newman's approach to reducing infant mortality was not original. It was based on the accumulated work of his fellow MOHs to which he made frequent reference throughout his book. In particular the work of Benjamin Broadbent and S.G.H. Moore, authors of the so-called 'Huddersfield scheme', had already sought to raise awareness amongst mothers in their fight against infant mortality (Moore, 1904; Newman, 1906: 264-265, 271-276, 280; Marland, 1993). However, Newman emphasised how mothers, with appropriate care and attention, could alleviate the various threats to their infants posed by living in a harsh environment. In his 'Preface' Newman stressed that the problem of high infant mortality was 'intimately related to the social life of the people' and that the solution likewise needed to be a social one

---

17 Walker-Smith (1998: 357) examined Newman's book and showed that it was quoted extensively in a number of important paediatric textbooks.

18 For instance, topics discussed at the first National Conference on Infant Mortality included, education, health visiting, milk, premature birth, alcoholism, registration of births, infant life insurance, women's employment, infant foods and midwives (McCleary, 1933: 106-107).

(Newman, 1906: vi). He realised that both the home and public environments were important and that public health officials could do much to bring about improvements there, but it was only with the active cooperation of all mothers that IMRs would be lowered significantly. Writing in 1931 Newman argued that, 'Now it is perfectly true that the State cannot itself save the child, *but it can help the mother to save it*; and all over England there has now grown up a network of organisations for so helping her' (Newman, 1931: 131). Newman never wavered on this point and throughout all his writings he was consistent both about the causes of high infant mortality and about how the problem needed to be tackled (Newman, 1907a: 108-133; 1920: 24-25; 1924: 7-8; 1928: 121-123; 1931: 129-132; 1939: 281-321; 1941: 19-21). Once infant mortality had begun to fall he also gave full credit to the mothers for the work they had done, arguing that the fight against infant mortality could be viewed as an excellent example of 'modern collective humanism at work' (Newman, 1931: 129). In one of his final statements on this subject he again reiterated his absolute and unshakable belief in the value of preventive medicine,

> Yet it was not Acts of Parliament but the mothers themselves who solved this great problem. From 1910-1930 they saved their own children at an average rate of 40,000 infant lives every year as compared with 1901-1910. It seems incredible, but there it is, set out year by year by the Registrar General. In 1891-1900 there were on average 140,000 infant deaths each year. In 1940 that terrible national loss was reduced to less than 35,000 per annum. The infant mortality rate thus fell from 150 per thousand births in 1900 to 50 per 1,000 by 1939. There has been nothing comparable in the history of Preventive Medicine in England with this great triumph. The State and the doctors have no doubt done their part, but this is the achievement of the mothers of England themselves, aided by a zealous army of devoted and skilled helpers (Newman, 1941: 19-20).

The centenary of the publication of *Infant Mortality* has afforded us an opportunity to look afresh at this book, which is perhaps best viewed as the culmination of the various attempts to address the issue of infant mortality in the late Victorian and Edwardian periods. Written by the 'world's most prominent health official' it provides a useful analysis of the causes of infant mortality and says much about how contemporaries viewed the problem (Hammer, 1995: viii).[19] By implication it also proposes an interesting thesis concerning the course of secular decline in infant mortality during the twentieth century.[20] Moreover, were he alive today Newman, being a passionate advocate of medical history, might well argue that his book is also relevant to those researching infant welfare issues at the beginning of the twenty-first century. Certainly his emphasis on the complex social influences on infant mortality would appear to resonate with current debates concerning the causes of modern health inequalities. Thus, for these reasons alone *Infant Mortality* is still relevant to the study of infant mortality in the past, the present and perhaps even the future.

---

19 This description was given to Newman by the New York Academy when it notified him of his election as an honourary fellow in 1916.

20 See Woods *et al.* (1989: 113-121) which examines contemporary approaches to solving the infant mortality problem, especially those put forward by Newman and Newsholme.

Chapter 3

Newman's *Infant Mortality* as an Agenda for Research

Robert Woods

George Newman made clear in the Preface to *Infant Mortality: a Social Problem* that, first and foremost, we should consider his book an introduction to the facts; a careful study influenced by the experience of five years work as Medical Officer of Health (MOH) for Finsbury, London. But this was not all since:

> …a high infant mortality rate almost necessarily denotes a prevalence of those causes and conditions which in the long run determine a degeneration of race; and further, that a high death-rate of infants is an indication of the existence of evil conditions in the homes of the people – which are, after all, the vitals of the nation. Poverty is not alone responsible, for in many poor communities the infant mortality is low. Housing and external environment alone do not cause it, for under some of the worst external conditions in the world the evil is absent. It is difficult to escape the conclusion that this loss of infant life is in some way intimately related to the social life of the people (Newman, 1906 [hereafter '1906']: v-vi).

In these few words there are some important indicators as to aims and objectives: the facts do not speak for themselves; infant mortality is a social evil; the factors that influence its high levels are rather complicated; the problem is one of particular national concern, and so on. In today's language we might call this 'evidence-based research directed to policy formulation and practical implementation'. As we shall see, Newman was most forceful in his stress on one particular aspect of the problem – motherhood –but this is not so apparent in the early pages of his study. His approach involved the use of a wide variety of sources with a heavy reliance on official reports by, for example, the Registrar General, the Medical Officer to the Local Government Board, the Interdepartmental Committee on Physical Deterioration and especially the annual reports of a wide collection of local MOHs, including his own for Finsbury. There are also many references to the findings reported in scientific and medical journals. England and its large towns provide the most important case studies, but Scotland, Ireland and France are also well represented. The problem identified, the solution provided; *Infant Mortality* proved to be a thorough and timely survey, one with a powerful message. The emphasis on motherhood, rather than sanitation or poverty, has continued to be controversial. It appears to give responsibility to just one parent, and not to the State, nor to employers, society or the medical profession, and thereby places mothers in a position to be blamed for their failures if infant mortality remained high. This chapter considers *Infant Mortality* as a text. It is interested in content, use of language, authorial intentions, as well as its context: how infant mortality came to be problematized, what was known of its causes and how that knowledge was acquired, what assumptions were made, and silences guarded. Via Newman and *Infant Mortality* we should be able to establish not only the state of contemporary thinking on infant mortality in 1906, but how those conclusions were derived, what 'research methods', as they might now be called, were employed and how effective they have proved against the test of subsequent work during the twentieth century. This chapter builds on Chapter 2, which was mainly concerned with Newman's background and experiences, as well as the reception *Infant Mortality* received. The two chapters need to be read in sequence.

Infant Mortality was organized into eleven chapters with the following titles:

Present position and incidence of infant mortality (Chapter I)[1]
Distribution of infant mortality in Great Britain (II)
The fatal diseases of infancy (III)
Ante-natal influences on infant mortality (IV)
The occupation of women and infant mortality (V)
Epidemic diarrhoea (VI)
The influence of domestic and social conditions (VII)
Infant feeding and management (VIII)
Preventive methods: (a) The mother (IX)
Preventive methods: (b) The child (X)
Preventive methods: (c) The environment (XI)

Although some chapters demand more attention than others, it will be convenient to consider them here under four broad headings: influences, causes, mother and home, prevention. Other sections in this chapter focus on Newman's sources, the shortcomings of his survey and some examples of areas in which he was especially insightful.

Influences on the Infant Mortality Rate

Chapters I and II report a number of 'facts' concerning the current (1905) level of mortality in Britain, its recent trends and comparative position. These can be listed quite simply in the form of ten points.

1. Although the general death rate was declining, the infant mortality rate (IMR) was not. In 1905 there were 120,000 infant deaths in England and Wales, a quarter of all deaths that year.
2. The birth rate was in decline.
3. The general death rate was clearly in decline.
4. The IMR was higher in foreign countries 'as a rule' than in Britain, but Ireland, New Zealand and Norway were exceptions. Infant mortality in Austria, Russia

---

1 Roman numerals are used to indicate Newman's chapters and differentiate them from chapters in the present volume which are given Arabic numerals.

and Chile were especially high.

5. New Zealand might be used as a special low mortality standard since in 1904 its infant mortality rate had reached 'an almost ideal figure of 71' (1906: 10).
6. During the first twelve months after live birth there was a distinctive distribution of deaths with the greatest proportion occurring in the first trimester while proportions in later months declined.
7. Infant mortality among illegitimate births was 'enormously greater' than among legitimate ones. In London it was twice as high, for example. As Newman put it:

> The infant mortality rate ... is not declining, and this is the broad fact which constitutes the problem to be considered. Children under twelve months of age die in England to-day, in spite of all our boasted progress and in spite of an immense improvement in the social and physical life of the people, as greatly as they did seventy years ago. Nor is England alone. Over all the continent of Europe there are but four States with a lower infant mortality rate than that of the British Isles. To find healthy rates elsewhere we must take the wings of the morning to the new lands of the Australian Commonwealth and the island of New Zealand. But even so we shall not wholly escape our problem.
>
> Further, we have seen that the incidence of death in infancy is not equally or uniformly distributed, but falls in exceptional measure on the first weeks and months of life, having a tendency to fall earlier rather than later, and carrying off more boys than girls. That there is here something in operation outside the laws of nature becomes certain when we find that infant mortality is enormously increased in illegitimately born infants as compared with legitimate.
>
> These are the ground facts of the problem (1906: 18).

8. Although the national IMR had remained constant over decades, there were strong differences between the industrial counties of the North and the agricultural ones in the South (a line from The Wash to the Severn marked the division). Even so, there was a 'general similarity of distribution' (1906: 22) among the counties and regions.
9. 'No mere physical or geological conditions' (1906: 25), or topography or pauperism explained the differences, but high population density and degree of manufacturing industry were important.
10. The effect of the towns was most significant and this could be demonstrated in Scotland, in the case of Bedfordshire, in London, and by comparing selected urban districts with rural counties:

> On the whole, Urban England, or the more densely populated districts, suffers a higher death rate from all causes and a higher infant mortality rate than Rural England, or the less densely populated districts (1906: 29).
>
> All diseases of infancy are heavier in the towns than the counties; but immaturity is twice as fatal and epidemic diarrhoea seven times as fatal in the towns (1906: 42).

This list of ten points gives a clear picture of turn of the century understanding. It emphasizes the principal characteristics of level, trend and variation, which subsequent studies have, with one important exception, largely confirmed and elaborated upon (Woods and Shelton, 1997; Woods, 2000). There was a marked association between infant mortality and population density that picked out the large urban centres, but not all rural districts were equally well favoured and several large cities were no more disadvantaged than some smaller towns. It was also the case that early childhood mortality (ages 0 to 4 in completed years) was even more sensitive to urban-rural differences than the IMR. Births to unmarried mothers were at much greater risk than legitimate offspring, but then so were the first-born regardless of their mother's marital status (although most illegitimates were first-borns) and those with third and higher birth orders. Risk of death also followed a distinct pattern during the first year after birth, a pattern that Newman and others were beginning to recognize. Today we distinguish between perinatal (stillbirths plus first-week deaths), neonatal (first four weeks) and post-neonatal mortality (remainder of first year). Of the three, post-neonatal mortality was the most sensitive to sanitary conditions and the effects of changed diet at weaning, it was also the element of infant mortality that began to decline first during the nineteenth century (along with or shortly after early childhood mortality). The idea that an IMR of about 70 might be considered 'almost ideal' (1906: 10) will appear strange today, but Newman was used to thinking in terms or 150 or 200 in the towns. New Zealand illustrated what was possible. It is also true that national levels of marital fertility had begun to decline since the mid-1870s and that life expectancy at birth had started to increase from the 1860s and was now at an unprecedentedly high level. It was clear to Newman and many professional observers that Edwardian Britain showed signs of having a new demographic structure: declining birth rates combined with declining death rates. Yet the IMR remained unchanged, or nearly so. Of course, we now know that national IMRs went into a long phase of secular decline after 1900, but this was not apparent to Newman. His focus on the causes of persistently high levels was understandable.

Causes of Death Among Infants

Newman was well aware that the official registration and reporting of causes of death was fraught with difficulty and that the problem was especially important among infants where the recognition of cause and its simple description were particularly troublesome matters on which to be precise (1906: 59). Nevertheless, he also appreciated that cause of death data should provide valuable insights on the reasons why infant mortality remained at a high level. Let us begin with his conclusions to Chapter III and then work backwards through his arguments and the evidence he marshaled to support them:

> First, the causes of infant mortality are composite. It has been well said that every effect has an ancestry of causes. Pre-eminently is this the case in regard to infant mortality, which is due to a combination of factors closely inter-related.
>
> Secondly, nearly one half (about 48 per cent.) of the infant deaths in towns occur in the first three months of life;
>
> Thirdly, the chief fatality in these first three months is caused by *prematurity* and *immaturity*;

> Fourthly, by far the greatest fatality in the remainder of the first year of life is due to inflammatory conditions of the lungs and to *epidemic diarrhoea*; and
>
> Lastly, infant mortality is not declining owing to the fact that while certain diseases have enormously decreased, *prematurity, pneumonia*, and *epidemic diarrhoea* have, in spite of all advance in science, steadily increased, particularly in the towns and where the lamp of social life burns low (1906: 60).

Newman defined immaturity by combining deaths due to prematurity and congenital conditions. Immaturity was credited with contributing 30 per cent of all infant deaths, and was ascribed to an 'ante-natal origin':

> These children are simply born in such poor physical condition that they are unfit to live, and find a few hours or days of extra-uterine life too much for them. They are not so much diseased as merely unfit, and either not ready or not equipped for separate existence (1906: 47).

A further 20 per cent of infant deaths were due to epidemic diarrhoea, which was the result of 'bad housing conditions, poverty, artificial feeding (and) domestic insanitation' especially in urban districts where rates were higher. The third corner of the triangle – prematurity, pneumonia, diarrhoea – was occupied by the respiratory diseases which were identified as being particularly prominent after the first trimester. Newman was also much exercised by the fact that these three categories were increasing in importance while some of the infectious diseases of childhood were declining. Chapter VI was entirely devoted to epidemic diarrhoea while Chapter IV dealt with ante-natal influences on immaturity. The former was much influenced by the work of Dr Arthur Newsholme, the then MOH for Brighton and an established medical statistician (Newsholme, 1889, 1891; Eyler, 1997), while the latter depended on the research of Dr John William Ballantyne, the distinguished pathologist (Ballantyne, 1891, 1892, 1902, 1904a, 1904b, 1905). We shall consider the role of epidemic diarrhoea here and return to prematurity etc. in the final section, which reviews Newman's particular insights.

It was clear to Newman (Walker-Smith, 1998a, 1998b), as it was to Newsholme (1899, 1902, 1902-03, 1906b) and most other observers, that epidemic diarrhoea was a disease especially of late summer and of the towns; that its incidence was at least conditioned by meteorological circumstances and that it was directly due to 'contaminated or injurious food', especially milk. Whether the milk supply was contaminated on the farms, in transit or in homes was an issue for debate, as was the combined influence of temperature and rainfall. On the latter, Newman observed that the old rule of thumb, by which epidemic diarrhoea would be triggered once the four-foot earth thermometer read over 56°F., was not a perfect guide. Certainly temperature was important, but it also needed to be dry. On the former, Newman was apparently convinced by Newsholme's argument that domestic conditions were of most critical importance in the contamination of milk, water and food given to infants; that flies could play a significant role as vectors, and that the 'ailing condition of infants when attacked' may hasten their demise. In short, epidemic diarrhoea was a most important cause of death especially among urban infants and its prominence was the result of the combined effects of meteorology; contaminated or injurious food; social conditions and domestic insanitation.

Mother and Home

Chapters V, VII and VIII of *Infant Mortality* dealt with the influence of the mother and the home. They provide the distinctive core of Newman's analysis, and they are controversial. For example, here is a concluding paragraph from Chapter V on 'The occupation of women and infant mortality':

> We had previously seen that town life is disadvantageous to the health and life of infants. That is a commonplace and obvious fact. But after the present chapter we must now add that in towns where women are largely employed in factories, away from home, the disadvantages are enormously increased. Broadly, this is the inevitable conclusion. There are reservations and differences, but the fact remains. And the operations of this fact are threefold. First, there are the ordinary injuries and diseases to which women and girls in factories are liable; secondly, there is the strain and stress of long hours and hard work to the pregnant woman; and thirdly, there is the absence from home of the mother of the infant. It cannot be doubted that these are the factors in the relation between factory occupation of women and a high infant mortality (1906: 131).

A close reading of Chapter V reveals that these 'reservations and differences' are far from trivial.

While Newman was inclined to follow Sir John Simon in the belief that 'infants perish under the neglect and mismanagement *which their mothers' occupation implies*' (1906: 93), he found it difficult to establish the case in a way that proved completely convincing. Certainly, it appeared that those districts with relatively high levels of female factory employment also had higher IMRs, but Newman was concerned to ensure that only married women in the child-bearing ages who were engaged in factory employment should be included in his comparative analysis. Once this correction had been made, he affirmed that it was 'beyond all question, that there is an intimacy – if not a relationship at least a correspondence – existing between the occupation of married women and a high infant mortality' (1906: 106). But the paragraphs that followed noted the possible inverse association between infant mortality and the proportion of female domestic servants in a district's population, as well as the point that infant mortality could still be high even where women were mainly employed at home. The appreciation of these complications led Newman to make the following observations by way of conclusion.

> No doubt the factory plays a part, but the home plays a vastly greater part, in the causation of infant mortality in the towns where women are employed at the mills. There are two influences at work: first, the direct injury to the physique and character of the individual caused by much of the factory employment of women; and secondly, the indirect and reflex injury to the home and social life of the worker (1906: 137).

But there is also, Newman acknowledges, a 'vicious circle': women work in the mills because their husbands are poorly paid, house rents are high and opportunities exist for non-domestic employment. Good hygiene is not enough in itself, the effects of poverty also need to be considered.

The scene is thus set for the discussions of, respectively, domestic and social conditions, and infant feeding and management in Chapters VII and VIII:

> It is a well-known fact that communities in which there is a large measure of poverty have a higher mortality from all causes at all ages than communities better circumstanced. This, of course, is not due only to poverty per se, but to all that poverty involves—heredity, upbringing, education, food, housing, overcrowding, &c (1906: 177).

Newman cited Charles Ansell's 1874 survey to the effect that among the professional and well-to-do classes infant mortality was 1 in 10 (i.e. an infant mortality rate of 100); among the families of tradesmen, 1 in 6 (166); but among the working classes it was 1 in 4 (250). Alongside these statistics he placed the differentials between groups of London districts: infant mortality was less than 140 in the better off and over 180 in the poorest districts. The conditions responsible for these differences were four in number: urbanization, housing and poor social life, alcoholism, and birth rates.

As far as urbanization was concerned:

> It is rather the habits and customs of town life, which militate against healthy infancy, especially in the artisan classes, to whom infant mortality is almost confined. Life in a large city means for the artisan limited accommodation (not necessarily overcrowding), late hours, short nights, manufactured foods, stress and strain, alcoholism, small excesses, and an almost total absence of restfulness, leisure, and home life. The homelessness of the people is one of the worst features of town life, and is operating injuriously on infancy (1906: 180).

Newman's discussion of housing and poor social life relied heavily on Charles Booth's great survey *Life and Labour of the People in London* (Booth, 1902: final volume) which he used in conjunction with Seebohm Rowntree's work on York (Rowntree, 1901) and a number of MOH reports, including his own for Finsbury, and especially that by Dr Robertson on Birmingham (Robertson, 1904). From these cases – London, York, Birmingham – Newman built a comprehensive and conclusive account of poverty's multiple effects.

Although there can be no doubt that infant mortality was adversely affected by poverty in all its forms, Newman was on less safe ground when it came to his third factor, alcoholism. Establishing the extent of alcohol consumption among women proved something of a problem. One source used was Dr Arthur Shadwell's *Industrial Efficiency* (Shadwell, 1906). Here is Shadwell in full cry as quoted by Newman:

> We certainly have a class of women already numerous and probably increasing who are a source of great national weakness. They are ignorant, idle, extravagant, and self-indulgent. They neglect their children and their homes, they drink and bet, and they exist in all ranks of society. The wretched appearance of so many working-class homes and children, which constantly horrifies visitors to this country, is quite as much due to this type of woman as to the self-indulgent man who matches her. Neither in America nor Germany nor in any other country that I have seen do women drink and bet as they do here (1906: 209):

Although Newman's language is more temperate, he appears to accept the contribution of alcohol consumption to the deaths of infants. He quotes from 'a body of unpleasant figures', which showed that 1,899 infants were overlain in bed in England and Wales in 1904 (584 in London), and that of the equivalent number in 1890 (2,020), 28 per cent – nearly twice the expected proportion given an equal distribution among the days of the week – died on a Sunday (1906: 211). Neglect or alcoholism was inferred.

Finally, Newman turned to birth rates and illegitimacy. As far as the former was concerned, he accepted, mentioning the example of Ireland, that there was no necessary causal link, but illegitimacy was a well-established and highly-damaging influence on infant mortality even though its rate had declined substantially since the middle of the nineteenth century. Infant mortality among illegitimate births was 1.9 to 2.8 times higher than that among legitimates. To illustrate his point, Newman mentioned the case of Glasgow in 1873 for which city William Farr found that the legitimate IMR was 154 while the illegitimate rate was 293; and although only 4 per cent of total births were illegitimate they contributed 25-30 per cent of total infant deaths (1906: 214). Newman drew the following conclusion:

> The causes of death in illegitimate infants are the causes of death in legitimate infants *plus* maternal indifference and social and economic disabilities of the mothers, the separation of infants from mothers, the secret adoptions of infants for gain, the various rapid or slow forms of infanticide. The chief of these in actual terms of disease are diarrhoea, atrophy and inanition, and syphilis (1906: 215).

The conclusion to Chapter VII on domestic and social conditions was very bleak. Housing conditions were certainly important, but poor housing alone was not the true cause of high infant mortality; rather the results of overcrowding were of special significance. And yet 'it is clear that it is not external environment which only, or in fact mainly, affects the problem' of infant mortality since the external environment had 'enormously improved' during the preceding twenty-five years while infant mortality remained 'as grave a problem as ever' (1906: 210).

By Chapter VIII Newman had come to the key issue in his survey: infant feeding and management, noting that:

> ...more than any other single agency, infant mortality depends upon infant rearing. Ignorance and carelessness as to the physical necessities of life may mean little to the adult, but they are unpardonable where infancy is concerned, and may as likely as not result in death to the new-born child. And, expressed bluntly, it is the ignorance and carelessness of mothers that directly causes a large proportion of infant mortality, which sweeps away every year in England and Wales alone 120,000 children under twelve months of age. This ignorance reveals itself in many ways, but chiefly, perhaps, in feeding, uncleanliness, and exposure (1906: 221).

Most of the chapter was concerned with a detailed review of the evidence on local infant feeding practices gathered by the MOHs for Derby, Finsbury and Brighton, although it was introduced by an international comparison which served to illustrate the benefits of breastfeeding and the evils of artificial and bottle feeding. But Newman began by quoting the work of Dr E. W. Hope, the MOH for Liverpool, on the state of infant mortality in that city. Hope had noticed that among working class families the IMR was substantially lower among the Irish than the English. This difference could be ascribed directly to the fact that Irish mothers suckled their own infants giving them thereby the 'proper food at proper temperatures, the supply being made when wanted' (1906: 226). English mothers were more likely to use bottles and artificial foods. The resort to 'proper feeding' was noted among members of the Jewish communities in London and New York.

The international comparison focused on the cases of Scotland, Norway and Sweden, which were said to have low infant mortality, and France and Austria, where infant mortality was higher. The exercise repeated the one reported by William Farr in an appendix to his paper in the *Journal of the Statistical Society* of London for 1866 (Farr, 1866). Dr Stark, Farr's Scottish correspondent, affirmed that 80-85 per cent of infants were breast-fed by their mothers, that in general they were weaned at nine months, and that solids were not given until after weaning. In Norway and Sweden it was also normal for infants to be breast-fed for at least the first six months with examples of suckling up to two or three years being not uncommon. In France, the practice of sending town-born infants away to the countryside to be wet-nursed was held to account for the persistently high infant mortality, but it was also assisted by the use of artificial feeds, by high illegitimacy and by child abandonment. As for Austria, there the habit of using wet-nurses, especially by the Viennese, was also blamed, along with the ignorant midwives. Newman drew the obvious conclusions from these observations: the survival chances of infants were much enhanced if mothers breast-fed their own infants for at least nine months and avoided giving them the feeding bottle, artificial foods or solids in general during that period.

However, the most telling sections of Chapter VIII were provided by the new survey data collected by the English MOHs: Dr Howarth in Derby, Dr Newman himself in Finsbury, and Dr Newsholme in Brighton. The largely rural county of Dorset was also used as a comparator. Newman was most impressed by the evidence from Derby where Howarth had attempted to record the experiences of 8,343 infants born during the 36 months up to November 1903 (see Reid in this volume). While the total IMR was 99, that for breast-fed infants (368 deaths) was 70; for those given a mixture of foods (142), 99; but for the hand-fed infants (321) it was 198. Among those who had been hand-fed, it emerged that condensed milk had an especially poor record: 255 deaths per 1,000 infants so fed. The role of condensed milk was also picked-up in the Finsbury and Brighton studies. Newsholme, in particular, emphasized the point that condensed milk, even more than cow's milk, was liable to be poor in nutrition and to be subject to contamination in the home environment. These three studies were, in combination, especially effective in drawing attention to the perils of infant feeding together with the special risks faced by those born in the four months July to October when epidemic diarrhoea could be at its most lethal among infants who were not being breast-fed.

The final paragraph in Chapter VIII again strikes a portentous note and is worth quoting in full since it sets the tone for the book's final part, which deals with preventive methods:

> The present chapter has but added to the cumulative evidence as to the absolutely vital importance of suitable infant feeding. It is not everything, but it may be said that it is a greater factor than any other single thing. The problem of infant mortality is after all one of those elementary problems which depend more upon instinct and the physical faculties and functions which nature has provided in the mother than upon external environment. It is so with the young of all animals, and the human species is not an exception. And so it comes about that many of the facts set forth in the preceding pages have importance chiefly as they concern a fulfillment of the primitive needs of food and warmth and cleanliness (1906: 256).

Prevention

The last three chapters of *Infant Mortality: a Social Problem* dealt with preventive methods. They focused on the mother, the child and the environment, but it was 'physical motherhood' that held the key. Once again, it is important to let Newman speak for himself:

> This book will have been written in vain if it does not lay the emphasis of this problem upon the vital importance to the nation of its motherhood. Wherever we turn, and to whatever issue, in this question of infant mortality, we are faced with one all-pervading primary need—the need of a high standard of physical motherhood. Infant mortality in the early weeks of life is evidently due in large measure to the physical conditions of the mother, leading to prematurity and debility of the infant; and in the later months of the first year infant mortality appears to be due to unsatisfactory feeding of the infant. But from either point of view it becomes clear that the problem of infant mortality is not one of sanitation alone, or housing, or indeed of poverty as such, *but is mainly a question of motherhood.* No doubt external conditions as those named are influencing maternity, but they are, in the main, affecting the mother, and not the child. They exert their influence upon the infant indirectly through the mother. Improved sanitation, better housing, cheap and good food, domestic education, a healthy life of body and mind—these are the conditions, which lead to efficient motherhood from the point of view of child-rearing. They exert but an indirect effect on the child itself, who depends for its life in the first twelve months, not upon the State or the municipality, nor yet upon this or that system of *crèche* or milk-feeding, but upon the health, the intelligence, the devotion and maternal instinct of the mother. And if we would solve the great problem of infant mortality, it would appear that we must first obtain a high standard of physical motherhood (1906: 257-58).

The content of the preventive methods chapters is summarized in Table 3.1. We shall consider each section briefly here, although the general tenor will already be apparent from the above.

Newman set great store by programmes to correct the 'ignorance and carelessness of mothers with respect to infant management' especially 'improper feeding and ill-timed weaning', along with schemes to educate the mothers of the future by bringing domestic hygiene and maternal skills into the classroom. He was also concerned to reduce as far as possible the 'necessary evil' of married women being employed outside the home. Here Newman proposed restricting the employment of women in factories during the final stages of their pregnancies which would involve, at minimum, enforcing the 'four-weeks rule' and making provision, where possible, for

Table 3.1 Preventive methods for the reduction of infant mortality as discussed by George Newman in *Infant Mortality: a Social Problem*

I. The mother

(1) The reorganization of existing agencies

(2) Education of the mother as to infant management

a. The instruction of mothers

b. The appointment of lady health visitors

c. The education of girls in domestic hygiene

(3) The occupation of the mother

II. The child

(1) Birth registration

(2) Protection

a. The Midwives Act, 1902

b. The Infant Life Protection Act, 1897

c. The Prevention of Cruelty to Children Act, 1904

(3) The artificial feeding of infants

III. The environment

'the external sanitary conditions surrounding both mother and infant'

(1) Improved sanitation in the factory

(2) Improved sanitation in the home

(3) Urban cleanliness

(4) Control of the milk supply

---

longer periods of rest before and after birth. The appointment of lady health visitors to assist in the provision of post-natal care, advice and instruction along the lines of the scheme pioneered in Huddersfield was also advocated (Marland, 1993). The entire package of schemes and interventions outlined in Chapter IX and targeted on the mother seemed obvious and consistent with international best practice.

In Chapter X Newman turned his attention to the child. He thought it desirable that the registration of births be tightened-up. Live births should be registered within 48 hours of delivery and not six weeks as per the 1874 Act since although two-thirds of births were registered within the legal time limit, a third of infant deaths occurred within the first four weeks. Existing special schemes in Glasgow and Huddersfield were to be recommended in this regard. Three other existing Acts could assist in prevention if their powers were enforced or extended. The registration of midwives, the regulation of infant nursing and adoption, and the limitation of cruelty to children all offered important measures which could, in combination, lead to an 'immense improvement' in infant mortality. But it was to infant feeding that Newman returned to give most emphasis. He championed the role of the infant milk depot, like the ones established in Paris in 1892, St Helens in 1899, and Finsbury in 1904. These establishments could provide not only pure milk suitable for infant consumption, but also advice to mothers concerning infant care in general. Whilst not 'the answer', their presence and effective use, in conjunction with greater maternal education and monitoring, should provide the essential policy interventions to resolve the problem of persistently high infant mortality.

Finally, Chapter XI turned to the environment, but this was not the external sanitary environment of the public health pioneers and water engineers rather Newman was concerned with those environmental issues that had a direct bearing on the health of the mother and thereby on the infant and child. The factory, the home, the wider urban environment and, yet again, the milk supply were the chief targets, but they were dealt with briefly, almost as an afterthought.

Sources

Chapter 2 has already mentioned some of the sources Newman used for his study, here we shall take another look at some specific pieces of research and report writing. Newman's account is heavily dependent on four distinct, although not entirely independent, sets of materials. The first, and probably the most important, is comprised of the regular annual reports and special one-off surveys prepared by fellow local MOHs. The most important of these related to the following areas: Barnsley, Bedfordshire, Birmingham, Blackburn, Brighton, Bury, Croydon, Derby, Dundee, Glasgow, Huddersfield, Hull, Liverpool, London County, London-Battersea, London-Finsbury, London-St Pancras, Manchester, Nottingham, Preston, Salford, Sheffield, Staffordshire, Stockport. Although the geographical coverage is impressive, several of the reports must have been selected because the authors were prominent figures in the public health movement and were known to Newman. Drs Niven (Manchester), Robertson (Birmingham), Chalmers (Glasgow), Sykes (St Pancras), Newsholme (Brighton), Hope (Liverpool), and Sir Shirley Murphy (London County) were among the leaders in their field. This was true partly because they were concerned to move beyond their routine activities; to undertake new, research-based studies founded on both medical statistics and social survey methods. The reports of the MOHs to the Local Government Board (LGB) were also used regularly as were other more general reports prepared for the LGB and the Chief Inspector of

Factories. Of the special reports the one by Dr John Robertson, *Special Report of the Medical Officer of Health on Infant Mortality in the City of Birmingham* (1904), was particularly timely. It furnished Newman with a substantial amount of material on the importance of domestic and social conditions from which he quoted at length in Chapter VII (1906: 190-196), but it also provided evidence on infant feeding for Chapter VIII (1906: 254).

The second set consists of the various reports, annual, decennial and special, prepared by successive Statistical Superintendents to the General Register Office, London. Although these publications normally appeared under the name of the Registrar General of the time, his involvement in their production can have been little more than token in nature. Drs Farr, Ogle and Tatham wrote the reports and supervised the statistical work on which they were founded. Each of these authorities was frequently mentioned by name in *Infant Mortality*. Newman also used some of the reports prepared by Dr Jacques Bertillon who, as head of the Municipal Statistical Department of Paris, performed a similar function in France. Of these, the work of William Farr certainly had the most lasting impact both through his role as one of the first statistical civil servants and through his scientific papers (Farr, 1885; Newsholme, 1923; Eyler, 1979).

Newman was writing at a time when several important social surveys were being published. These 'quality of life' reports gave *Infant Mortality* substantial evidence for the social divisions within British society, and especially the extent and forms of poverty experienced by members of the working classes. The most important of these were Charles Booth's *London Life and Labour* and Seebohm Rowntree's *Poverty* (1901), which documented conditions in York. Considerable use was also made of Charles Ansell's (1874) postal questionnaire directed at families in the 'upper and professional classes'.

The fourth set of sources involved papers in scientific and especially medical journals, as well as larger medical treatises. *The Lancet* and the *British Medical Journal* were well represented, but the following were also used: *Public Health, Journal of Hygiene, The Practitioner, Transactions of the Epidemiological Society of London, Journal of Mental Science, Clinical Journal, Transactions of the American Pediatric Society, Johns Hopkins Hospital Reports, Revue d'Hygiène et de Mèdicine Infantiles, L'Obstetrique* and the *Journal of the Royal Statistical Society*. Of the book-length studies, the most important were Thomas Oliver's *Dangerous Trades* (1902), Arthur Shadwell's *Drink, Temperance and Legislation* (1902), George Frederick McCleary's *Infantile Mortality and Infants' Milk Depôts* (1905), William Smoult Playfair's *A Treatise on the Science and Practice of Midwifery* (1886), Pierre Budin's *Le nourrisson* (1900) and John Ballantyne's manuals of foetal pathology (1902, 1904).

Finally, and in a category of its own, comes the *Report of the Inter-Departmental Committee on Physical Deterioration* (1904). This important government report had a substantial impact on the thinking of health professionals during the early years of the twentieth century. It symbolized a change of emphasis away from cities and sewers to the physical health of individuals and what could be done to improve its quality. An emphasis on the role of mothers, whether prospective or actual, emerged as a fundamental part of this changed perspective, one that Newman's work clearly reflected.

A list in this form makes *Infant Mortality* appear like a compendium, an encyclopaedia of Edwardian infant mortality, but such an impression would be misleading. Newman had a particular story to tell, one with a very strong message. He was careful in his selection of evidence and he was certainly uncritical or critical of his sources by turns as they conformed to the case he wished to make.

Limitations

As we have already noted, Newman was writing at a particularly interesting time for the history of infant mortality in Britain. While the IMR stayed on a relatively high plateau during the nineteenth century (150 infant deaths per 1000 live births provides a rough guide), in the twentieth century it passed into a sustained decline. *Infant Mortality* appeared at or just after the key turning point in national rates, but it was not possible for Newman to see the turn with the evidence available. His entire approach was directed to the question 'why has infant mortality remained at a high level when the general death rate has begun to decline?' Newman's answer focuses on motherhood and the policy implications he draws involve the education of mothers and future mothers in the practices of superior childcare, including the proper feeding of infants. In the absence of major schemes to alleviate poverty and improve housing conditions, neither of which formed part of the preventive response, training, advising, inspecting and supporting mothers would undoubtedly have had a beneficial impact on the survival chances of the newly born. However, even if widely implemented to 'best-practice' standards such policies could only have made a small, yet important, contribution to the secular decline of infant mortality in the twentieth century. What else was required?

There are a number of important points that Newman failed to emphasize or was unable to see. Some of these may usefully be touched upon here. First, the decline of infant mortality during the twentieth century was a Europe-wide phenomenon, one that can also be discerned more generally in the industrialized West (Schofield *et al.*, 1991; Bideau *et al.*, 1997; Woods, 2000). While there were differences in level and trend between European countries, and especially regions, many populations experienced a downturn (or accelerated decline) in rates during the early years of the twentieth century. Britain was not alone in this regard, therefore. Policy shifts unique to one country will not help to reconstruct the broader picture. One other demographic development affected most European societies during the last quarter of the nineteenth century: the decline in marital fertility and thus the reduction in family size. Demographic theory consistently argues for a causal link between fertility and early-age mortality. A prior decline in the latter motivates the former while a reduction in the former limits the numbers of vulnerable high-parity births and extends the intervals between births. The widespread decline in marital fertility in Europe between 1870 and 1940 undoubtedly had a beneficial effect on the improvement in IMRs, but it is very difficult to quantify this relationship and to separate it from other effects. Newman and his contemporaries reacted to the combination of lower fertility and unchanged infant mortality with fears for the future of the race. The full, and mutually beneficial, consequences did not become apparent until the 1930s when there were new concerns over below replacement-level fertility.

Second, Newman was strangely silent about the role of medical science, the medical professions (practitioners, nurses and midwives) and hospitals. Certainly, the Midwives Act of 1902 is mentioned (1906: 261), but the potential for advances in obstetrics, gynaecology and paediatrics was not considered, nor was the importance of the lying-in hospital and the contribution it could make via safe delivery, ante- and post-natal care, as well as advice for mothers. Although Newman's principal professional interest was in public health, this should not absolve him from an appreciation of other possibilities and responsibilities beyond an emphasis on maternal duties. The case of maternal mortality offers some interesting side-lights on this matter. Irvine Loudon (1991, 2000) has shown that maternal mortality in England and Wales did not begin to decline until the late 1930s; that the fall from a plateau of around 40 deaths per 10,000 births to less than 10 by 1950 was directly attributable to the introduction and effective use of the sulphonamides, which were directed against puerperal fever. This example clearly illustrates the important contribution medical advances could make in the twentieth century.

Third, the politics of motherhood and gender relations in general have become far more important issues for debate and academic study in the twentieth century. For example, Davin (1978), Dyhouse (1978), Lewis (1980, 1986), Oakley (1984), Ross (1993) and Apple (1995), among many others, have explored the various ways in which the accusative finger of attention was pointed by commentators like Newman at women and especially mothers, giving them thereby special responsibilities for the health of the nation without empowering them. Infant mortality did not decline during the twentieth century because women became better mothers, more caring, more skilled, and more responsible. Rather, it was reduced via a mutually-reinforcing combination of medical interventions (including hospitalization, improved practices, effective drug therapy, immunization and vaccination programmes); better diet and living conditions in general (electricity and the refrigerator); lower fertility; a new focus on ante- and post-natal care by the State. Clearly mothers played their parts, but they were greatly helped by new knowledge and social investment (Preston and Haines, 1991; Millward and Bell, 2001). The insult of parental, especially maternal indifference, has not been supported by empirical evidence from the nineteenth century. Ross's observations have been re-confirmed (Strange, 2005; Woods, 2006a):

> Mothers, especially among the poor, did not see a long, healthy life as the birthright of each of their children, for they knew that many forces that parents could not control could thwart their development. At birth, infants' survival was still tentative, and as the babies grew, their mothers knew that attempting to save them from accidents or diseases would be among their most taxing and heartbreaking jobs.
>
> This 'selective neglect', which anthropologists have noted among some Third World shantytown dwellers today, seems rare among married pre-World War I London mothers,

though it may be more accurate to say that it remained almost unremarked and unnamed (Ross, 1993: 179).

Insights

Despite its limitations, *Infant Mortality* offered some very important insights on the health problems of Edwardian Britain. Newman marked the shift in emphasis away from public health, especially the sanitation and water works variety, towards the health and responsibility of the individual (personal health) and this for a local Medical Officer represented a significant re-orientation in thinking. But Newman also appreciated, in a way that most non-specialists did not at the time, the particular significance of fetal health and ante-natal care. After the 'broad survey' and the emphasis on personal health via 'physical motherhood', Newman's most important contribution was probably to stress the role of ante-natal environment.

Chapter IV on 'Ante-natal influences' relied heavily on the work of John William Ballantyne (1861-1923). Ballantyne was an Edinburgh-based physician specializing in gynaecology, obstetrics and foetal pathology. His two volume study *Manual of Antenatal Pathology and Hygiene* (1902, 1904a) defined the field of morbid pathology as it then stood, while his other studies and textbooks helped to summarize current knowledge on the wider area (Ballantyne, 1891, 1892, 1904b, 1905, and 1914). Newman also made extensive use of the multi-volume work on midwifery and obstetrics by Tarnier *et al.* (1888-1901). Newman (1906: 62), following Ballantyne, identified three periods of life before birth: the germinal, before conception; embryonic, beginning in the first six weeks of intra-uterine life; and the foetal, from six to eight weeks to the point of birth during weeks 36 to 40. During the germinal period hereditary influences were most important while the embryonic and fetal stages were influenced by infections and toxaemias. Following Ballantyne, Newman listed certain diseases which they believed could be passed on in-utero and might lead to miscarriage or premature birth: smallpox, malaria, measles, scarlet fever, erysipelas, influenza, whooping cough, typhoid, tuberculosis, anthrax and syphilis. In terms of the toxaemias, metal poisoning (e.g. lead, mercury, phosphorus) and alcohol abuse were singled out. The risks of lead poisoning for miscarriages, stillbirths and prematurity in general were emphasized and much was made by Newman of the dangers to the health of the unborn of alcoholism among parents. While it had to be acknowledged that it was impossible to measure the impact of such infections and toxaemias on foetal health and morbidity, the stress placed on them by Ballantyne and Newman was entirely appropriate and is only now coming to be fully appreciated, although their list of infectious diseases is clearly contentious.

Newman also considered the causes of prematurity and immaturity in Chapter IV. Apart from the infections and poisons, the other influences on prematurity were said to be 'pathological states' which 'lie in some measure outside control' and the 'environment of the pregnant woman or her general habits', but also 'profound emotion', strain, stress, physical labour, and poor or insufficient nourishment (1906: 80). Immaturity, defined rather confusingly as 'all conditions of congenital disability other than prematurity' (1906: 83), was due particularly to nutrition since, quite simply, 'the size of the child may be reduced by restricting the diet of the mother' (1906: 83). All-in-all the general conclusion to ante-natal influences was that 'in spite of the tendency of nature on behalf of the new-born child, poor physique and ill-nutrition of the mother exerts, in a considerable percentage of cases, an important effect upon the infant' (1906: 89).

Clearly, Newman and his contemporaries faced several substantial problems in any consideration of fetal health. For example, they were unable to say with any certainty what the level of late-fetal or stillbirth mortality actually was during the Victorian and Edwardian eras (Mooney, 1999; Woods, 2005). Knowing that 5.5 per cent of all recorded cemetery interments were stillborn infants (quoted 1906: 83) or that 20-25 per cent of pregnancies might end in abortion or that 20 per cent of hospital births might have been premature does not provide a sufficiently secure basis for understanding the scale of fetal loss. It only became possible to analyze these important issues in detail after the registration of stillbirths was introduced (1927 in England and Wales, 1939 in Scotland) and then new research methods were required emphasizing epidemiology rather than pathology, and developing what came to be known as 'social obstetrics' (Baird, 1960).

Infant Mortality: a Social Problem should be regarded as a signpost publication in that it pointed both forwards to better health, lower infant mortality, and back to an even more dangerous past. In his review for the *Journal of the Royal Statistical Society* Newsholme (1906a) described it as extremely well written, full of useful information, presented in an orderly and masterly manner. But there is no mention of Newman's stress on 'physical motherhood', or of his playing down the role of public health measures.

PART II

Chapter 4

Place and Status as Determinants of Infant Mortality in England c. 1550-1837

Richard Smith and Jim Oeppen

In the first two chapters of *Infant Mortality: a Social Problem* George Newman drew attention to two features of infant mortality in Britain in the very early twentieth century that he thought were noteworthy: the increase of infant mortality in the previous half century relative to mortality within the population at large and the increase of mortality within the first three months and especially the first week and month of life which he suggested was a function of the 'increased immaturity of infants at birth' (Newman, 1906: 15). He also noted that in the urban population, or in counties with high urban proportions in their populations, mortality in the later months of the first year of life tended to be high relative to levels found at that phase in infancy among rural populations (Newman, 1906: 20-42). This chapter will consider the same set of parameters that exercised Newman's attention in so far as they are identifiable and measurable within datasets that are spatially more restricted than those available to Newman, but extending over a 250-year period prior to the onset of civil registration in 1837. The aim of the chapter is to exploit changes in those parameters, particularly using trends in the component parts of infant mortality to make greater sense of the ways in which mortality regimes as a whole changed across the early modern period.

The dataset initially under consideration in this chapter in part concerns the amalgamated results of 26 parish register-based family reconstitutions that were assembled at the Cambridge Group for the History of Population and Social Structure (Wrigley *et al.*, 1997). These parishes form a mixture of rural and small town populations, although drawn from an array of environments that for much of the period capture the structure of the national economy as a whole. However, they possess too small a number of urban populations and in the subsequent discussion it will be necessary to consider evidence from such settings separately. In addition, certain elements in the parish register-based dataset will be compared with those retrievable from another body of demographic evidence constructed from genealogies relating to the British peerage extending from the early seventeenth century. We begin with an assessment of the accuracy of these parish register-based datasets since they derive from sources that were never intended to be systematically created records of vital events.

Much ink has been spilled in the assessment of the accuracy of English parish registers as reliable data sources for demographic purposes. It would be correct to state that any parish register randomly selected from those that survive in their thousands would display considerable data deficiencies as a result of damage, changing customs relating to baptism, the growth of nonconformity and laziness or incompetence of the parish incumbents who maintained them. However, those that have been chosen to constitute the 26-parish sample have been subjected to some very stringent quality controls and it can be shown that the estimates of mortality obtained are likely to capture relatively accurately actual demographic patterns and trends. In addition, continuity of registration over long periods is a central requirement for the effective identification of individuals through nominative linkages which is a fundamental requirement of the technique of family reconstitution – a factor increasing the confidence in the demographic rates so derived (Wrigley, 1997; Wrigley, 2005: chapter 15).

In marked contrast to the calculation of many demographic parameters the nominative linkage procedures that are required to define the presence of an infant in observation throughout his or her first year of life are such that a relatively high proportion of all births – around 80 per cent – can be used to estimate infant mortality rates (IMRs) (Schofield and Wrigley, 1983: 158). However, infant deaths have frequently been found to be those most at risk to be under-reported in European parish registers. This risk is of course highest for those individuals whose deaths occurred in the first days or weeks of life. Infants who died before baptism constituted a category especially likely to lead to a shortfall in recorded deaths. This likelihood was increased in England as a result of the tendency of the interval between birth and baptism to be extended over the course of the eighteenth century (Berry and Schofield, 1971).

There are various ways in which the adequacy of birth coverage can be tested using baptismal registrations in parish registers. There are certain physiological processes that create distinctive birth patternings which should be evident in well recorded populations not engaging in family limitation. Long birth intervals should be rare in such populations. By using fecundability models it is possible to show that in such a population birth intervals greater than 60 months would not be expected to exceed 2 per cent of the total (Bongaarts, 1975; Wilson, 1982). If births were being omitted it would be expected that this would result in a far larger proportion of longer intervals. In fact the English data extending from 1570-1837 reveal a proportion of such long intervals slightly lower than 1.8 per cent (Wrigley *et al.*, 1997: 474-5). Such a close fit with model estimates suggests that few births may have been missed. Furthermore it is known that sterility measured by estimating the percentage of married women categorised by marriage age who reach their fiftieth birthday without giving birth follows a straight line, rising from about 3 per cent among those who marry between ages 15 and 19 to nearly 70 per cent for those who marry between 40 and 44. These are characteristics widely encountered in non-contracepting societies and the patterns displayed by the evidence from English parish registers accord with the biological 'norm' (Pittenger, 1973). If births were omitted these percentages would all be greater and highly unlikely to assume a straight line when plotted logarithmically.

Louis Henry (1967: 22-25) devised a remarkably powerful means of estimating the adequacy of birth registration and death coverage in infancy by dividing baptisms into three categories according to the subsequent fate of the child: those cases where it is known that the child died before his or her first birthday; those where the child is known to have survived his or her first birthday; and those where the fate of the child is unknown. It is possible with the 26-parish dataset to investigate many tens of thousands of such intervals categorised in the fashion stated previously and this enables a rigorous statistical analysis of these patterns to be completed.

If registration was complete, all the children whose fate was unknown might be legitimately supposed to have migrated rather than dying in the parish unrecorded. If that were the case the intervals between their birth and the next birth to their mother might be supposed to be similar to that of those children who are known to have lived through their first year because another event in the parish registers identified him or her at a later date. If a significant proportion of those in the 'fate unknown' category had indeed died in infancy but their deaths went unrecorded, we would expect the mean and the shape of the distribution of such intervals to differ markedly from those whose fates were known. In the parish register data in the event of an early death we find that the mean birth interval was about eight months, whereas if the child survived through into the second year of life because of the widespread presence of relatively long-duration breastfeeding the mean interval was close to 30-31 months (Wrigley *et al.*, 1997: 430-449). It has been found that the shape and mean of birth interval patterns after 1600 was such that no statistical difference could be observed between birth intervals where the child survived and where the child's fate was unknown. This is powerful evidence supporting the conclusion that death registration was very complete. It is especially reassuring that the statistical patterns are present both in the periods 1600-1749 and 1750-1837. In the latter period, with a widening of the birth-baptism interval as a result of a changing christening customs, it might be expected that a rising number of children would have died unbaptised thereby distorting the distribution of observed intervals. However the interval data from registers that were well kept suggest that this problem did not arise to undermine accurate measurement of infant death rates (Wrigley *et al.*, 1997: 101-106). This confidence is strengthened still further when a comparison is made between the estimates of infant mortality from the parish reconstitutions from 1825-37 and estimates based on recorded births and deaths for the registration districts within which the parishes fell in the 1840s. While registration districts extended over an area larger than that of the parish, a comparison of rates shows that in almost all cases there was very close agreement. In a comparison of eight parish family reconstitutions and eight registration districts the mean absolute discrepancy between the two rates was only 8 per 1,000 (Wrigley *et al.*, 1997: 92-7). Once again it might be supposed that as IMRs were the measurement most susceptible to under-recording the closeness of the two sets of values suggests that much confidence can be placed in the estimates of infant death rates that derive from parish registers.

Of course, family reconstitution is based upon births occurring in marriage and those deemed to be legitimate so that any estimate of infant mortality based solely on legitimate births will carry some inaccuracy. We do have an estimate of the bastardy ratio (Adair, 1996; Laslett, 1977) from a sample of English parish registers extending from the mid-sixteenth century through to 1837. We have good reason to believe that infants born outside of wedlock suffered far higher rates of mortality in their first year of life than those born within marriage. Using the proportion of all infants born outside of wedlock we can inflate the IMRs obtained from family reconstitutions by assuming that the mortality rate of illegitimate births was double that of legitimate

Table 4.1 Illegitimacy ratios, legitimate and overall infant mortality rates: 26 English parishes, 1580-1837

Date	Illegitimacy Ratio	Legitimate infant mortality rate	Overall infant mortality rate
1580-9	3.46	168.4	174.2
1590-9	4.05	173.0	180.0
1600-9	4.02	165.4	172.0
1610-9	3.46	166.7	172.5
1620-9	2.89	153.4	157.8
1630-9	2.87	160.3	164.9
1640-9	2.27	150.2	153.6
1650-9	1.17	164.3	166.2
1660-9	1.37	169.3	171.8
1670-9	1.68	168.2	171.0
1680-9	1.70	201.9	205.3
1690-9	2.06	174.9	178.5
1700-9	2.31	174.1	178.2
1710-9	2.23	203.3	207.9
1720-9	2.49	193.3	198.1
1730-9	2.78	195.1	200.5
1740-9	3.25	186.6	192.5
1750-9	4.02	159.6	166.0
1760-9	4.98	165.6	173.8
1770-9	5.34	156.8	167.3
1780-9	6.00	163.4	173.2
1790-9	6.12	151.3	160.2
1800-9	6.36	136.6	145.3
1810-9	5.82	133.1	140.8
1820-9	6.06	144.7	153.5
1830-7	6.00	140.4	148.8

Source: Wrigley et al. (1997: 224)

births (Glass, 1973: 197; Kitson, 2004: 229-236). The resulting estimates of infant mortality are shown in Table 4.1 (Wrigley *et al.*, 1997: 224). One striking feature that is evident from these data is that levels of infant mortality in rural and small town English settings were by pre-industrial European standards relatively low. While there was significant geographic variation there was also a noticeable temporal shift in rates. Rates certainly rose in the later seventeenth century to reach a relatively high level through the period 1680-1749. In fact this phase contained the only two decades throughout the whole period from 1580 when IMRs exceeded 200 per 1,000, but even in these decades the estimated rates were significantly lower than

Table 4.2 Infant and early childhood mortality (1,000q_x) in 26 English parishes, 1580-1837

	Days							Years	
Date	**60-89**	**90-179**	**180-273**	**274-365**	**366-457**	**458-548**	**549-730**	${}_4q_1$	${}_5q_5$
a **1580-99**	12.2	25.0	11.9	12.3	10.0	11.3	15.2	84.5	46.3
b **1600-24**	10.1	18.6	11.5	9.5	10.2	9.9	16.3	81.6	36.1
c **1625-49**	8.4	18.7	12.3	13.8	13.0	10.5	17.8	100.0	48.0
d **1650-74**	9.3	19.1	14.1	13.0	13.4	10.8	20.4	111.1	50.9
e **1675-99**	13.1	24.6	16.0	16.6	15.2	13.0	20.0	107.6	45.9
f **1700-24**	14.1	26.1	21.2	17.5	16.6	12.1	19.9	107.9	46.4
g **1725-49**	13.8	26.9	22.5	20.4	22.1	14.1	24.8	121.0	50.1
h **1750-74**	11.6	25.7	22.6	19.1	17.6	13.5	22.6	107.3	41.1
i **1775-99**	11.0	26.6	23.0	17.7	16.1	13.5	23.2	107.7	34.7
j **1800-24**	10.9	23.6	21.0	16.8	15.7	13.1	19.7	98.0	25.7
k **1825-37**	10.7	27.1	27.2	22.3	14.6	15.0	19.7	98.3	34.7
Indexed data[a]									
a-d **1580-1674**	71.6	76.9	56.9	64.2	60.1	80.9	78.2	82.4	93.9
h-k **1750-1837**	79.3	97.1	107.2	100.1	82.6	105.0	95.5	89.8	70.6
Ratio (h-k):(a-d)	1.108	1.264	1.886	1.560	1.375	1.299	1.221	1.090	0.752

Notes: [a] *Index of 100 calculated as ((data for1700-24) + (data for 1725-49))/2*

Source: Wrigley et al. (1997: 252)

those found in most parts of Europe where rates in excess of 250 per 1,000 were relatively common. After 1750 IMRs within the 26-parish sample fell markedly to the lowest levels found in the parish register period in the first two decades of the nineteenth century, after which the final decade before the onset of civil registration revealed an upward drift.

Although infant mortality was such an important component of overall mortality and hence a central factor determining life expectancy at birth it would be unwise to assume that it provided a relatively straightforward predictor of overall survivorship trends. In fact there were periods in which movements in IMRs and other age-specific mortality rates were at best only loosely positively correlated. Early childhood mortality moved in a fashion mirroring that displayed by infants (see Table 4.2, also Figure 1.2 in Chapter 1). However, rates rose earlier for children from 1640 and by the 1680s attained levels that were 50 per cent above those to be found in 1580-9. Despite falling abruptly in the 1690s they recovered to remain through much of the period prior to 1750 approximately 40 per cent above those that held at the close of the sixteenth century. After the mid-eighteenth century childhood mortality rates fell, although in 1820 they were still 20 per cent higher than those to be found in 1600. While infant mortality even in the two most hazardous decades of 1680-9 and 1710-9 never rose more than 20 per cent above the levels found around 1600, infants after 1740 began to exhibit a significant improvement in their survival chances so that by the first decade of the nineteenth century their death rate was 20 per cent lower than that to be found at the end of Elizabeth I's reign.

A more finely tuned analysis (Table 4.2) helps to disaggregate and thereby better understand early childhood mortality trends (Wrigley *et al.*, 1997: 252). In the second year of life the rates changed dramatically between 1580-99 and 1725-49, rising by 60 per cent. The second quarter of the eighteenth century was a period of high mortality in all of the individual years in this age group, although falling to lower levels by the early nineteenth century. In the second year of life the fall after 1750 was slight. A similar pattern was exhibited by rates applying to those in the second half of the first year of life. Rates for this latter age group doubled by 1725-49 and thereafter fell only marginally by the early nineteenth century compared with those applying to infants in the first six months of life. Children and older infants therefore exhibited a deterioration in their life chances that started sooner in the seventeenth century than younger infants and benefited from far smaller improvements over the second half of the eighteenth century.

Adults like infants and children experienced a fall in their survivorship chances over the seventeenth century (Figure 4.1). From about 1700 expectations of life at age 25 began to rise and sustained that improvement through into the early nineteenth century, although this was more marked in the half century before 1750 than the half century afterwards (Wrigley *et al.*, 1997: 281). In fact the period between 1680 and 1749 was decidedly unusual in that infants and children were displaying a significant worsening of their life chances while adults were beginning a significant improvement which was sustained throughout the eighteenth and early nineteenth century (Wrigley *et al.*, 1997: 282-4). Only in the late eighteenth century did infant and adult mortality rates move in an evidently synchronised fashion. Although this association was closer for adults and younger infants in this period it is noteworthy

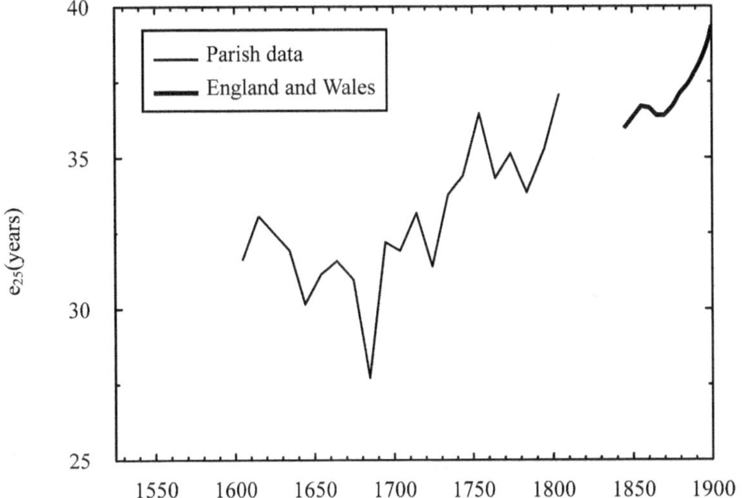

Figure 4.1 English life expectancy at age 25: sexes combined, 1600-1900

that older infants and young children showed a slower and more muted gain in survivorship than did adults. Having noted the significant contrasts displayed by trends in infant mortality in the first and last sixth months of the first year of life we proceed to consider in more detail the changing levels of exogenous and endogenous infant mortality. We know that infant mortality in the first month of life was high and showed a tendency to remain close to 100 per 1,000 live births for most of the period from 1580 to 1750. Thereafter it fell uninterruptedly through time so that by the onset of civil registration in 1837 it was below 50 per 1,000. Neonatal mortality showed an even greater decline over the same period.

The analysis of these developments in early infant mortality is assisted by use of the method of biometric analysis that was first developed by Bourgeois-Pichat (1951). This has been widely exploited by historical demographers since it purports to distinguish between causes of deaths that are only rarely directly provided in the sources. The technique makes use of the distinction that is conventionally drawn between endogenous and exogenous causes of infant deaths. Endogenous causes include the influence of prematurity, the birth trauma itself or inherited genetic defects and factors that may owe much to intra-uterine conditions and the health of the mother. In contrast exogenous causes are those which result from the invasion of the body by external agents, principally infectious disease which could also be a function of the environmental circumstances in which the child lived as it grew older.

Bourgeois-Pichat plotted the rising totals of deaths within the first year of life on a graph in which the horizontal axis representing days over the first year of life was converted to a logarithmic scale.[1] The point of the cumulative totals of deaths from month one to month twelve normally assumes a straight line. Assuming that the overwhelming majority of all deaths from month one are due to exogenous causes

1 $[\log(x+1]^3$ where x is age in days.

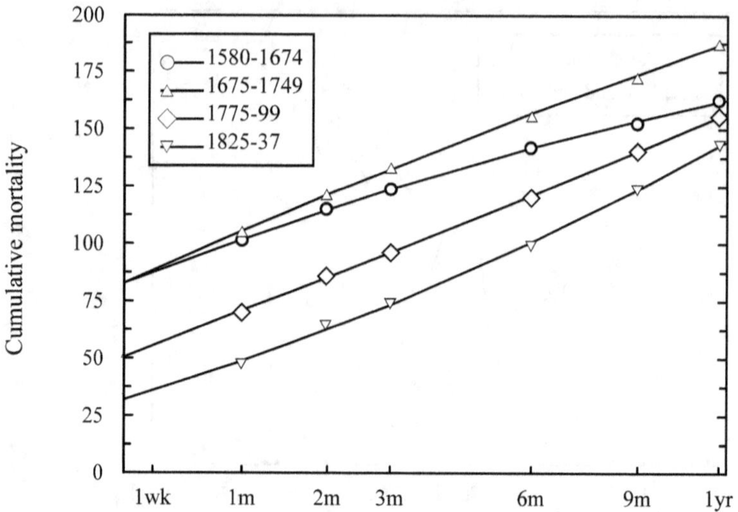

Figure 4.2 Biometric analysis of infant mortality: 26 English parishes, 1580-1837

Source: Wrigley et al. (1997; 227)

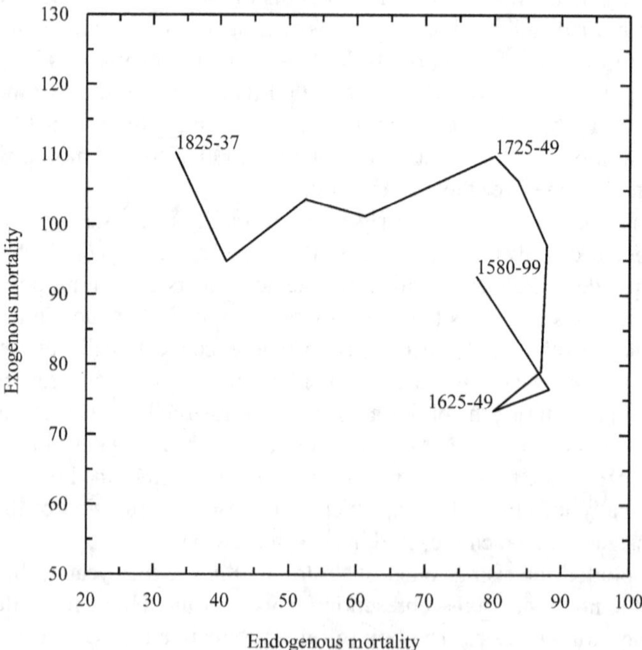

Figure 4.3 The relative movements of endogenous and exogenous infant mortality: 26 English parishes, 1580-1837

Source: Wrigley et al. (1997; 229)

extending the line back to the point at which it crosses the vertical axis suggests that the height of the intercept represents the level of endogenous infant mortality and the remainder of the first month mortality can be allocated to exogenous causes. Using this method and plotting the results in Figure 4.2 reveals some remarkable changes from the late sixteenth to early nineteenth centuries (Wrigley *et al.*, 1997: 227). It shows that there was little change in endogenous infant mortality from 1580 to 1750 although exogenous causes rose and were almost exclusively the reason for the infant mortality rise that occurred over that period. It shows, however, a striking fall in endogenous mortality over the next 80 years by almost two-thirds. Another way of considering the relative movements of exogenous and endogenous infant mortality is shown in Figure 4.3 and it appears that the levels of each category showed a two-phase pattern of shift. Prior to 1750 all change was focused upon exogenous mortality and thereafter hardly any decline in exogenous mortality took place. All the substantial decline in infant mortality that was a marked feature of the period after 1750 can be attributed to a large decline in the rate attributable to endogenous causes (Wrigley *et al.*, 1997: 228-9).

Certain developments in the pattern of infant birth and burial registration that have been frequently discussed by critics of parish registers as historical demographic sources suggest that their potential fallibility needs careful consideration in assessing the likelihood that any changes identified were real rather than the artefact of changing registration practices. A delay in the date of baptism relative to birth might mean that infants deaths could go unreported and those individuals being baptised and exposed to mortality might be more than a month or two in age when they first come into observation, leading to the possibility that changes in death patterns occurring later in the first year of life might be spuriously located in the earliest weeks of life. We have already stressed that the pattern of registration of events in the parish sample used in this discussion was not susceptible in any major fashion to these influences and hence interpretational problems. We have focused previously on certain features of the birth intervals and levels of sterility which suggest that patterns assume a form that might *a priori* be expected in a population that was not yet engaging in family limitation. Furthermore we have used the technique advocated by Louis Henry to plot the distribution of birth intervals differentiated on the basis of the subsequent fate of the child and once again we have discovered patterns that are fully compatible with a belief in the accuracy of the data. Perhaps the strongest body of evidence which suggests that the pattern of declining endogenous infant mortality was real comes from data that the Registrar General collected and reported in the earliest decades following the onset of a system of civil registration in 1837 for a few scattered years in the 1840s (Wrigley *et al.*, 1997: 232). We have used these data to suggest that the similarity of infant deaths in registration districts in which individual parishes in the sample were located leads to the conclusion that few infant deaths were escaping the clutches of the system of parochial registration and therefore made it likely that the overall rates of infant death were accurately measured for the period 1825-37. However if the data assembled by the Registrar General are also analysed using the biometric method they reveal a remarkable similarity with that calculated from the last period analysed using the technique of family reconstitution. The national level of endogenous infant mortality was 27 per 1,000 and that from the family reconstitution sample 33 per 1,000 (Figure 4.4). Assuming that the endogenous infant mortality levels were accurately estimated prior to 1750 it would seem clear that this component of infant mortality had experienced a dramatic decline over the previous 75 to 100 years. What is more, the Registrar General's data for a number of very different communities exhibiting very different overall infant mortality all suggest that whether infant mortality was high or modest or indeed uncharacteristically low there was a level of endogenous infant mortality that was remarkably similar in all of them. Whether the community in question was an urban area such as Liverpool or an unhealthy low-lying registration district such as Ely, in Cambridgeshire, with IMRs in excess of 200 per 1,000, or the Healthy Districts as defined by William Farr with IMRs slightly below 100, or the Devon registration district of Bideford with overall infant mortality below 60, endogenous rates tended to fall within the range of 25 to 35 per 1,000. Indeed the same levels of endogenous infant mortality were to be found among those unfortunate infants born outside of wedlock whose overall infant mortality reached levels twice the national rate.

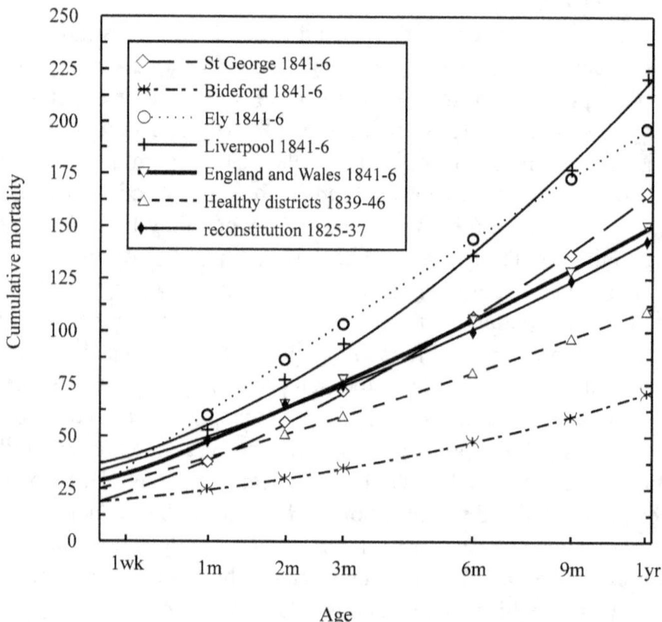

Figure 4.4 A comparison of endogenous and exogenous infant mortality in the last years of reconstitution data with data from the Registrar General's returns for the 1840s

Source: Wrigley et al. (1997: 232)

These striking geographical contrasts in overall infant mortality and hence exogenous infant mortality but decidedly invariant levels of endogenous infant mortality constitute a very important finding. Likewise the discovery that illegitimate infants were in no sense disadvantaged in their early existence by a higher incidence of endogenous causes of death than those born in wedlock is *a priori* a surprising result of evidence collected by the Registrar General (Wrigley, 1977: 306). Environmental and social status or material conditions may indeed have served to influence the levels of exogenous infant mortality but would appear to have a rather limited impact upon the level of endogenous infant mortality.

To better understand these developments a fuller discussion of geographical and social status differences in infant mortality is ideally necessary. In the remainder of this chapter an assessment of the still limited quantity of such evidence that is available will be attempted, beginning with some further interrogation of the 26-parish sample of family reconstitutions. Changes in individual parish register estimates over the periods from 1675 to 1844 have been assessed by utilising estimates from the family reconstitutions and those obtained from the earliest years of civil registration relating to the registration districts within which the parishes were located (Wrigley *et al.*, 1997: 268-73). While this is not an ideal procedure for capturing the geographically finely grained patterns of mortality change, some suggestive results do emerge.

Table 4.3 Infant mortality in the 26 parishes in 1675-1749 and in the registration districts in which they were located in the 1840s

Parish	Reconstitution parishes 1675-1749	Registration districts 1838-44
March	311	195
Gainsborough	270	141
Great Oakley	269	123
Lowestoft	246	132
Alcester	236	133
Willingham	222	138
Banbury	209	139
Bottesford	167	144
Earsdon	163	146
Terling	176	139
Southill	172	154
Reigate	147	103
Shepshed	139	178
Methley	134	164
Morchard Bishop	131	87
Aldenham	130	154
Birstall	128	170
Colyton	125	105
Ipplepen	125	98
Ash	121	107
Austrey	118	131
Gedling	112	159
Dawlish	108	98
Odiham	96	101
Hartland	94	80
Bridford	92	112
Agricultural	124	113
Manufacturing	126	169
Trade & handicraft	240	136
Other	139	127

Source: Wrigley et al. (1997: 270-271, 274)

In Table 4.3 parishes have been ranked according to their IMRs for the period from 1675-1749. These ranged from the highest rates in the Fenland parish of March, in northern Cambridgeshire, to Bridford, which was one of a group of Devon parishes with exceptionally low infant mortality for the period. It is clear that the highest rates of infant mortality prior to 1749 were displayed by the low-lying marshland parishes – March, Willingham and Great Oakley – and the market towns – Gainsborough, Lowestoft, Alcester and Banbury – each one a location with rates significantly in excess of 200 per 1,000. Some rural parishes, particularly those not located in low-lying localities had IMRs that by the standards of most communities in much of Europe at the beginning of the nineteenth century were unusually healthy with IMRs below 120 per 1,000. Wrigley *et al.* (1997: 275) have introduced some order into these parochial data by excluding the three low-lying marshland parishes and grouping the remaining 26 parishes into four categories based upon their occupational structure as revealed by the 1831 census: those in which the proportions of the adult male labour force engaged in agriculture exceeded 60 per cent in the 1831; those in which the proportions in manufacturing exceeded 30 per cent; those in which those employed in the retail trades and handicrafts exceeded 40 per cent; and all remaining parishes not falling into any of the other three categories.

The manufacturing group, unfortunately small in number for the purposes of the analysis, nonetheless reveals a noteworthy deterioration in IMRs over time. The development of domestic industry and the population growth associated with it drove down infant survivorship chances. In the settlements with a sizeable retail and handicraft sector, as might be expected of market towns, there was a remarkable decline in mortality rates among infants. Such places in the late seventeenth century had been among the most unhealthy settings of early modern England, but by the late Georgian period were achieving mortality rates for those in their first year of life that were characteristic of the majority of agricultural parishes that were not low-lying a century earlier. Agricultural parishes and those that were not industrialising or market towns showed little net change in infant mortality or in the probability of dying between birth and age 15.

Currently we lack a detailed analysis of trends in endogenous and exogenous mortality across time in all of these sample parishes although there is highly suggestive evidence in the data collected by the Registrar General for the whole of England and Wales for the 1840s that the level of endogenous mortality had fallen markedly across the preceding century. As we have already demonstrated this decline is mirrored in the aggregated dataset relating to the parish sample. We can be reasonably confident therefore that endogenous mortality had fallen in each of the sample parishes. The differing trends revealed by those parishes might be regarded principally as a function of ways in which exogenous IMRs had changed in the different parish groupings. We might therefore propose that exogenous infant mortality had risen markedly in the manufacturing parishes. These were often associated with increased population densities and overcrowding. The most intriguing case concerns the urban centres which experienced what would have been a significant decline in exogenous infant mortality. This intensified the overall fall in infant mortality, assisted by the declining levels of endogenous mortality.

The late eighteenth century fall in exogenous infant mortality occurring in urban centres remains one of the unsolved mysteries of urban population history. Improved physical environments, investment in the housing stock and greater attention to public amenities may to varying degrees provide part of the explanation. A possible rise in exogenous mortality in rural settings is also a likely development that our evidence reveals and is not easily explained. There may be reasons for supposing that living standards in the rural sector of the economy were under greater pressure, given what is known of wages paid to farm labourers and expenditure on poor relief, but the exact link between IMRs and living standards remains far from clear since at a national level throughout the period an inverse correlation between mortality at all ages and real wages is so decidedly absent.

A geographical approach to these issues is potentially important since it should also be stressed that towns and cities may have been settings in which the seventeenth- and early eighteenth-century rise in exogenous infant mortality (especially mortality in the latter part of the first year of life) and very early childhood were most marked so that there may have been more potential for subsequent improvement in those age groups in such environments. Developments in London are especially germane to this issue. Family reconstitution is a technique that does not adapt well to use on the parish registers of larger urban centres. A city of the size of early modern London made up of a large number of geographically relatively small parishes containing populations characterised by high rates of turnover creates a setting in which it is very hard to hold individuals in observation for sufficient time to generate demographic measures relating to them. As a result work on London's demography has made relatively limited use of nominative linkage techniques. Nonetheless the relatively small amount of research that has been successfully undertaken has yielded findings that are helpful for the present discussion.

The most reliable family reconstitution-based study of metropolitan infant mortality levels over the relevant period has been conducted by John Landers (1993) and it makes use of the records of the Quaker meetings of Southwark and 'Peel'. Southwark covered the built-up area south of the river Thames including Rotherhithe and Bermondsey as well as Southwark itself and the parishes of Lambeth, Camberwell and Newington Butts. Peel meeting covered a less well-defined area consisting of parishes lying in a north-western quadrant outside the city walls and incorporating Clerkenwell. In combined populations from these two meetings Landers (1993: 136) showed that infant mortality worsened from a rate of 251 per 1,000 in the period 1650-74 to 263 in 1675-1699 and reaching 342 in 1700-24 (see Table 4.4). IMRs were 36 per cent higher in the quarter century between 1725 and 1749 than in that between 1650 and 1674. The rise in infant mortality among the London Quakers was therefore proportionately greater than that exhibited by the non-metropolitan sample represented by the 26 parishes. Early childhood mortality rose relatively further. However infant mortality fell quite sharply after 1750 and by 1825-49 was only 60 per cent of the level prevailing in 1650-74. Mortality in the second year of life also fell but not to the same extent as for infants. Exogenous infant mortality rose to high levels for much of the period up to 1774 when it was 50 to 60 per cent higher than in 1650-74. There was little increase in endogenous infant mortality but after 1750 it declined sharply so that by 1824-49 the rate was only 18 per cent

Table 4.4 **Age-specific mortality for specific age groups 0-9 years, plus endogenous and exogenous infant mortality among London Quakers, 1650-1849; and values indexed against 1650-1674**

	Mortality rates by age									
	Months				Years					
Cohort	0	1-2	3-5	6-11	0	1	2-4	5-9	Endogenous	Exogenous
1650-74	108	152	51	70	251	103	190	66	76	175
1675-99	115	158	46	82	263	113	132	69	80	183
1700-24	125	197	59	130	342	145	177	89	75	267
1725-49	112	204	58	121	341	143	186	109	81	260
1750-74	96	168	82	119	327	150	159	91	43	284
1775-99	81	114	38	80	231	101	141	32	48	183
1800-24	40	53	41	95	194	93	85	79	27	167
1825-49	33	33	37	76	151	77	93		14	137
	Index values									
1650-74	100	100	100	100	100	100	100	100	100	100
1675-99	106	104	90	117	104	110	69	105	105	105
1700-24	116	129	115	186	136	141	93	135	99	153
1725-49	104	134	114	173	136	139	98	165	107	149
1750-74	99	111	157	170	130	146	84	138	53	162
1775-99	75	75	75	114	92	98	74	48	63	105
1800-24	37	35	80	136	77	90	45	120	35	95
1825-49	31	22	63	109	60	74	49		18	78

Source: Landers (1993: 136, 140)

of that found in the mid-seventeenth century. It would appear that London Quakers exhibited trends in infant and early childhood that were quite similar to those found elsewhere but in certain respects the patterns assumed a more extreme version of those generally found. Mortality of Quaker infants worsened more markedly as a result of exogenous mortality rises, and fell more markedly as a result of a very sharp decline in endogenous mortality, assisted by some fall in the level of exogenous mortality. Exogenous infant mortality fell by 22 per cent from the level prevailing in 1650-74, whereas nationally there had been a rise of comparable proportions within this component of infant mortality.

In presenting these findings Landers commented on a somewhat less robustly determined set calculated by Roger Finlay (1978) from six London parishes. These indicated a significant rise in infant mortality over the second half of the seventeenth century, although the aggregate figure that he obtained for the 1690s was 185, having risen from 157 per 1,000. The higher rate is still below 200 per 1,000 and might not seem to be so striking when set alongside estimates for areas outside the metropolis which reached values closer to 190 per 1,000 at this time. At a comparable date Landers noted, however, that Finlay's aggregate figures may have been an underestimate since the two poorest parishes in his sample, St Mary Somerset and St Botolph Bishopsgate displayed an apparent improvement against the general trend. Had the rates for these parishes remained as they were at mid-century the rate for Finlay's sample in the 1690s might have reached 250 per 1,000 – a rise of nearly 50 per cent over the rates found around 1630. A recent attempt (Smith and Newton, forthcoming) to reconstitute the parish registers of a group of wealthy Cheapside parishes – Allhallows Honey Lane, St Mary le Bow, St Pancras Soper Lane, St Martin Ironmonger Lane and St Mary Colechurch – reveals that infant mortality rose from 195 per 1,000 from 1670-99 to 299 per 1,000 between 1700 and 1721. This rise of 53 per cent is more marked than the increase displayed by the London Quakers. It is worth stressing too that this rise took place in parishes that had undergone very substantial rebuilding and general refurbishment following the severe damage they experienced as a result of the Great Fire. When account is taken of the environmental characteristics of the parishes and the fact that Cheapside was a wealthy district and that the decline of infant survivorship chances was on a par with or actually exceeded those observed by Landers for the London Quakers – a relatively well-off, but certainly less well positioned group than the bulk of the parishioners in the 26-parish sample – we may have assembled evidence which would seem to suggest that environmental factors *per se* were not principally responsible for the infant mortality rise that took place over the late seventeenth and early eighteenth centuries.

It is useful to situate Landers' findings regarding infant mortality among the London Quakers alongside an analysis by Laxton and Williams (1989: 124-6) of the London bills of mortality from 1728 to 1842. The London Bills of Mortality state the number of deaths of children in the first two years of life and any estimate of infant mortality based upon them must therefore make an assumption about the proportion of total deaths under two which referred to infants under one (see Woods, 2006b). Laxton and Williams made maximum and minimum assumptions about this proportion and thus maximum and minimum estimates of the IMR. The rates calculated suggest an infant mortality that was as high as 350-400 per 1,000 in 1728-

42 when Quakers rates were at their highest and also close to 350 per 1,000. Recent analysis (Smith and Newton, forthcoming) of the large extra-mural but rapidly growing suburban parish of Clerkenwell indicate a rate there of 380 per 1,000 in 1734-1753. There seems good reason to suppose that Laxton and Williams' estimates are plausible. They can at least be favourably compared with the Registrar General's data for London in the 1840s to which they were very similar (Wrigley *et al.*, 1997: 256-7). At that time London's IMR was 160 per 1,000 but had recently risen from a low point in the second decade of the nineteenth century when rates may have been as low as 120 per 1,000. The improvement in London's infant mortality by a factor of more than three over the preceding three-quarters of a century was a remarkable development. We cannot be sure that the changes in endogenous and exogenous infant mortality that are found among the Quakers was equally characteristic of the larger London population of which they were part, but we also have stray pieces of evidence that seem to be supportive of similar developments which were themselves not unduly influenced by factors to do with wealth or social status. We have noted the sharp decline in endogenous infant mortality among the London Quakers after 1750. The biometric plots of infant deaths in Figure 4.4 reveal that in 1841-46 in St George's in the East, which was a poor area of East London, infant mortality was around 150 per 1,000 and the endogenous rate no greater than 30 per 1,000. We may suppose that this implied fall in endogenous infant mortality was a significant contributor to the decline in infant mortality although insufficient by itself to account for the whole of the improvement in infant life chances, since a significant fall in exogenous infant mortality also occurred. That this fall in exogenous infant mortality was not restricted to those who were wealthy or resided disproportionately in the wealthier areas of the city is important to stress since in this respect the period of declining infant mortality resembles those years during which infant mortality rose so sharply after 1675 when wealth and high status offered little protection against these damaging trends. That the foundlings who were admitted to the London Foundling Hospital after 1740 should also reveal, once the high mortality of the General Reception period from1756 to 1760 is excluded, a fall in their death rates in the later eighteenth on a scale not disimilar to the falls shown by the Quakers and the population captured by the Bills of Mortality is indicative of the social reach of the improvements in infant mortality that characterised the 80 year period from the mid-eighteenth century in the capital (Levene, 2005).

Clues to an explanation of some of these developments might reside in a closer consideration of the links between early infant deaths and maternal mortality. As noted a sizeable proportion of endogenous infant deaths were related to the birth trauma itself and the days immediately following the birth. One might expect there to be some association between levels of maternal mortality and endogenous infant mortality. There is no space to discuss the complex issues that surround the estimation of maternal mortality rates when parish registers are used but Roger Schofield (1986) has pioneered this investigation using techniques that both enable adjustments for missing events as well as making use of information from other European societies during the eighteenth century. It is noteworthy that, over the same period during which endogenous infant mortality dropped so sharply, maternal mortality was also falling. Table 4.5 shows the maternal mortality rate for successive

Table 4.5 Maternal mortality, early infant mortality and endogenous infant mortality: 26 English parishes, 1580-1837

Cohort	Maternal mortality	Infant mortality		Endogenous mortality
		0-6 days	0-29 days	
1580-99	12.3	63.8	101.9	77.6
1600-24	12.8	75.1	108.2	88.5
1625-49	14.0	68.5	94.5	80.0
1650-74	17.0	76.3	104.2	87.3
1675-99	15.6	78.3	109.7	88.3
1700-24	13.4	64.5	106.3	84.0
1725-49	12.3	62.7	101.6	80.5
1750-74	9.5	50.2	78.5	61.3
1775-99	9.0	47.2	71.3	52.6
1800-24	6.3	33.6	57.3	41.0
1825-37	4.7	22.6	48.7	33.3

Source: Wrigley et al. (1997: 236)

quarter centuries beginning in 1580 together with the endogenous IMR as well as the rate for the first month and the first week of life (Wrigley *et al.*, 1997: 236). There is a strikingly good fit between the maternal mortality rate and the mortality rate amongst infants in the first week of life. The endogenous rate is also strongly correlated. Wrigley *et al.* (1997: 236-7) have noted that the similarity between trends in maternal mortality and neonatal mortality suggests the possibility of a common influence. They ruled out shifts in prematurity/low birth weight or inherited defects as factors responsible for these changes but wondered whether more effective midwifery practices may have worked to the benefit of the mother and the child. In a subsequent study Wrigley (1998) extended this analysis to make better sense of an apparent rise in marital fertility in the eighteenth century which he concluded was a likely consequence of falling stillbirth rates. Since stillbirths would not, in the vast majority of cases, have been recorded in parish registers, the increasing registration of live births would appear as a rise in fertility rather than as a decline in the stillbirth rate. In this study Wrigley is willing to entertain the possibility that since birth weight is a vital determinant of pre-natal infant mortality the improvement in stillbirth rates was evidently a consequence of a rise in maternal net nutrition. He suggests that this may also be reflected in a general improvement in nutritional status that correlates well with Fogel's (1989) findings regarding an increase in adult heights in both men and women. While Wrigley is cautious not to overstate this emphasis, if it were likely to provide a framework within which to assess not only falls in the stillbirth rate but also in the neontal and maternal mortality, we will need later in this chapter to subject this argument to detailed social and geographical scrutiny.

We have somewhat limited evidence currently available to engage in a comprehensive assessment of social status difference relating to many of the issues we have discussed so far, but there are aspects of the demographic patterns revealed by the British peerage that provide some helpful pointers in this debate. Demographic data relating to the British peerage were first published by Hollingsworth in his classic work of 1964 (see also Hollingsworth, 1977) from which valuable findings were unearthed concerning the peers' mortality experience. In the forty years since the peerage data were first reported there have been massive advances in both computing power and techniques for mortality estimation from incomplete data. Dr Hollingsworth has given us access to his original data and these have been entered into a database as part of a wider study of historic mortality trends. As a result it has proved possible to make a number of re-estimations as well as to generate new measurements that would have been difficult to generate with the data in their original state. There is insufficient space in the current discussion to detail these procedures but two points can be made. Firstly, Hollingsworth only used individuals who were born into the dataset. We have used techniques which allow individuals to enter the risk-set at any age, without biasing the results. Secondly, we have taken advantage of major advances concerning survival estimation techniques. In 1976 Turnbull published a method that allowed deaths to be interval-censored on the right – that is, it was possible to specify a 'window' defined by two dates within which a death must have occurred if it was not known exactly, and to make efficient and consistent use of this information. The more recent surge of research into incomplete data estimation caused by the emergence of AIDS offered a means of determining an equivalent window on the left that sets the dates between which a death must have occurred (Heisey and Nordheim, 1995; Sun, 1995). Datasets with 'windows' on both ends of the event-history and people joining at different ages are described as 'double-censored and left-truncated'. In AIDS studies, not knowing exactly when a person was first infected is equivalent to not knowing exactly when a peer was born. Hollingsworth was aware of the possibilities of tackling these issues but did not implement any adjustments when undertaking his study. The evidence reported here incorporates the results of applying these techniques. It should also be stressed that all violent deaths have been treated as if they are right-censored, or died at some unknown point after the actual death, which would be equivalent to assuming that the individuals concerned migrated in a conventional study. This problematic assumption is made to facilitate comparison with populations in which this category of death would be minimal.

As far as possible comparisons will be made between the peerage and the 'population' created by amalgamating the 26 parish family reconstitutions. Figure 4.5 shows married female life expectancy at age 25 for the peerage. Data from the parish family reconstitutions are plotted in the same figure and relate only to married persons since the technique and the data available do not reveal information on the mortality experiences of the unmarried. The slight decline in adult female life expectancy among the married population of both status groups is closely synchronised before 1700 as is the steady improvement that occurs in an almost unbroken fashion after

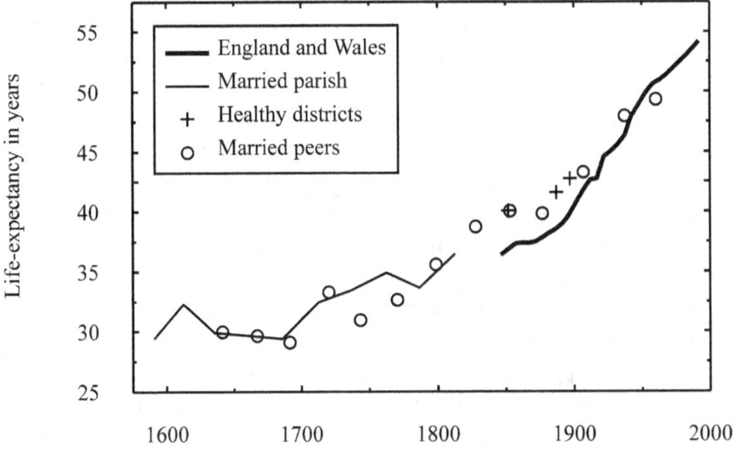

Figure 4.5 **Female life expectancy at age 25: cohort data plotted at mean age of death for British Peerage and English parish populations and England and Wales**

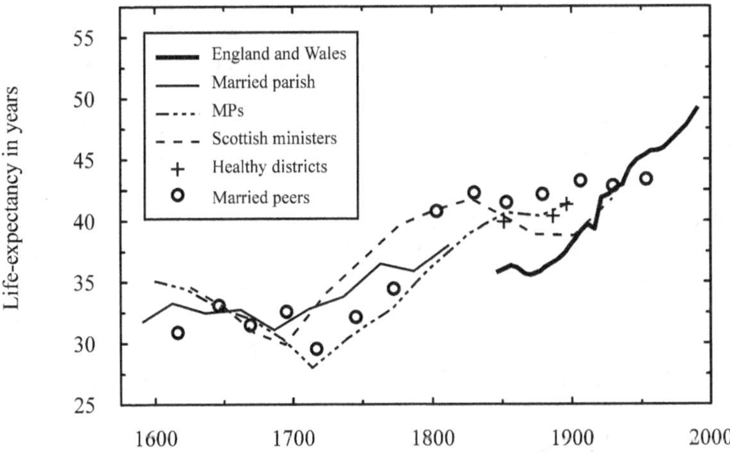

Figure 4.6 **Male life-expectancy at age 25: cohort data plotted at mean age of death for British Peerage, MPs and Scottish Ministers, English parish populations and England and Wales**

1700. There is an emerging gap as peeresses pull away slightly from the general population from the third decade of the nineteenth century, although at no point do married peeresses outperform women of a similar age for the total population resident in the Healthy Districts identified by William Farr to be unusually salubrious later in the nineteenth century (Woods, 2000: 183). Married male peers also follow the course of mortality change revealed by the general population of married males but clearly trends are less closely paralleled than was the case with the wives of both status groups (Figure 4.6). For comparative purposes adult life expectancy measures for Ministers of the Church of Scotland and Members of Parliament are also shown in Figure 4.6, although neither of these groups is distinguished by marital status. The males of all of the status groups in Figure 4.6 show an improvement in survivorship after 1700 but from some point in the early nineteenth century the peers, ministers and MPs begin to outperform married males in the general population. But all status groups show a slow down in the pace of improvement, even a slight decline in the mid-nineteenth century. The gap between adult mortality in the Healthy Districts and the national figures and the gap between married males in both status groups suggest that the urban/industrial penalty was higher for men than women in the nineteenth century. However, the patterns, particularly when account is taken of the evidence from the Healthy Districts, suggest that these trends for the elites are more readily explained by arguments of epidemiological externality since we cannot invoke the usual factors of personal poverty, occupational risk, or poor housing in their case. If there was an improvement in nutritional status that was driving these developments then it might be more appropriate to focus on the claims on body energy as a result of a common exposure to epidemic disease working its influence across the whole social spectrum than an increased energy input arising from better diets or improved physical environments.

Such an interpretational position may seem more acceptable when the pattern of maternal mortality is considered. Although some cause of death data are recorded on the peeress forms created by Hollingsworth, it is sporadic and generally restricted to violent death and death in childbed. However there is little reason to suppose that death in childbed was fully reported. As with the parish sample we are forced into using the indirect evidence presented by the association between a child's birth and the death of a mother. The difficulties this presents for estimating maternal mortality in the English parish data have been rehearsed elsewhere (Schofield, 1986). The problems with the peerage data are compounded by the likelihood that births associated with infant deaths are probably under-recorded before the mid-eighteenth century and that some dates are indirectly known. In the current analysis therefore only women with exactly known dates of death and exactly known childbirth dates are selected. It is hard to know how, if at all, this restriction may bias the results obtained.

Figure 4.7 contrasts the peerage results with the parish population, indicating two standard deviation limits. After 1700 it is clear that the peerage experienced higher maternal mortality than the parish women and this is true over the whole period. One widely promoted interpretation of the fall in maternal mortality has been seen to be the result of improved obstetric practice (Wilson, 1995). However, Figure 4.8 shows the relative risk for the mother compared with the risk of death for the father in the 60 days after the childbirth in the parish data. Using the risk for the father as a denominator is an attempt to control for non-maternal risk within the family and for the background level of adult mortality, which changed over time. The results suggest that while the risk is indeed higher at about 6 times the 'paternal' risk, there is little evidence of a trend, which indicates that maternal mortality is simply background mortality 'writ large'. The alternative conclusion is that any improvements simply allowed maternal mortality to keep pace with the general decline in adult mortality.

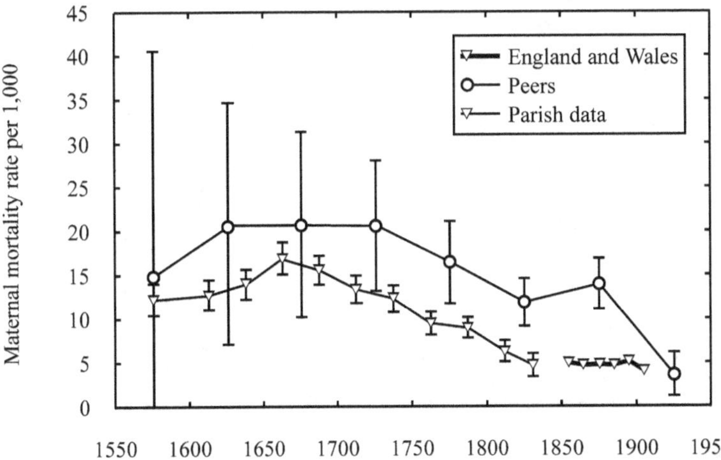

Figure 4.7 Maternal mortality among English parishioners, British Peers and England and Wales

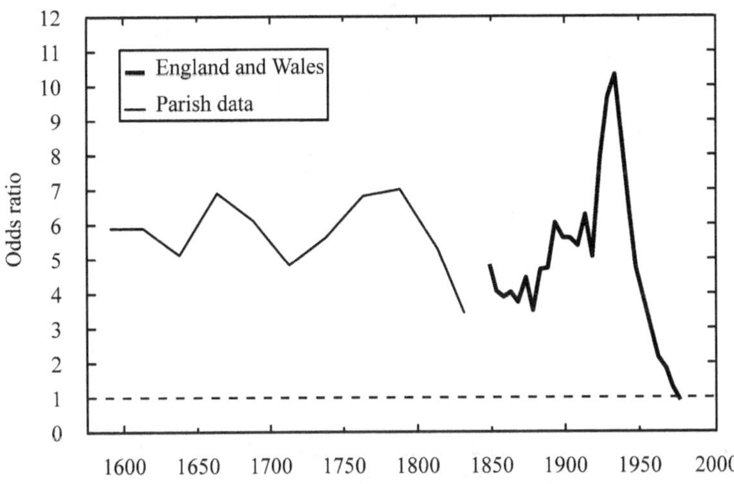

Figure 4.8 Relative risk of mortality: maternal over paternal among English parishioners and England and Wales

Unfortunately, the numbers in the peerage dataset are too small to repeat this calculation. However it is possible to compare the risks between the two groups of women. Figure 4.9 shows a rising relative risk for the peeresses over the general population across the periods only returning to parity in the twentieth century. If it is true that elite women were more likely than the general population to be assisted with their births – perhaps we would say interfered with – and that this was an

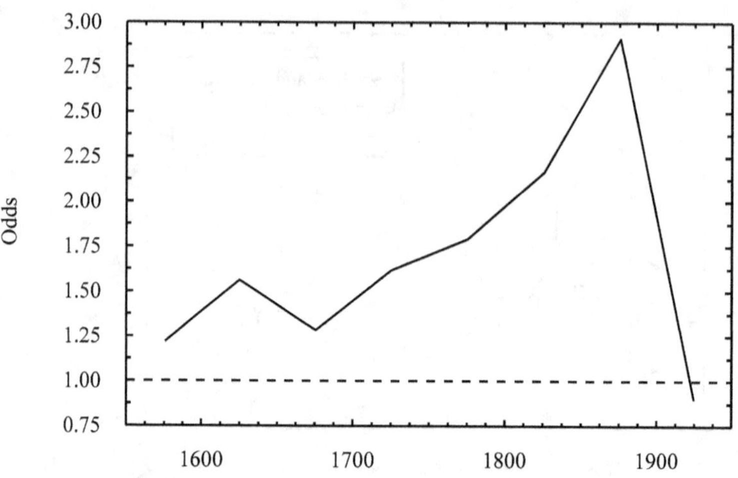

Figure 4.9 The relative risk of maternal mortality for British Peers

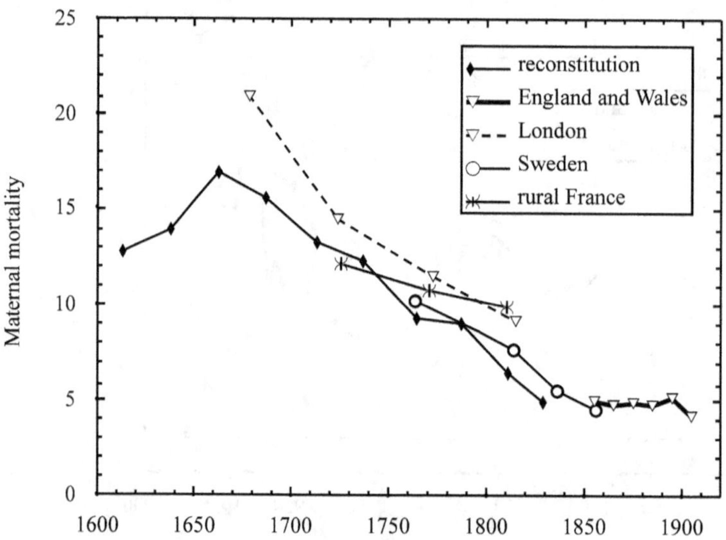

Figure 4.10 Maternal mortality: English data and international comparisons

increasing tendency over time, then these results support Loudon's (1992) view that assistance has a negative impact on maternal survivorship in childbed in the past. Contemporary interpretations seem to have focused on the weakness of the elite mothers (Lewis, 1998). However, we would stress the similarities in the trends exhibited by elite and non-elite rates of maternal mortality. Both exhibit a tendency to rise over the seventeenth century before embarking on a decline through the 'long eighteenth century'. In that respect they reflect trends that are detectable in other populations shown in Figure 4.10. Maternal mortality in the London population recorded in the Bills of Mortality shows higher rates throughout, but over the period when they can be compared maternal mortality rates are quite close to those displayed by the British peeresses. Another distinctive feature of Figure 4.10 is that trends and levels in maternal mortality for the English parochial population, Sweden and rates derived from a sample of French parish reconstitutions are strikingly similar. It might therefore be important to stress the similarity of level and simultaneity of trend regarding maternal mortality rates in these populations notwithstanding their very different environmental settings – that is their levels of industrialisation and urbanisation – and levels of per capita economic well being.

Unfortunately we lack maternal mortality time series as long as those now available for the English parish and peerage populations but it is important to note not only the fall in rates over the eighteenth and early nineteenth centuries but also the rise during the seventeenth century. There is no reason to believe that obstetrical practices deteriorated in the seventeenth century. As we have noted previously the general level of mortality deteriorated, with particularly that of infants and young children suffering a decided worsening. It is now reasonably well known that women in their third trimester of pregnancy, or newly delivered, are several times more likely than others of the same age to become infected by, and to die of, diseases that were common such as tuberculosis, smallpox and influenza (Schofield, 1986: 254). This vulnerability arises from the fact that pregnancy suppresses cell-mediated immunity in a way which permits foetal retention, but interferes with resistance to specific pathogens. Such pregnancy associated immune deficiency could help to explain the contrasts between maternal mortality rates in a metropolitan centre such as London, as well as the greater likelihood of exposure to streptococcal infection that married women in the peerage experienced as a result of interference by midwives and general practitioners. However, we might also suppose that the pathogenic load to which mothers were exposed may have declined over time after peaking in the early eighteenth century. Furthermore we might also entertain the possibility that mothers in both status groups may have achieved a greater immunity to those pathogens that advantaged them during the third trimester of their pregnancy.

In support of this viewpoint we may add just one final substantive consideration which is stimulated by recent work on past stillbirth rates by Robert Woods (2005). Woods has subjected Wrigley's (1998) pioneering attempt to estimate stillbirth rates to critical scrutiny and has carefully considered datasets relating to late-foetal mortality from certain Scandinavian countries. In addition he has re-estimated English stillbirth rates using data from the 1930s to recalibrate the relations between stillbirth rates, early neonatal and maternal mortality. Wrigley, by ingeniously using neonatal/endogenous IMRs as predictors, had concluded that there was a dramatic fall in stillbirth rates over the 150 years after 1670 dropping from rates between 100 and 120 per 1,000 to levels closer to 40 per 1,000 and thereby enabling him to make sense of an apparent rise in marital fertility rates occurring over the same time period. Woods' new estimates suggest a more muted degree of improvement of stillbirth rates from the seventeenth to early nineteenth century than was proposed by Wrigley. In addition the new estimates locate the English stillbirth rates and patterns of change more firmly within what Woods (2005: 157) views as the 'temporal and geographical experience of northern, Protestant Europe in the past'. Woods proposes a decline in stillbirth rates from levels of 70-75 to 40-45 per 1,000 between the late seventeenth and early nineteenth centuries. We have already seen that in the matter of maternal mortality in the eighteenth and early nineteenth century levels and trends in England and Sweden mirrored each other very closely. One might therefore propose that a close fit between stillbirth rates in these societies would therefore be expected. Furthermore, in concluding comments to this analysis Woods (2005: 160) states that 'changes in disease environment were more important than variation in the nutritional status of mothers in influencing the pattern of early age mortality'. Such a sentiment accords well with much of the evidence presented in this chapter relating both to the life chances of infants and their mothers.

We are handicapped in further developing our discussion by the far from secure estimates of infant mortality that we can derive from the genealogical data from which our estimates for the British peerage are derived. Genealogical data are often prone to inaccuracy and bias. A recording system concerned with male primogeniture is likely to have paid less attention to females and non-heir males and children who died young. Hollingsworth (1977) nonetheless regarded these data as superior to those extracted from parish registers, but concluded that before about 1750 a certain and probably significant number of children who died young were not recorded. Before 1750 his estimates of infant and child mortality under age 5 are raised markedly by the estimated numbers of missing births and infant burials that he added. These inflated estimates are in broad agreement with the levels from the 26 parish sample, although they suggest an earlier fall in infant and early childhood mortality following a period of rising age-specific death rates. From the late eighteenth century they diverge far more markedly when the peerage rates fall more sharply. The peerage estimates for ages 5 to 15 are a remarkably close match with the parish estimates over the whole period. We are as a consequence of data lacunae and uncertainties unable to subject the peerage data to biometric analysis. However, given what we have found from other sources there would be reason to suppose that some of the mortality fall in the first year of life among the peers would owe a good deal to improvements in early infant mortality in the late eighteenth and early nineteenth century. We might also suppose that they are far less likely than other groups to have been exposed to rising exogenous mortality after 1750. Indeed the probable resort to maternal breast feeding by many members of this group who had previously shunned the practice (Lewis, 1998) will have also gained them an added boost to their infant survivorship – as such a behavioural change may well have done to London populations at the same time (Landers, 1993: 357) – which would have helped still further to lower their mortality relative to a general population where we have no reason to believe there were significant changes in maternal feeding practices. Whether the peerage were also able to benefit from some of the advantages that were also accruing to urban residents more broadly in the later eighteenth century is difficult to detect, but it may well be the case that as a group they spent a significant amount of time in London as the eighteenth century progressed, and as we have seen, infant mortality in the capital went through a decline far exceeding that found in other non-urban or industrialising settings.

In reflecting on the course of infant mortality changes we have invoked changes in exposure to infection as a primary influence eventually enabling infant peers to secure a greater improvement in their survival chances than was found to be the case within the general population, although never giving them an advantage over those of the general population resident in the Healthy Districts. However, in many respects both status groups were likely to have been subject to similar influences bearing upon their exposure to infection. We should also give serious attention to the fact that adult mortality among both aristocratic and parochial males and females fell at approximately the same time from the early eighteenth century. The fact that this coincides with a phase when in the general population infant mortality and early childhood mortality remained high or rose still further suggests that infants were increasingly exposed to infections that were taking a toll of their early lives. However those who survived may well have acquired immunities that were a factor in sustaining improved longevity subsequently through their teens and into their adult years. It has been suggested that as a result of an increasingly spatially integrated economy (Kunitz, 1983), in part revolving around the dominant role played in the articulation of goods, peoples and diseases by London, within not only a national but also an international economy, a growing share of the population was likely to have acquired immunity to infection as a result of prior exposure (Landers 1993: chapter 9). Adults, of course, would be the first to experience this advantage, followed by older children and very young children and infants last of all.

It would be in this context that mothers would have become less vulnerable to infections and hence transmitted this benefit to their conceptions both at a foetal and early infant stage. Consequently we might expect this benefit to have contributed most markedly to the lowering of neonatal and endogenous infant mortality and to some extent to the stillbirth rates. Some interpretations would wish to give a prime place in this story to an immunity-granting disease such as smallpox and possibly also some role to inoculation (Razzell, 1977). Undoubtedly these aforementioned influences may have been contributory factors of note but it would be unwise to become too focused on any one influence, given the broad geographical extent and temporal simultaneity of many of the changes we have observed. The decline of endogenous infant mortality would appear to have been geographically remarkably widespread within England before 1830 and some of the changes in stillbirth and maternal mortality may have occurred more generally over a large tract of northern and north-western Europe.

While this chapter has assumed the appearance of a *jeu d'esprit* in covering such a large sweep of time and also inclined at times to mobilise speculatively material from a broad array of contexts, it ends by stressing the need to give far greater attention to fine-grained analysis which differentiates more carefully between infant age groups and those in early childhood as well as looking more systematically at the links between maternal mortality and the component parts of foetal and infant mortality across a broad spectrum of social status groups and environmental settings. Furthermore, our arguments in this area of mortality analysis cannot continue to be confined to interpretations that rely exclusively on English data, since England was part of a larger and geographically specific epidemiological regime that extended over a substantial part of northern or north-western Europe. Our knowledge of these issues is still far from adequate for the period between 1550 and 1850 both in England and in her closely located European neighbours. Perhaps the example set by George Newman a century ago can still serve to inspire far more attention to these matters than they have received hitherto.[2]

2 The authors wish to thank Max Satchell for his help in producing the figures for this chapter.

Chapter 5

A Double Penalty? Infant Mortality in the Lincolnshire Fens, 1870-1900

Sam Sneddon

Introduction

Nineteenth-century infant mortality is a subject that has attracted a great deal of attention, both among contemporary and modern researchers. Reflecting its distinctive epidemiological regime, it has been conducted essentially as a specialist subject in its own right (Garrett *et al.*, 2001; Woods, 1995; 1997).

Contemporaries' burgeoning interest in infant mortality as a key indicator of the nation's health was prompted by the economic and political climate of Victorian England, which had invoked a keen interest in the size and health of the nation's population (see Chapter 2 in this volume). In contrast with Newman, who examined local variations in infant mortality, many contemporaries concentrated on national trends. In part, despite the existence of a vast array of Victorian demographic material, full exploitation was limited due to a lack of sufficiently advanced computerised methods capable of coping with the magnitude of data produced during this period.

Although the calculation of national trends and averages of infant mortality can provide a useful guide, subsequent researchers have often found them to be 'misleading' (Johansson and Kasakoff, 2000: 56-58). In turn, this has prompted inaccurate explanations of the trends and patterns of mortality experienced within the nation, where averages and trends 'mask as much as they reveal' (Smith, 1979: 65). Bearing this in mind, the most important aspect of nineteenth-century infant mortality that has to be appreciated, especially for this investigation, is the extent of its spatial variability. Even Newman himself pointed out that infant mortality varied so greatly that it was 'impossible to form any reliable opinion except on broad lines' (Newman, 1906: 20). He therefore advised that it was 'necessary to take counties and, where possible, even broader areas as a basis of observation' (Newman, 1906: 20).

In contrast, modern research now has the tools available to deal with the plethora of data produced in the nineteenth century, so Newman's advice has become outdated. Now, smaller scale data, below county level, are analytically accessible. Although the larger geographical units of analysis, such as registration counties or divisions, are an important starting point in any examination of nineteenth-century infant mortality in England and Wales (Newman, 1906; Lee, 1991), in this chapter the value of geographical analysis of mortality at a smaller scale is emphasised.

Woods and Shelton (1997) have already successfully demonstrated the value of analysing mortality by registration district, the 'optimum geographical unit' for their analysis of age-related mortality in Victorian England and Wales. Since the 'locality of mortality was of particular importance' in determining the effects of a disease environment on the inhabitants, registration districts were considered the 'optimum' unit for 'capturing local rather than very broad regional variations' (Woods and Shelton, 1997: 16). Woods and Shelton considered, but discarded, the use of registration sub-districts, claiming that only crude birth and death rates could be calculated for this smaller geographical unit (1997: 16).

However, thanks to the work of Williams (1992) and Mooney (1994a, 1994b), we have been alerted to the existence and utility of Quarterly Returns, another of the government-aggregations of civil registration data.[1] This under-used source contains several pieces of information that can be used to analyse certain aspects of mortality for these smaller sub-district units. Most importantly for this piece of research, infant mortality rates (IMRs) can be calculated easily, providing a more spatially detailed picture of this measure of mortality than previously attained through registration district analysis. Williams' (1992) and Mooney's (1994a, 1994b) studies of Sheffield and London demonstrate the benefits of using sub-districts when investigating infant mortality within large urban centres rather than using time-consuming family reconstitution methods, more appropriate for smaller communities (Wrigley *et al.*, 1997). As an extension of this, this investigation of infant mortality in the Fens of Lincolnshire also advocates the use of spatial analysis at the sub-district level for broader rural areas. It demonstrates that sub-district studies can be used to gain rapid understanding of the picture of infant mortality across the variety of different communities and disease environments that may exist within a broad rural region.

The majority of researchers rightly associated high levels of infant mortality with many urban or industrialized areas in England and Wales (Woods, Watterson and Woodward, 1988, 1989; Woods, 1978, 1984, 1991; Williams, 1992; Mooney, 1994a, 1994b; Williams and Mooney, 1994). In contrast, rural areas were generally thought to provide a comparatively healthy environment for infant life. As a consequence, the demographic study of mortality among rural populations has been neglected, and generally urban factors have been identified as being responsible for high levels of mortality.

Nevertheless, one mid-nineteenth century observer, a Dr H. J. Hunter, certainly noted that high IMRs were not exclusive to urban areas. In a parliamentary report, Hunter made the striking observation that 'some few of the thoroughly rural districts, having no extraordinary mortality at all ages, and no factories whatever, still have an infantile mortality which is equalled only by a few of the large towns' (Hunter, 1864: 454-55). Hunter located five hazardous rural areas, generally situated on, or near, the estuaries of rivers flowing into the North Sea: the Humber Estuary, the Fens, Yarmouth, Norwich and Thetford, and the Medway Estuary.

The largest of these was the Fens (a region renowned for its flat, watery landscape and unhealthy atmosphere) surrounding the Wash in Cambridgeshire, Lincolnshire and Norfolk. As one contemporary writer described:

1 The Quarterly Returns report on demographic events occurring between January and March (Winter), April and June (Spring); July and September (Summer); October and December (Autumn).

The moory soil, the watry atmosphere,
With damp, unhealthy moisture chills the air
Thick stinking fogs, and noxious vapours fall,
Agues and coughs are epidemicall,
Hence every face presented to our view
Looks of a pallid or sallow hue.

(Anon., cited in Darby, 1940: 117)

The high mortality among the youngest fenland inhabitants, observed by Hunter, has since also been commented upon by Woods and Shelton (1997), Galley and Shelton (2001) and Woods (2000). However, no in-depth investigation of these abnormally high rural IMRs within this region has yet been published. In fact there have been no previous examinations of late nineteenth-century rural fenland communities.[2] This chapter argues that the comparatively high IMRs prevalent in the Fens in the nineteenth century are a subject worthy of substantial interest. In addition to providing a model for research into rural infant mortality using sub-district level material, the research presented here is intended to fill a gap in our knowledge of rural IMRs and thus, further our understanding of the geographical variability of infant mortality across England and Wales during the nineteenth century.

Spatial Variability of Infant Mortality and the Gap in Rural Mortality Research

'Where one lived in Victorian England critically affected not only one's life chances, but also the manner in which death might occur' (Woods and Shelton, 2000: 142). Even in a small, relatively homogenous nation, like England and Wales, there was well-defined differentiation between the mortality experiences of different localities. This statement was especially appropriate for infants, as demonstrated by the mass of research produced regarding this subject.

Even contemporaries such as Hunter (1864), Newman (1906), Ashby (1922) and Newsholme (1923), whose analytical techniques were limited in comparison with those of researchers today, did not fail to notice the magnitude of the spatial variations in infant mortality. However, apart from Newman's examination (1906) of the distribution of infant mortality by registration county, 'surprisingly little inroad into these spatial and environmental issues' of mortality was made (Dobson, 1992: 81), until work was carried out by Woods (1982).

Nonetheless, since the late 1980s, the spatial dimension of infant mortality has received growing attention (Williams and Galley, 1995). Over the last two decades, the level of spatial detail has become progressively finer: moving from registration county (Lee, 1991) and registration district level (Woods, 1982; Woods, Watterson and Woodward, 1988, 1989; Woods and Shelton, 1997; Woods, 2000) to registration

2 With the exception of fen-edge parishes such as Melbourn (Mills, 1978, 1984) and Waterbeach (Ravensdale, 1974), only market towns in the heart of the Fens, such as March, have been examined (Reynolds, 1987; Bevis, 1980).

sub-district level in certain areas of the country (Williams, 1992; Mooney, 1994a, 1994b; Sneddon, 2002).

The most notable geographical feature of infant mortality in these studies was the dramatic difference between urban and rural areas, with the unfavourable nature of towns being first recognized by the likes of Graunt (1662) and Malthus (1798). Others also swiftly realised the dangers of these urban 'graves of mankind' during the nineteenth century as urban and rural environments became more diverse in their mortality and health experiences (Farr, 1837: 572; Chadwick, 1842).

Gradually contemporaries compiled series of statistics supporting the notion of an urban-rural dichotomy in infant mortality (Newsholme, 1923). The most prolific researcher investigating this dichotomy was Woods (1982, 1991, 1995, 1997, 2000), and a series of collaborators (Woods, Watterson and Woodward, 1988, 1989; Woods and Woodward, 1984; Woods and Shelton, 1997; Woods and Hinde, 1987; Woods, Williams and Galley, 1997). All of these studies confirmed a strong 'urban penalty', although some rural exceptions were also noted. Evidence against a simple urban-rural dichotomy was presented by Huck (1994), and to a lesser extent Wrigley (1985), West (1974), Woods and Shelton (1997), Woods (2000) and Galley and Shelton (2001). For infants, the urban-rural differential was not necessarily clear-cut, especially in the 1850s, 1860s and 1870s, since the supposed relative health of rural regions in England and Wales was not always guaranteed (Woods, 2000). Huck (1994) asserted that localised variations within rural districts became less pronounced after 1861, while Woods and Shelton (1997) and Woods (2000) pointed out that some rural districts in the Eastern counties of Norfolk, Cambridgeshire, Lincolnshire and the East Riding of Yorkshire continued to display unusually high IMRs throughout much of the nineteenth century. Thus, it was clear that there was more to the spatial picture of infant mortality, than a mere urban-rural divide.

Williams (1992), also investigating the nature of the 'urban penalty', chose to take a more localised approach. Using registration sub-districts, she found that the spatial variations in infant mortality within the city of Sheffield operated on a much smaller, more complex scale, and still existed within entirely urban communities. Working with Galley, Williams continued along this research avenue, covering a wider spectrum of different environments across England and Wales, and concluded that mortality varied dramatically even within essentially rural areas (Williams and Galley, 1995).

Given that most registration districts were very large, with an average acreage of 59,240 statute acres (Census Reports, 1861, 1871, 1881, 1891, and 1901) the majority of registration districts encompassed a wide array of different environments and communities. Many rural districts contained one or two market towns, inevitably displaying urban characteristics, thus including pockets of urban mortality levels. Within these same registration districts substantial variation in the environment of the rural hinterland surrounding these market towns was also found. All of these different physical, social, economic and cultural environments naturally produced an extensive array of demographic experiences, which were combined and averaged out over these substantial areas (Hardy, 1994; Lawton, 1978).

Thus, although researchers have now firmly established that mortality was exceedingly spatially variable across England and Wales, even when the picture of infant mortality is broken down into smaller geographical units, such as registration districts (Williams, 1992), further exploration at a smaller spatial scale is still required to decipher the patterns and trends within rural registration districts. Thus far, internal variations within rural registration districts have seldom been addressed, since urban mortality has received a disproportionate amount of attention in the research conducted. Consequently, our understanding of late nineteenth-century rural demography may be considered inadequate. However, it is now time to place the Fens under closer scrutiny, using the smaller geographical tool of registration sub-districts to explore the complexities of infant mortality within rural areas.

Sources and Methods

Officials at the General Register Office compile aggregated tabulations and reports, using individual level demographic data (Szreter, 1991a, 1991b; Drake and Finnegan, 1994). During the nineteenth century these aggregated figures were collated for various geographical units – registration divisions, counties, districts and sub-districts – and were supplied for a number of different intervals: quarter, year and decade.[3] The computerisation of some of these data (Woods and Shelton, 1997), and their subsequent incorporation into the Great Britain Historical Database has provided researchers with several very powerful tools for analysing national data both statistically and visually through maps (Sneddon, 2002; Southall, 2006).[4] Two of these government-compiled sources are used in this chapter; the Decennial Supplements, which provide data at the registration district level and the Quarterly Returns, which supply registration sub-district level data.

Absolute totals of deaths of children aged under one year for each registration district, given in the Decennial Supplements, were transformed into IMRs using the conventional formula – infant deaths divided by births multiplied by 1,000 (Shryock and Siegel *et al.*, 1976: 235). This allowed the differences between districts within the three Eastern counties of Norfolk, Lincolnshire and Cambridgeshire to be compared, and patterns and trends to be established. In order to examine levels of high infant mortality in the Eastern region, the average IMR for these three counties was compared graphically with the overall national average infant mortality rate at the registration district level (Figure 5.1). This comparison admirably displayed the 'Eastern County penalty' upon infant mortality which operated over the 1850s, 1860s and 1870s.

3 Various legislative improvements put in place over the course of the nineteenth century, were designed to increase the accuracy, completeness and therefore reliability of sources derived from this registration system. However inaccuracies continued to appear in the individual records (Glass, 1973; Nissel, 1987). Researchers have to be aware that such errors were inevitably incorporated into the aggregated reports, although little can be done to rectify the problem.

4 Further information on the Great Britian Historical Database and associated research can be found at http://www.port.ac.uk/research/gbhgis/aboutthegbhistoricalgis/database and http://www.VisionofBritain.org.uk

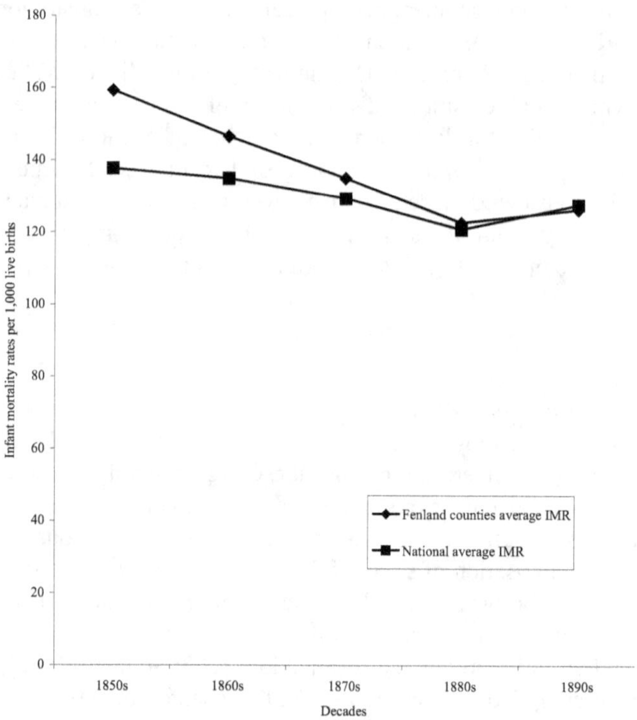

Figure 5.1 Average infant mortality rate in the three Fenland counties, per decade 1850s-1890s, compared with the national average infant mortality rate

The IMRs were then mapped using a classification of four nested means (Gregory, 1998) to facilitate comparison of the IMRs across the three counties.[5] The maps were used to identify spatial patterns and to establish which particular registration districts were contributing the lion's share of the high mortality levels found in these three counties.[6]

---

5 Classes using 'nested means' are determined by balancing the frequency distribution of the set of mortality rates for each map about the mean to give two classes and then each of these classes is subdivided around its own mean. This allows any particularly high or low outlying values, which may have skewed the distribution of the data, to be accounted for (Scripter, 1970), and gives four classes which may be viewed as being 'well above', 'above', 'below' and 'well below' average for each time period.

6 The mapping process was not flawless. One of the most serious problems was the frequent alteration of boundaries during the nineteenth century. Woods and Shelton (1997), like others before them (Howe, 1970; Lawton, 1968), avoided this problem by arbitrarily choosing one set of boundaries to be used for the entire period. However the Great Britain Historical Database allows tracking of the continual boundary changes occurring during the nineteenth and twentieth centuries and permits the construction of maps with exact boundaries correct for any date (Gregory, 1998). This resulted in a more accurate series of maps than

Despite their apparent value, the Quarterly Returns have been substantially under-used because of certain limitations, namely their absence prior to 1869, and their lack of age-specific data for those over the age of one year.[7] Although mentioned by several researchers (Williams, 1992; Lewes, 1991; Woods and Shelton, 1997), the only other demographer to use them systematically was Mooney (1994a, 1994b), who demonstrated that the restrictions imposed by the structure of the Quarterly Returns were not insurmountable. Therefore, as long as they are used with caution the Quarterly Returns, and their sub-district material, can at least add further depth to our understanding of infant mortality.

In order to examine the possibility of a 'fen penalty' IMRs were calculated for each registration sub-district within Lincolnshire, again employing the 'conventional' IMR formula.[8] To compensate for the small number of events in some of the sub-districts it was necessary to aggregate the data in order that the calculation of mortality rates could be considered more robust. To preserve the greater spatial detail provided by sub-district units the data was aggregated temporally; the number of births and deaths occurring in the four quarters of each year was collected, and then the annual data was combined to calculate quinquennial IMRs from 1870 to 1899.

Temporal changes to infant mortality across Lincolnshire were analysed by comparing them graphically (see Figure 5.4). The quinquennial IMRs were then analysed spatially through a series of maps, enabling verification of any associations between certain types of environment – such as 'fen' or 'urban' settings – and the infant mortality level (see Figure 5.5). Further assessment of these 'fen and urban penalties' was carried out by sub-categorising the sub-districts into four groups - Urban Fen, Urban Non-Fen, Rural Fen and Rural Non-Fen – and calculating and comparing the average infant mortality rates for each of these sub-groups.[9]

---

previously obtained (Sneddon, 2002; Southall, 2006). However, boundary changes over the fifty-year period (1861-1901) can present problems if comparisons of mortality are drawn between the beginning and end of the period.

7 Unfortunately apart from infant deaths and those over 60 years, no other age subdivision of deaths was included in the Quarterly Returns. Therefore, it is not possible to examine other age-based mortality measures using this source.

8 Lincolnshire alone was selected due to time and space constraints. This county displayed a good mixture of upland and lowland areas, which enabled the possibility of a 'fen penalty', mentioned earlier in discussions of the three eastern counties, to be explored.

9 These four groups were determined using two sets of criteria. Firstly, population density was used to determine whether the sub-district displayed urban or rural characteristics. If the density measure exceeded 0.5 persons per acre then the sub-district was classed as urban (Williams and Galley, 1995: 406). The second criterion was a much less easily quantified factor and assessed whether the district lay within the fenland region or not. Although it may have been possible to use height above sea-level as the determinant criteria, this was ruled out because it altered substantially over the course of the nineteenth century as the Fens became more permanently drained and the processes of 'shrinkage and wastage' lowered the surface level of the ground (Godwin, 1978). Instead, inclusion within the fen or non-fen categories was determined by comparing each sub-district with Waller's geological map of the Fens (1994), see Figure 5.3. On the basis of these two criteria each sub-district was assigned to one of the Urban Fen, Urban Non-Fen, Rural Fen or Rural Non-Fen categories.

High Infant Mortality in the Eastern Counties, 1850-1900: A New Type of Penalty

Corroboration for the 'penalty' in the Eastern Counties is first observed in Figure 5.1, which clearly shows that the average rates of infant mortality in the three Eastern Counties were above or, at least in later decades, equal to the national average.

However, given the counties predominantly rural nature, one might have expected their IMRs to be somewhat below the national average (Hunter, 1864; Whitely, 1864; Wheeler, 1898; Miller and Skertchley, 1878; Horn, 1976; Taylor, 1973; Higgs, 1995). The majority of the Eastern registration districts displayed typical rural population densities of 0.2-0.3 persons per statute acre throughout the second half of the nineteenth century. Similarly, industrial activity, commonly known to inflate levels of mortality, was very limited in this area (Godwin, 1978; Thirsk, 1957; Darby, 1940, 1983). As a result various researchers have stated that this rural region was almost entirely governed by an agrarian economy (Thirsk, 1957; Hills, 1967; Darby, 1983), and noted that any light industry existing in the region was strongly connected with agriculture, such as woad or flax production (Godwin, 1978; Wills, 1970).

Figure 5.1 also indicates that the impact of the 'eastern county penalty' declined somewhat after the 1870s, allowing IMRs to decline at a faster pace than the national average. This, therefore, supports the notion that very different factors were at work in this region, in comparison with other highly dangerous mortality regimes, such as urban or industrialized areas. Contrary to expectation, therefore, this rural-agricultural region suffered from an infant mortality regime that was more akin to that of urban-industrial regions and bore little resemblance to that of similar rural-agricultural regions elsewhere in England and Wales. What factors were operating upon infant health to bring about this unusual picture? One method of exploring this phenomenon further is to map the IMRs for each registration district in the three Eastern counties by decade (Figure 5.2).

One prominent feature of the five maps in Figure 5.2 is the concentration of above-average IMRs in the low-lying region surrounding the Wash, commonly known as the Fens. In order to show which districts lay specifically within the Fens a copy of a geological survey of land in this region (Waller, 1994) was superimposed onto a boundary map of the three counties with the shaded areas indicating the Fenland districts (Figure 5.3). Using Figures 5.2 and 5.3 a 'fen penalty' becomes evident. Pockets of high mortality also occurred around Norwich and in Guiltcross and Wayland, an area to the south of Norwich, which displayed a similar marshy, low-lying character to other fenland districts surrounding the Wash. However, from the 1880s, as might be expected from Figure 5.1, the detrimental impact upon infant life in the low-lying fenland regions lessened somewhat as the areas with above-average infant mortality in these areas of the three counties reduced in size.

In the later decades of the century other increasingly urban districts, such as Yarmouth, Kings Lynn and Grimsby joined the already highly urbanised districts, such as Norwich and Lincoln, in producing high levels of infant mortality and increasing the average IMR of the three counties as a whole. It seems, therefore, as the century progressed, that, in contrast with the 'fen penalty', the 'urban penalty' played an increasingly dominant role in the spatial patterns of infant mortality within the Fenland counties. This was especially visible in the north of Lincolnshire, near Grimsby, where some of the highest rates in the area began to appear from the 1870s onwards, thanks to increasing levels of industrialization and, consequently, urbanization.

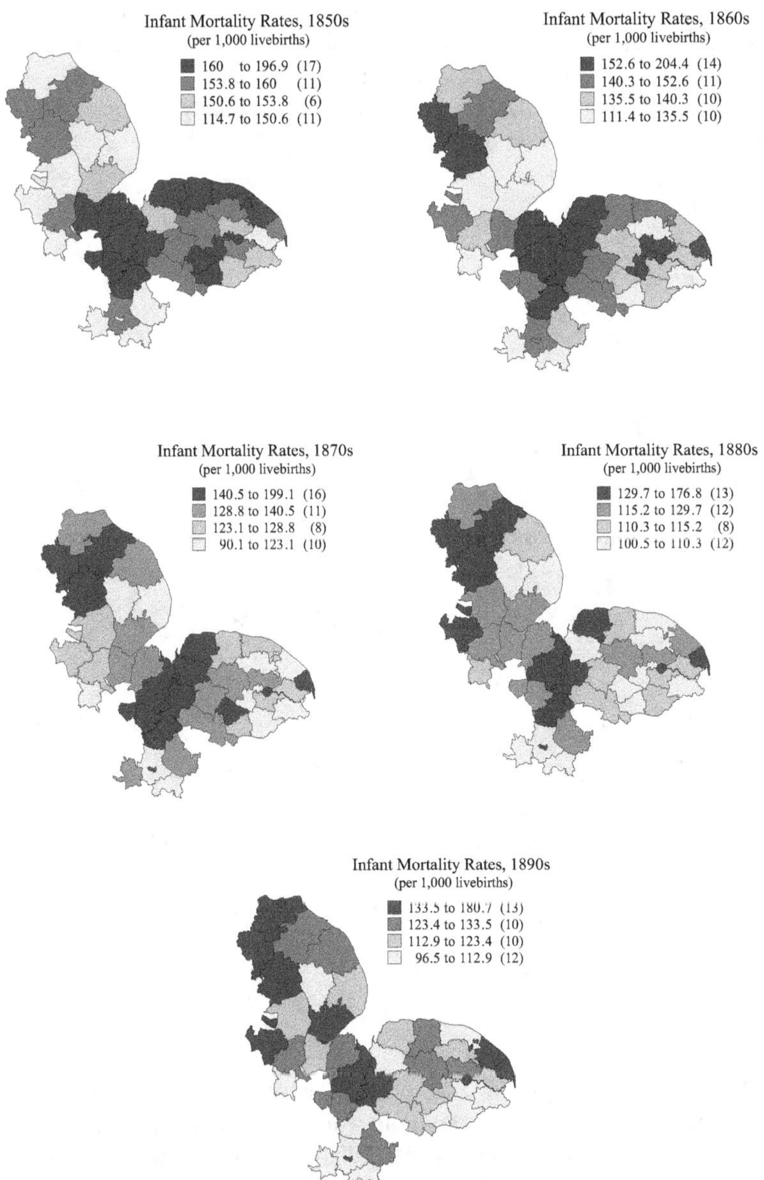

Figure 5.2 Infant mortality rates for the registration districts within the three Fenland counties, mapped according to four nested means, 1850s-1890s

On the basis of this analysis, it is clear that there were two common characteristics that tended to produce a high mortality regime for infants in this region: – a fenland location or an urban profile. This suggests that in addition to the more frequently studied, and, therefore, better understood, 'urban penalty' upon infant life, some element of fenland life – physical, economic, social or cultural – also contributed to high levels of infant mortality. Furthermore, over the course of the century it became clear that the effects of that factor or factors gradually ceased to have such an impact. Instead, the 'urban penalty' came to dominate patterns of mortality among the young inhabitants of fenland counties. The effects of the 'fenland penalty' can be pinpointed more accurately using the smaller registration sub-district units.

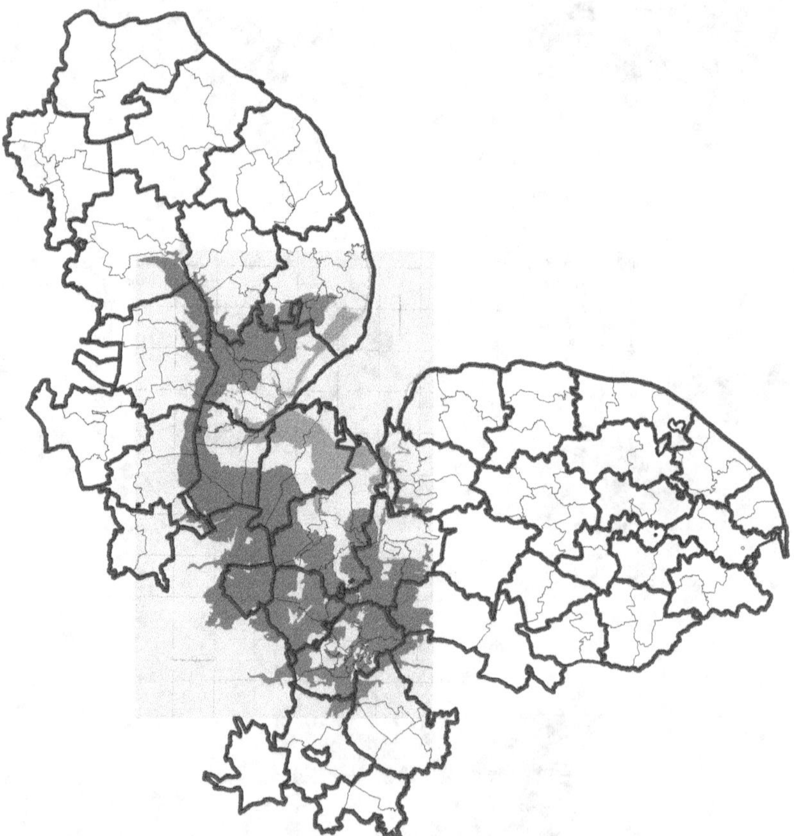

Figure 5.3 Boundary maps showing both registration districts and sub-districts of the three Fenland counties, superimposed with a geological survey map (Waller, 1994) to indicate which particular districts lay within the Fens

The Regional Picture of Infant Mortality, 1870-1899 – The 'Fen Penalty'

Whilst the smaller sub-district units are still an aggregation, the resulting spatial blurring of the mortality experience within each unit is greatly reduced compared with that of registration district studies. Due to the large number of sub-districts across these three counties, this part of the study focuses purely on Lincolnshire. The temporal pattern of infant mortality in Lincolnshire is best viewed by plotting the mean quinquennial IMRs for all the registration sub-districts in Lincolnshire (Figure 5.4). The most obvious point to note from this graph is the steady decline in infant mortality over the 1870s and 1880s, then a reversal with a substantial increase occurring with the mean IMR hitting a peak in the late 1890s. This may have been produced by the same series of hot summers that caused high national levels of infant mortality in the 1890s, or possibly by a sudden shift in the growing impact of the 'urban penalty' affecting the north of the county (Garrett *et al.*, 2001).[10]

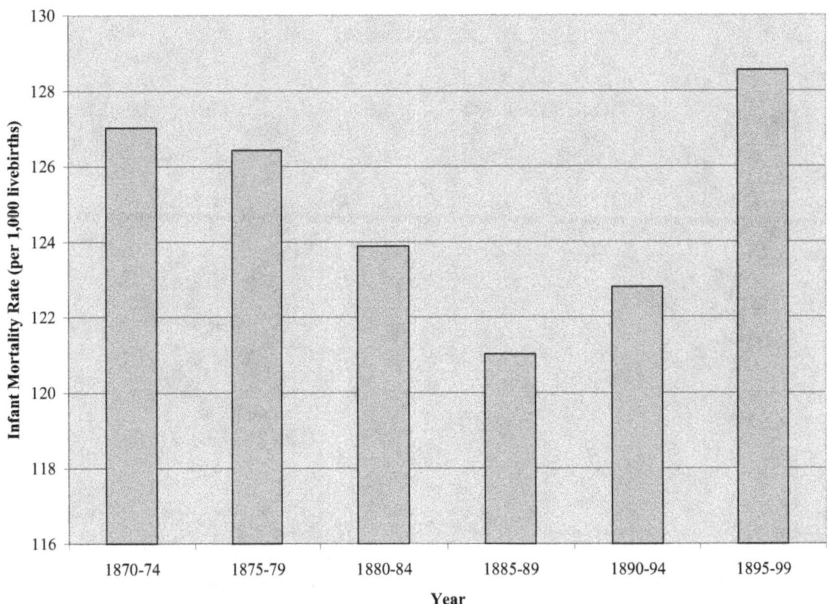

Figure 5.4 The mean quinquennial infant mortality rate for the registration sub-districts of Lincolnshire, 1870-1899

A more complete understanding of infant mortality within these Lincolnshire sub-districts can be obtained by mapping the data for the six quinquennial periods between 1870 and 1899 with the data again being categorised according to four

10 Unfortunately, further examination of the rates in the subsequent century, which cannot be pursued here, is required, to ascertain whether either of these scenarios are accurate explanations of the dramatic rise experienced in the late 1890s.

nested means (Figure 5.5).[11] Figure 5.5 clearly illustrates a much more complex spatial picture of infant mortality within Lincolnshire than that previously exposed by the registration district analysis. In spite of the seemingly uniform rural and agricultural nature of Lincolnshire, it is clear that there was a great deal of variation between the registration sub-districts of the county.

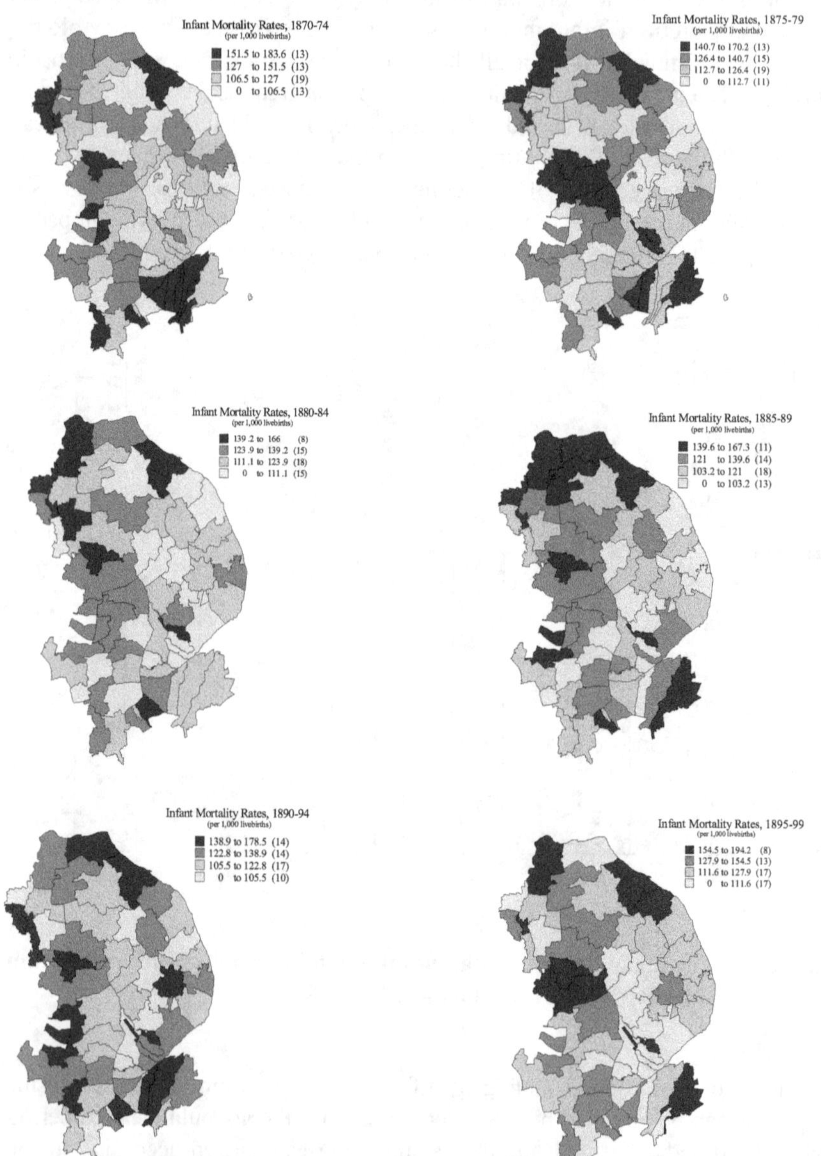

Figure 5.5 Quinquennial infant mortality rates for registration sub-districts in Lincolnshire, 1870-1899, mapped using four nested means

11 For explanation of nested means see footnote 5.

However, if Figure 5.5 is viewed in conjunction with Figure 5.6 and Table 5.1, which give the names of the various sub-districts, it is possible to see three main regions where high infant mortality was principally concentrated – in the midwest surrounding Lincoln (sub-district 18 on Figure 5.6), in the Fens around Boston (36) and Spalding (53), and in the north of the county near Grimsby (5). The urban sub-districts – Grantham (29), Gainsborough (14), Great Grimsby (5), Lincoln (18), Boston (36), Spalding (53) and Stamford (49) – also displayed above average IMRs at various points throughout the period.

Therefore, these maps verify that there were certain hazardous environments within Lincolnshire and confirm that both the 'urban' and 'fen' penalties, detected at the registration district level of analysis, are both also strongly visible at this smaller spatial scale.

However, when the maps in Figure 5.5 are viewed chronologically, it appears that the effect of the 'urban penalty' increased over time, both within certain sub-districts and across the whole county, with a growing number of increasingly urbanized sub-districts displaying above average IMRs. Rising levels of urbanization in this rural county, were most discernible in the expansion of the urban centre of Lincoln into parts of Lincoln North (17) and Lincoln South sub-districts (19), inevitably generating elevated IMRs from 1885. Similarly, mounting industrialization in the north of the county in later decades was directly responsible for coincidental urbanization in the sub-districts surrounding Grimsby and Immingham (1, 2 and 5), and the resultant heightened IMRs in the northern sub-districts.

Equally, however, visual analysis of Figure 5.5 shows the impact of a low-lying fenland setting upon infant mortality. When the maps in Figure 5.5 are examined, bearing Figures 5.3 and 5.6 in mind, the detrimental impact of the 'fen penalty', in the large number of districts surrounding Boston and Spalding (24, 33-38, 40, 43-46 and 52-57), becomes apparent. Chronological examination of this phenomenon over the six five-year periods also reinforces the conclusion that the detrimental impact of this 'fen penalty' on infant life was not uniformly sustained throughout the period. By the 1880s the effect of the 'fen penalty' had begun to wane, though it did appear to have re-emerged to some degree in the late 1890s, possibly mirroring the national rise in infant mortality in this decade.

To explore this in more depth, the registration sub-districts were divided into four sub-categories – urban fen, urban non-fen, rural fen and rural non-fen. To estimate the different effects of these four categories upon infant mortality, the average quinquennial infant mortality rates for each sub-group were calculated and compared (Table 5.2). Table 5.2 shows that the most distinctive disparity lay between the urban and rural rates, the former being from 18 to 33 infant deaths per 1,000 live births greater than the latter in each of the quinquennial periods, confirming the operation of an 'urban penalty' upon infant life in Lincolnshire throughout the last three decades of the nineteenth century.

Generally, the disparity between fen and non-fen sub-districts was of a rather smaller order, 8 deaths per 1,000 births at most, apart from 1895-99. In this quinquennium the two urban fen districts experienced an unusually high average infant mortality rate of 160.3 infant deaths per 1,000 births whereas, in previous quinquennia their IMRs rates had declined steadily from 155 to 140 per 1,000 births.

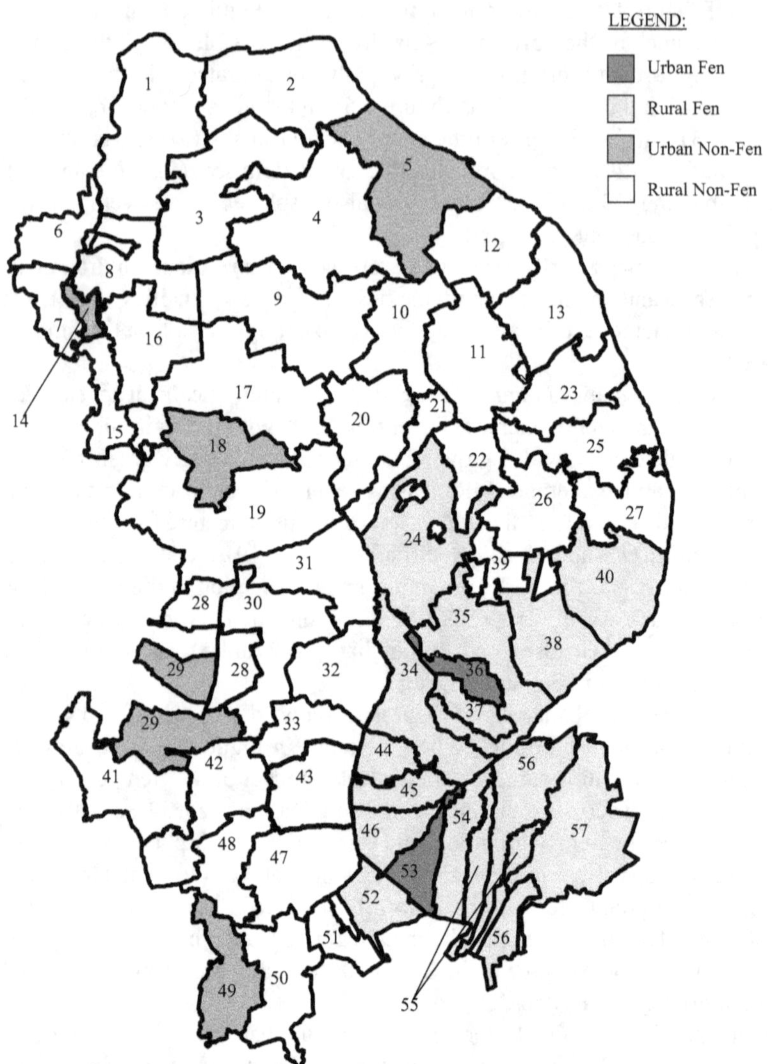

Figure 5.6 The location of the registration sub-districts in Lincolnshire, indicating whether urban fen, rural fen, urban non-fen or rural non-fen

However there were only two sub-districts in the urban fen category, and if we concentrate on the more numerous rural sub-districts then a clearer pattern of decline in the disparity between the fen and non-fen categories becomes visible. By the 1880s, the difference between the average IMRs of the two groups is minimal, and by the early 1890s virtually identical. By 1895-99 the rural fen sub-districts appears to have a slight advantage over rural non-fen. Thus, regrettably, the registration sub-district analysis in Table 5.2 would seem to capture merely the tail-end, or final decade, of the 'fen penalty' in the 1870s.

Table 5.1 Key for Figure 5.6, indicating the identifying number and name of each sub-district and whether rural fen, rural non-fen, urban fen and urban non-fen

Rural Non-Fen		Rural Fen		Urban Non-Fen		Urban Fen	
ID	Name	ID	Name	ID	Name	ID	Name
1	Winterton	24	Tattershall	5	Great Grimsby[a]	36	Boston
2	Barton[b]	33	Aswarby[b]	14	Gainsborough	53	Spalding
3	Brigg	34	Swineshead	18	Lincoln Home		
4	Caistor	35	Sibsey	29	Grantham		
6	Owston[b]	37	Kirton in Holland	49	Stamford		
7	Misterton	38	Bennington				
8	Scotter	40	Wainfleet				
9	Market Rasen	43	Aslackby				
10	Binbrook	44	Donington				
11	Louth	45	Gosberton				
12	Tetney	46	Pinchbeck				
13	Saltfleet	52	Deeping St Nicholas				
15	Marton	54	Moulton				
16	Willingham[b]	55	Holbeach				
17	Lincoln North East	56	Gedney Hill				
19	Lincoln SouthWest	57	Long Sutton				
20	Wragby						
21	Horncastle						
22	Tetford[b]						
23	Withern						
25	Alford						
26	Splisby						
27	Burgh						
28	Leadenham						
30	Sleaford						
31	Billinghay						
32	Heckington[b]						
39	Stickney						
41	Denton/ Grantham South						
42	Colsterworth[b]						
47	Bourn						
48	Corby						
50	Barnack						
51	Deeping						

Notes: a) In 1870-74 Great Grimsby was in the Rural Non-Fen, but population growth put it in the Urban Non-Fen after this date. b) These sub-districts were absorbed by others in the district between 1870 and 1899

Table 5.2 Average quinquennial infant mortality rates per 1,000 live births for urban non-fen, urban fen, rural non-fen and rural fen registration sub-districts of Lincolnshire[a]

Sub-Category	1870-74 IMR	N	1875-79 IMR	N	1880-84 IMR	N	1885-89 IMR	N	1890-94 IMR	N	1895-99 IMR	N
Urban Non-Fen	158.2	4	150.1	5	144.1	5	135.9	5	146.9	5	136.7	5
Urban Fen	155.9	2	150.1	2	150.3	2	138.6	2	140.1	2	160.3	2
Rural Non-Fen	122.0	35	120.7	34	120.1	32	118.5	32	119.6	31	126.8	28
Rural Fen	126.7	16	128.5	16	121.8	16	119.3	16	119.3	16	124.6	15

Note: a) For reasons why the number of sub-registration districts included in the calculations varies, see notes to Table 5.1
Source: Quarterly Returns, Lincolnshire, 1870-1899

Conclusion

This chapter has explored the spatial variability of infant mortality over the latter half of the nineteenth century within one of the most seemingly uniform rural-agricultural regions – the Fens. The application of a 'regional magnifying glass' to the fenland county of Lincolnshire, bringing smaller spatial units into focus, has reinforced the operation of an 'urban penalty' and helped to establish the existence of a 'fen penalty'.

Space constraints have prevented further analysis of potential factors producing the 'fen penalty' being discussed here.[12] However, despite various limitations, such as a lack of specific data relating to female employment (Miller, 1984), household income and the health care of individual infants, it is possible to speculate on the causes of this phenomenon.

Some authors have alerted us to the existence of malaria or fenland ague in the fenland region (Whitley, 1864; Miller and Skertchley, 1878; Dobson, 1994) and to the common use of opiates by Fenlanders to ease the symptoms of the disease (Berridge, 1979; Berridge and Edwards, 1981). Whilst conceding that fenland ague was, indeed, in part, responsible for the 'chronic state of ill-health' found in this region (Nicholls, 2000: 526), the mechanisms producing high levels of infant mortality in the Victorian period were inevitably varied and multifaceted, and were, in fact, even in this seemingly homogenous rural-agricultural area, the result of a complex interaction of various physical, socio-economic, demographic and cultural factors (Williams and Galley, 1995).

The Fens have always possessed a unique physical environment. By the late nineteenth century they were distinguished from most rural areas of England and Wales by their low-lying, flat landscape and exceedingly nutrient-rich soil, a direct consequence of the successful draining of the region in the early nineteenth century (Hills, 1967; Darby, 1940, 1983). The resulting highly fertile land was farmed

12 Further discussion can be found in Sneddon (2002).

intensively, with fields producing several crop yields every year (Godwin, 1978; Thirsk, 1957), causing fenland agriculture to take on an almost industrial slant in terms of production levels.

Intensive farming of the Fens during the nineteenth century meant that men, women and children were all employed in great number as agricultural labourers, contrasting strongly with the more generally upheld picture presented in much of rural England and Wales as the nineteenth century progressed. Traditional views claim that changes in crop demand, agricultural techniques and increased mechanisation, particularly the use of the heavier scythe as opposed to the sickle (Perkins, 1976; Roberts, 1979), led to a fall in the demand for labour, particularly female agricultural workers (Snell, 1985). Female agricultural labourers became marginalized, and a 'sexual division' of agricultural tasks began to appear (Snell, 1985). However, much debate surrounds Snell's hypothesis and recently several researchers have reasserted Pinchbeck's original work (1930) that this picture was not as universally applicable across England and Wales as Snell's view implies (Bouquet, 1985; Armstrong, 1988; Howkins, 1991; Gielgud, 1992; Sharpe, 1996;. Burnette, 1999; Speechley, 1999; Verdon, 2001, 2002a, 2002b).

Research on the Fens of Lincolnshire (Sneddon, 2002) supports evidence found by Verdon in the Norfolk Fens (2001, 2002a, 2002b). As in Norfolk, the enormous demand for labour created by the intensive scale of farming found in this fenland region, coupled with the cultivation of highly labour-intensive crops (an array of garden vegetables, woad and flax) for which the Fens' nutrient-rich soil was ideally suited (Miller and Skertchley, 1878; Thirsk, 1957; Wills, 1970; Godwin, 1978), positively encouraged the employment of women and children in the fields due to the dexterity required in the cultivation of these crops. Consequently, these female and child labourers found employment on a casual basis in the large number of agricultural gangs that were scattered across the Fens of Lincolnshire and Norfolk (Pinchbeck, 1930; Kitteringham, 1975, Verdon, 2001).

The employment of married women, and more particularly, mothers, was fundamental to the issue of infant mortality in this area. Holding down a career and raising children has increasingly become the norm in modern times, and although still a difficult task, modern advances have made this easier. The employment of mothers in the nineteenth century in whatever capacity, be it on the land or in factories, inevitably had a detrimental impact on infant feeding and health care. Without today's legislative protection of maternity leave, and the array of modern childcare options available, mothers were faced with the difficult decision of staying at home to feed their young infant or going out to work to supplement the family income to ensure that there was sufficient food for the rest of the family (Verdon, 2002b). Rarely, some types of employment, particularly work in the fields (Hunter, 1864; Verdon, 2001, 2002b), allowed the infant to accompany the mother, however, being out in all weathers was hardly a suitable environment for a newborn. Without the benefits of modern artificial milk supplements – which did not begin to appear until the very end of the nineteenth century – Victorian mothers were forced to rely on breast-feeding or to wean the child early. The vast majority of mothers opted to take their children off the breast, often at less than a month of age, and leave them in the care of elderly relatives or older siblings, who frequently dosed these children on opiates in order to sedate them (Hunter, 1864; Berridge and Edwards, 1981; Nicholls, 2002; Verdon 2002b). Consequently, many were weaned early onto highly suspect and positively dangerous mixtures of 'sop' (a concoction of bread, sugar and water) (Hunter, 1864). Particularly dangerous for these infants was the use of potentially contaminated water in this mixture, which could easily promote typhoid and other enteric water-borne diseases that were so inimical to infants and young children. Like most rural areas, piped water supplies were practically unheard of until well into the twentieth century and there were frequent anecdotes of drinking water being taken from the same drainage ditches where human waste was deposited. Even in households where this was not the case it was also highly feasible that, due to the high water table found across most of the fenland region, contaminated material could seep into the communal wells and pumps in the region, again prompting outbreaks of typhoid and infantile diarrhoea.

As mentioned by Sayer (1995), by the 1860s, concern regarding this situation provoked sufficient adverse public opinion to instigate a parliamentary inquiry into this matter (Sixth Report of the Medical Officer of the Privy Council, 1863), which focused on women's employment in these agricultural gangs as a cause of infant death. Eventually this bad press resulted in the Gangs Act of 1867, prohibiting the employment of married women and children within these gangs, restricting, though not entirely preventing their employment in the fields *per se*. Verdon (2001, 2002b) points out that this legislation only referred to 'public' not 'private' gangs. Thus, although discouraged, gang labour persisted until the late 1890s on certain exceptional farms, such as Lodge Farm, Castle Acre, Norfolk. However, 'elsewhere the system … became redundant from the early 1870s as farmers began to abandon cleaning operations such as weeding and stone picking in response to the agricultural depression' (Verdon, 2001: 54).

Coincidentally (or perhaps not), as discussed in the body of this chapter, the 1870s also witnessed the beginning of the decline in infant mortality within the Fens. Taking into account a time lag of a few years, although this legislation may explain some of the decline in infant mortality, it is unlikely that it is wholly responsible. It is more likely that a whole host of factors, such as household income, housing, overcrowding, contaminated water supplies, family size, birth parity, availability of adult carers besides the mother, all combined to create an environment which exposed infants born into the Fens to rather more risks to their health than most rural newborns. For instance, successful and permanent drainage of the Fens in the 1820s triggered a sudden enormous demand for labour and a subsequent population explosion in this area in the first half of the nineteenth century. Unlike other rural areas, this demand for labour did not wane and continued throughout most of the nineteenth century. These population pressures, in turn, placed a great strain on housing provision. Poorer households, where infants and children were already at a health disadvantage, were frequently overcrowded, exposing the young inhabitants to respiratory disease, one of the main killers of the very young in the Fens.[13]

13 Examination of the seasonal patterns of infant mortality in the Fens (Sneddon, 2002), which can be derived at the registration sub-district level from the Quarterly Returns (Williams, 1992; Mooney, 1994a, 1994b, Sneddon, 2002) indicates a summer peak in the urban sub-districts, as expected, caused by enteric and diarrhoeal disease due to problems

Importantly, as both Williams and Galley (1995) and Newman (1906) maintained, no single factor is to blame for high IMRs. No matter whether a rural or an urban area is being investigated, infant mortality is always governed by a wide range of factors, determined by the physical, socio-economic, demographic and cultural environment of each community.

---

with water, drainage and sewage. However, within the fen sub-districts a strong late winter/spring peak is visible, suggesting respiratory diseases were one of the main killers within the rural Fens.

Chapter 6

Infant Mortality in Northamptonshire: A Vaccination Register Study

Tricia James

On page 44 of *Infant Mortality: a Social Problem* Newman, quoting Creighton (1903), notes in passing that in the latter half of the seventeenth century,

> … there were many market towns and country parishes whose registers showed an excess of burials over christenings, the special occasional causes having been widely prevailing fevers, epidemic agues, with influenzas and smallpox (Newman, 1906: 44).

Since that time, however, 'some changes have occurred'. Newman continues:

> Measles is probably less fatal, and smallpox, compared with its ravages before the time of Jenner, has almost vanished. Epidemic ague has also been driven out (Newman, 1906: 44).

For centuries smallpox had been the scourge of populations throughout the world, resulting in the death of one in every seven of those infected (Loosmore, 1996). Treatment was minimal and those who survived were often severely disfigured. As Table 6.1 shows, in the 1850s some 42,000 people, approximately 10,000 of them under one year of age, died from smallpox in England and Wales. Yet in the final decade of the century only 4,000 succumbed to this dreaded disease, and only 500 of these were infants. Newman noted that:

> … there has been a decline in smallpox in England and Wales, in which the whole population has shared, from an annual mortality of 576 per million persons in 1838-42 down to 13 per million persons in 1891-1900 (Newman, 1906: 59).

However, in concentrating on those conditions which continued to carry off much larger numbers of the nation's youngest citizens, he did not dwell on the fact that across the second half of the nineteenth century, in an attempt to ensure that every infant was brought before a medical man in order to be vaccinated against smallpox, an extensive monitoring system had been put in place under the auspices of the Medical Department of the Local Government Board (Drake, 2005: 40). This system generated a series of individual records which, as the civil records of birth, marriage and death of England and Wales remain inaccessible for the purposes of demographic study, are unparalleled. This chapter will consider the evolution of the vaccination system and its records, as the former very much influenced the latter, setting limits on the use of the records in research. It will also seek to illustrate the potential of this valuable source to our understanding of the decline in infant mortality by investigating infant mortality amongst the population of Higham Ferrers registration sub-district, a shoe-making district in the county of Northamptonshire in the final decades of the nineteenth century.

Smallpox Vaccination Registers

Variolation, inoculation of patients with lymph from someone already infected with smallpox, was the only form of protection until May 1796, when Dr Jenner presented his findings on human-to-human inoculation using live infectious material from individuals infected with cowpox, which he termed 'vaccine' (Wolfe and Sharp, 2002). The use of vaccination then became widespread, but as the vaccine could not be preserved, and was not readily available because cowpox infections occurred only rarely, the human-to-human method continued and children were used to transport lymph around the world by vaccinating them in succession, the lymph produced by one inoculated individual being transferred to the next.[1]

Originally variolation and then vaccination were promoted chiefly through charitable sources such as the first National Vaccine Establishment founded in 1808 but, following the first Vaccination Act in 1840 (British Parliamentary Papers [BPP], 1840 c.29), vaccination became the remit of the Boards of Guardians responsible for the administration of the Poor Law (Greenwood, 1901).[2] Under an Act of 1853 (BPP, 1853 c.100), vaccination became compulsory and to assist in compliance with the Act, an administrative system was set up based on the well-established structure whereby the civil registration of births, deaths and marriages took place. By law a child's birth had to be registered within 42 days. Each Registrar was required to give statutory notice of the need to have the child vaccinated to the person registering the birth, usually a parent. The notice indicated that the child should be taken for vaccination within three months of birth and gave places, dates and times of relevant sessions. Parents were then under a legal obligation to take the child for vaccination and return a week later for inspection and to provide lymph for the vaccination of another infant (Simon, 1872). A penalty of 20 shillings or a prison sentence could be imposed for failure to comply with the law; the fine or imprisonment to be repeated if non-compliance continued (Durbach, 2000).

Within the terms of the 1867 Vaccination Act (BPP, 1867 c.84), each Board of Guardians was ordered to contract with a registered medical practitioner who would serve as the Public Vaccinator. In addition to the qualifications of a District Medical Officer, the Public Vaccinator was required to have a certificate of proficiency in vaccination techniques but after 1886, vaccination skills were incorporated into general medical training.[3] The 1867 Act laid down that the Public Vaccinator was

1 On one such trip, organised by King Carlos IV of Spain in 1803, 230,000 vaccinations were carried out in Central America, the Philippines and – eventually – China (Loosmore, 1996).

2 As a result of which vaccination was bedevilled for generations through its association with the Poor Law (Durbach, 2000).

3 The qualifications for District Medical Officers (DMOs) were defined by Greenwood (1901) prior to 1886 as being registered under the Medical Officer's Orders December 10th

to be paid a minimum of 18 pence for each successful vaccination, with additional payments for re-vaccination should the first attempt not 'take'. An additional sum was also paid for a report, which was required to be sent annually to the General Board of Health by the Vaccinator (Greenwood, 1901).

The Guardians had to provide a venue for vaccinations to take place and, as this could not be the Public Vaccinator's surgery, except in exceptional circumstances, public vaccination stations were set up; often in inappropriate venues such as public houses (Durbach, 2000). In rural districts the Guardians were also empowered to arrange for the Public Vaccinator to visit less often than the statutory calendar; in many cases only twice a year. This meant that the age at which rural infants were vaccinated depended more on the month of their birth than was the case in the urban areas, where sessions were held on a much more frequent basis (Greenwood, 1901).

Vaccination Birth Registers

Under the terms of the 1871 Act (BPP, 1871 c.98), Vaccination Officers were to be appointed in addition to the Public Vaccinators. The Officers were to be responsible for the administration of the vaccination system although the Registrars continued to serve vaccination notices (Drake, 1997). With information supplied by the Registrar and the Public Vaccinator, the Vaccination Officer compiled infant birth registers. These 'vaccination birth registers' (VBRs) were originally compiled on a monthly basis on loose leaf sheets which were bound at the end of each year (Simon, 1872). Each volume was ordered by the date on which notification to vaccinate was given out, in other words on the date on which the births were registered, not the date on which they occurred, and thus spans parts of two years. In most volumes information was recorded on births from November in one year, to October or November the following year, depending on the time it took parents to register a birth. The VBRs provide detailed information for each individual infant including date and generally place of birth, name and gender of the child and name and occupation of the father, if he was present in the household. If the mother was not married, or had been widowed, then her occupation was given.

If a child died before it could be vaccinated the date of death was noted in the VBR, but not the cause. For each surviving child the VBR included a vaccination history giving the date of notification of the requirement to vaccinate, the name of the person to whom notification was given, the date of vaccination and the outcome. Outcomes could be either successful vaccination; postponement for a period of up to two months due to poor health; or 'insusceptibility'. A certificate of insusceptibility was issued if child had been unsuccessfully vaccinated three times (Greenwood, 1901).[4] If the child had had smallpox this was recorded. Finally the name of the vaccinator and the Registrar were appended.

1859, which qualified him to 'practise medicine and surgery'. From 1886 DMOs had to be registered under the Medical Act (BPP, 1886 c.48).

4 The vaccination process was deemed to have 'failed' if there was no reaction to the vaccine or development of a vesicle (Roitt, 1988).

The format of the registers changed following the revisions of the 1898 Vaccination Act (BPP, 1898 c.49) so that information on 'conscientious objection' could be introduced. As smallpox epidemics lessened and the anti-vaccination lobby gained ground, national rates of non-compliance with the Vaccination Acts had risen to worrying levels. As Greenwood put it:

> [with] the determined opposition of a comparatively small portion of the community, who lost no opportunity of traducing the principles of vaccination, and of dragging into publicity, and exaggerating, every accident that occurred in the working of the law ... popular opinion was slowly undermined and increasing difficulties arose in carrying out the law (Greenwood, 1901: 40).

Greenwood attributed the success of the anti-vaccination campaign to a small group of protesters who were able to infiltrate the Boards of Guardians - the overseers of the vaccination system. He added:

> As all prosecutions were thought to require the consent of the Guardians, the latter were practically able to paralyse the existing law (Greenwood, 1901: 40).

In 1889 a Royal Commission was set up to investigate the high levels of non-compliance. After numerous sittings the Commission published its final report in 1896, forming the basis of the 1898 Vaccination Act (Greenwood, 1901). Amongst its many revisions was the recommendation that animal lymph should be used instead of the previously used human lymph – a major issue for the anti-vaccination campaigners, many of whom believed the use of animal lymph to be against the laws of nature (Greenwood, 1901). The new Act also recommended that the time allowed for vaccination be increased from three to six months. Public vaccination stations were abolished in favour of the vaccinator visiting the child in his or her home, for which the vaccinator would be paid five shillings. Finally the penalty for non-compliance with the regulations could not be repeated until the child was four years of age (Greenwood, 1901). The major concession to the anti-vaccination element, however, was the introduction of a conscientious objection clause, which allowed parents to obtain a certificate of exemption from a stipendiary magistrate stating their belief 'that [vaccination] would be prejudicial to the child's health' (Greenwood, 1901: 42). Whilst these new regulations improved the uptake of vaccination initially, they failed to appease the anti-vaccination lobby, who urged parents to refuse to apply for a certificate of exemption, and the overall effect of the Act on levels of vaccinations was limited, particularly, according to Greenwood, as 'many of the Boards of Guardians throughout the country are still controlled by the anti-vaccinationist element, [and so] the action of the Vaccination Officers in prosecuting offenders is largely paralyzed' (Greenwood, 1901: 53).

The anti-vaccination campaign continued unabated until 1907, when a further amendment to the 1898 Act (BPP, 1907 c.31) reformed the conscientious objection clause so that only a statutory declaration was required instead of external arbitration by a magistrate (Durbach, 2000). By 1939 only about 34 per cent of Britain's children were vaccinated, although vaccination was still compulsory and remained so until the introduction of the National Health Service in 1948. The vaccination programme continued on a voluntary basis until 1971 (Baxby, 1999). After a further series of global campaigns aimed at eradicating the infection, the last naturally acquired cases occurred in Somalia in 1978 and the World Health Organisation finally declared smallpox to be extinct in 1980 (Walsh, 2002).

In England and Wales, as Table 6.1 shows, the decline in deaths from smallpox during the second half of the nineteenth century was dramatic, particularly in the two closing decades, making it 'one of the nineteenth century's most successful stories as far as preventative medicine is concerned' (Woods and Shelton, 1997: 73). After a peak in the 1870s, when more than 23 persons per 10,000 died from smallpox, mortality tumbled to fewer than 5 deaths per 10,000 in the 1880s and less than 2 deaths per 10,000 in the 1890s. Although the disease accounted for only a tiny fraction of all infant deaths at mid-century, this age group accounted for almost a quarter of all smallpox deaths. By the end of the nineteenth century children under the age of one accounted for one in eight smallpox deaths having been as low as 1 in 25 in the 1870s; an indication of the protection offered by the vaccination programme.

Table 6.1 The decline of smallpox in England and Wales, by decade 1851-1900

Decade	Total number of smallpox deaths	CDR from smallpox per 10,000 population	Percentage of smallpox deaths occurring among infants	Percentage of all infant deaths which were due to smallpox
1851-1860	42,071	22.1	24.6	1.04
1861-1870	34,786	16.3	23.6	0.71
1871-1880	57,422	23.6	4.3	0.19
1881-1890	12,280	4.5	11.3	0.11
1891-1900	4,058	1.3	12.4	0.04

Source: R.I. Woods: Study 3552, UK Data Archive 'Causes of Death in England and Wales 1851-60 to 1891-1900'

Vaccination Death Registers

Alongside the vaccination birth registers, registers of infant deaths (IDRs) were also compiled on a monthly basis, from information provided by the Registrars. Presumably designed to improve the efficiency with which infants requiring vaccination could be traced, these gave information on all infant deaths under the age of 12 months. In general, fewer of these registers appear to have survived than the VBRs (Drake, 1997). They provide much the same socio-biographical information as that seen in the VBRs; that is date and place of birth, name of child and father, father's occupation and birth registration number. Details such as date of death, age at death and death registration number were also included but, unfortunately, cause of death was not.

As mentioned above, deaths were only noted in the VBRs if the infant had died *before* vaccination. If the average time to vaccination changed, or the proportion of children who were not vaccinated altered, then the number of deaths in the birth registers could change, regardless of changes in infant mortality. However, when it is possible to link entries from the VDRs to the IDRs a fuller, more accurate account of infant mortality experience can be gleaned than is normally available from period measures. Using the registers it is possible to follow individual infants across their first year to observe whether or not they died before their first birthday, whereas period measures simply compare the number of infants dying in a particular year with the number born in that year. Indeed the view given of infant mortality by the VBRs and IDRs in combination is likely to remain unparalleled 'until the civil registers of births and deaths are made freely available' (Drake, 2005: 52).

For the Higham Ferrers registration sub-district VBRs are available for the years 1879-1909 and IDRs from 1871-1893.[5] The issue of changes in the proportion of infants being vaccinated is well illustrated when the number of deaths in the birth registers from Higham Ferrers over the decade 1880 -1889 is compared with that in the death registers (Table 6.2). In the majority of cases, infant deaths were recorded in both the vaccination birth registers and in the vaccination death registers. However, at the beginning of the decade, in 1880, over 90 per cent of infants born were being vaccinated, the vast majority by the age of three months. As the deaths occurring after vaccination were not recorded in the VBR, 11 (31 per cent) of the 35 deaths occurring amongst the 1880 birth cohort before they reached their first birthday, were only seen in the IDR. By 1889 the proportion of infants being vaccinated had dropped to less than 50 per cent thanks to the effectiveness of the anti-vaccination lobby, particularly among the shoe workers, and all infant deaths recorded in the IDR were also noted in the VBR.[6] Even then the records were not consistent. Between 1880 and 1889 a total of 544 infant deaths occurred in the registration sub-district. Of these 110 (20 per cent) occurred *after* vaccination, 58 (53 per cent) of which were recorded in the VBR. Why the vaccination officer chose to include some of the post-vaccination deaths in the birth register, but not others, is unclear. In addition, under the terms of the 1871 Vaccination Act (BPP, 1871 c.98), each Vaccination Officer was required to 'follow up' infants when they left the district before vaccination and record in the VBR when and where vaccination (or death if this occurred before vaccination) had occurred (Simon, 1872). The Higham Ferrers Vaccination Officer also appears on occasion to have followed infants beyond their first birthday, recording deaths of children up to 15 months old. These deaths do not, of course, appear in the infant death register as they fail to meet the criteria for inclusion. Infant deaths thus appear to have been inconsistently recorded in the VBR making the

5 Both sets of records may be found in Northampton Record Office; the VBRs under reference number PL12/378-403 and the IDRs under PL/161-162.

6 The decrease in the proportions vaccinated continued and by 1894 only one per cent of infants were vaccinated; a proportion which did not increase again until the introduction of certification in 1899.

Table 6.2 A comparison of annual deaths recorded in the Vaccination Birth Registers and Infant Death Registers for the Higham Ferrers sub-district, 1880-1890[a]

	Year									
	1880	1881	1882	1883	1884	1885	1886	1887	1888	1889[a]
Number of births in the VBR	356	409	441	447	440	438	470	507	485	552
Number of deaths in the IDR[b]	46	50	54	39	57	44	57	59	60	77
Number of deaths in the VBR[b]	35	40	45	32	51	40	48	59	60	77
Difference in number of deaths (IDR-VBR)	11	10	9	7	6	4	9	0	0	0
Percentage of infants vaccinated	90.2	87.0	79.5	81.8	73.6	78.9	74.1	68.2	60.0	48.1

Notes: a) Although births in the VBR and deaths in the IDR are given for each year listed, some deaths of infants registered in the VBR for 1889 actually occurred in 1890. b) Deaths include only those dying aged less than 1 year within the sub-district

IDRs a particularly valuable complementary resource and it is unfortunate that the latter are only available for Higham Ferrers up to 1893.

Infant Mortality in Higham Ferrers – A General Picture

When, in 1882, Dr Thomas took up the post of Medical Officer of Health (MOH) for Wellingborough Rural Sanitary Authority, in which Higham Ferrers registration sub-district lay, his first impression was that although he had charge of a rural area, 'the conditions of life are much what they are in the factory towns where the rate of infant mortality is always high' (Thomas, 1883). Figure 6.1 shows the IMRs which may be calculated for the sub-district from the Quarterly Registration Reports, 1871-1890.[7] The IMR for the year 1882 was a relatively high 134 deaths per 1,000 live births but this was significantly lower than the rates experienced over the preceding decade: in 1874 the IMR had stood at over 200 deaths per 1,000 live births.

The present study focuses on the 1880s, considering infant mortality amongst the annual birth cohorts from 1880 to 1889 inclusive, which may be calculated using the VBRs and IDRs for the sub-district. In addition, infants and their parents identified in the combined registers (CVRs) in these 10 cohorts can then be linked to the census enumerators' books (CEBs) covering the sub-district in both 1881 and 1891. This allows additional information on the infants' family circumstances to be gleaned, so that they can then be placed geographically and socially.

Of the 4,547 children born in the sub-district from 1st January 1880 to 31st December 1889, 4,003 are thought to have survived infancy and 544 were observed to

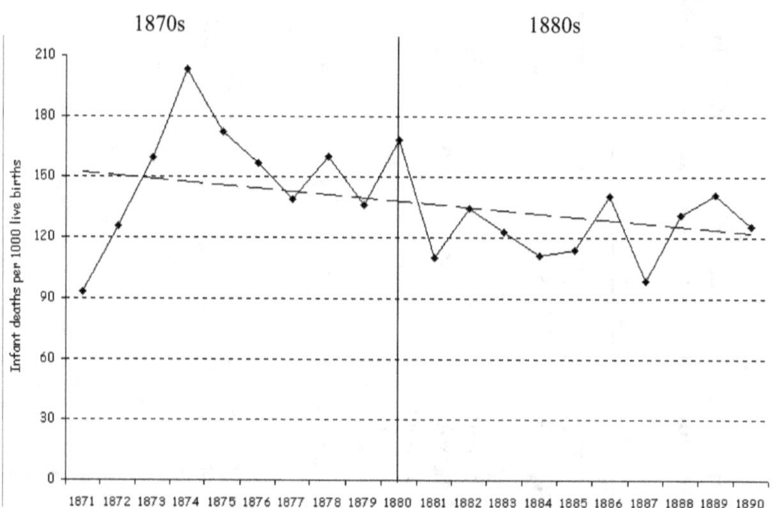

Figure 6.1 Annual infant mortality rates in the Higham Ferrers sub-district, 1871-1890

7 *Registration Reports (England) Quarterly and Weekly*, MRC, 1871-1888 and *Quarterly Reports of the Registrar General 1888-1910:* both located in the London School of Hygiene and Tropical Medicine.

die before their first birthday; an overall *cohort* IMR of 119.6. Note that this measure is calculated by following all infants born in a particular year to their first birth day, or to their death if this intervenes. The number of deaths observed is then divided by the number of births in the cohort. The deaths are therefore related exclusively to the birth cohort being observed. This differs from the more conventional measure of IMR which observes the number of deaths occurring in a calendar year and divides these by the number of births occurring in that year – as, for example, in Figure 6.1. In a year when epidemic disease was rife in the first half of the year many of the babies dying as post-neonates would have been born in the previous year. The conventional IMR thus gives a somewhat misleading impression of the relative life chances of infants born in the two years.

Of the infants in Higham Ferrers who survived their first year, 3,083 (77 per cent) could be located in the 1891 census returns, still resident in the sub-district. In all they belonged to some 1,417 family units. A further 352 of the infants who had died in infancy during the decade could also be identified as belonging to these family groups. An additional 33 married couples who had borne only one child and had lost it before its first birthday, could be located in the 1891 census.

Where parents of infants could not be located in the Higham Ferrers 1891 census enumerators' books, they were sought in books outside the sub-district and in the 1881 census returns. A further 148 family groups who had moved out of the sub-district prior to the 1891 census were identified. These families included 285 infants born between 1880 and 1889; 258 of whom were observed to survive to their first birthday whilst 27 died in infancy.

Higham Ferrers sub-district contained the town of Rushden, with a population of around 3,657 in 1881; three smaller towns: Higham, Irthlingborough and Irchester; and four villages: Farndish, Newton, Podington and Wymington. For the purpose of this study the sub-district was divided into three: Rushden; the 'small urban settlements' of Higham and Irthlingborough; and the 'rural areas' which included the four villages and Irchester, which displayed an occupational profile which was more 'rural' than 'urban'. Over the course of the 1880s there was an expansion in shoemaking, the area's principal industry, and the population of the sub-district grew by 40 per cent, with much of the expansion centred on the town of Rushden which grew by an impressive 103 per cent.

Despite this marked population growth, the IMR calculated for each cohort from the CVRs was relatively static across the 1880s, with two dips in 1883 and 1885 and a sharp rise amongst the 1889 cohort (Figure 6.2).

However, when the three divisions of the sub-district are considered, a rather different picture emerges. Over the 1880s as a whole there was a noticeable gradient in IMR from Rushden, at just under 140 deaths per 1,000 births, to the two smaller urban parishes Irthlingborough and Higham, where IMR was 120, down to the rural parts of the parish where it lay at 100. As the three-year moving averages across the decade in Figure 6.3 show, cohort infant mortality in Rushden followed an upward trend, while in the rural areas it saw a marked downward movement, but rates in Irthlingborough and Higham remained largely unchanged. The survival chances for infants born in Rushden thus worsened notably over the decade in comparison with those born elsewhere in the sub-district.

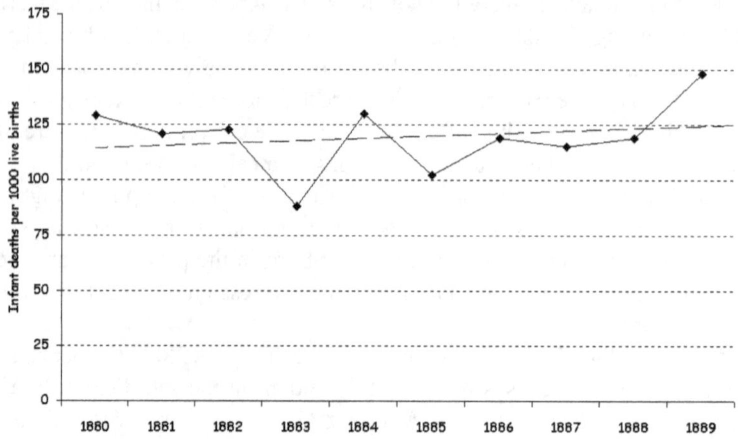

Figure 6.2 Annual cohort infant mortality rate in the Higham Ferrers sub-district, 1880-1889

Additional nuances are indicated in Figure 6.4. Because the VBRs supply exact dates of birth and death, it is possible to calculate an infant's age at death very accurately, thus distinction can be drawn between neonatal mortality and post-neonatal mortality. Decadal neonatal mortality rates (NMRs; deaths in the first 4 weeks, or 28 days of life) were virtually the same in the small urban and rural parishes, but markedly higher in Rushden. However, there is little difference in post-neonatal mortality rates (P-NMRs) between Rushden, Higham and Irthlingborough while the rural areas enjoyed considerably lower rates. The penalty of being resident in Rushden therefore seems to fall most heavily on infants less than a month old.

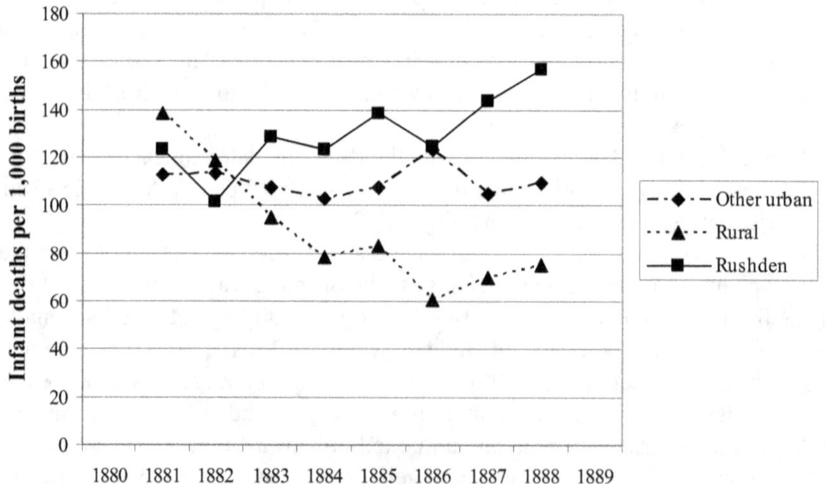

Figure 6.3 The 3-year moving average IMR for Rushden, other urban areas and rural areas: Higham Ferrers sub-district, 1880-1889

If the NMR and P-NMR are followed over the 1880s cohorts, it can be shown that the P-NMR for the sub-registration district remained more or less constant across the decade, but the NMR rose. The apparent constancy of the P-NMR for the sub-district obscured the fact that in Higham and Irthlingborough the rate moved downwards over the decade whereas the rate in Rushden rose steeply. The NMR rose in both Rushden and the small urban parishes, but more steeply in Rushden. By contrast, in the rural areas the trend in NMR was downward.

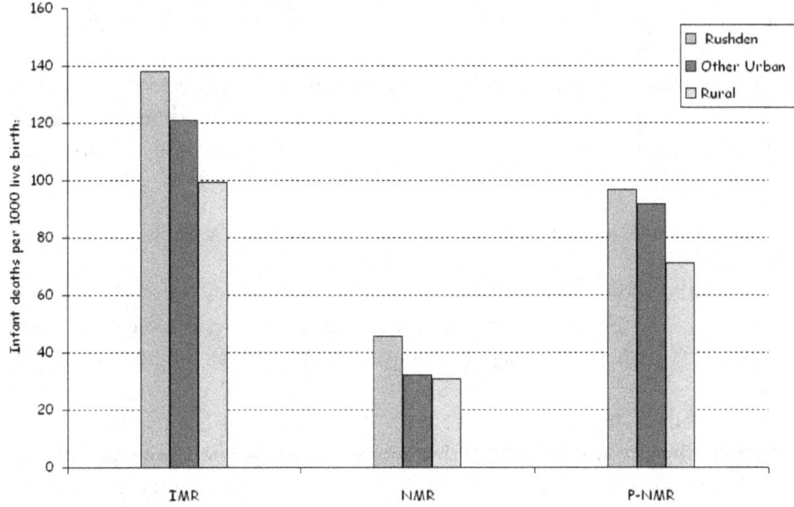

Figure 6.4 Comparison of infant mortality rates, neonatal mortality rates and post-neonatal mortality rates for the towns and villages of the Higham Ferrers sub-district, 1880-1889

Thus the apparently stable level of IMR within the registration sub-district hid three separate trends: in Rushden both NMR and P-NMR were rising across the 1880s; in Higham and Irthlingborough the NMR was rising, but the P-NMR was falling; and in the rural parishes both the NMR and P-NMR were falling. Infants in the urban centre of Rushden thus saw their life chances deteriorate markedly over the 1880s compared to those born elsewhere in the sub-district, particularly those in the rural areas, where survival chances, already good, appear to have improved dramatically during the decade.

In 1881 the Higham Ferrers sub-district held 10,788 persons. Ten years later this had risen to 15,072; a 40 per cent increase. The town of Rushden accounted for 88 per cent of the growth; its population rose from 3,657 to 7,445 persons in the course of the decade; a 103 per cent increase; in 1881 it held just over one third of the sub-district's population, by 1891 virtually one half. Rushden's explosive population growth in the 1880s may have underlain the deteriorating conditions for older infants. Increased population pressure meant that the environment into which infants were born grew potentially more overcrowded and unsanitary. If house building and improvements in the sanitary arrangements had kept pace with the increase in the number of residents, the effects of population increase may have been mitigated. There is some indication that the house building programme which increased housing stock by 102 per cent over the 1880s may at least have managed to maintain the status quo as average household size in Rushden remained at 5.1 persons from 1881 to 1891. However, population density within the town rose (from 1.3 to 2.7 persons per acre), and continued to rise, as the town expanded. Whilst population increased in the other parishes of the sub-district, at 7 per cent the growth was much more limited and appears to have had little impact on the mortality experience of infants in the post-neonatal period.

In contrast, the situation for infants in the first month of life deteriorated not only in Rushden, but in all parishes.[8] This would suggest that the detrimental influences on the infant experience at this age were generalised throughout the sub-district.

Infant Mortality in Higham Ferrers – From Partial Family Building Histories

Using the histories included in the combined VBRs and IDRs for Higham Ferrers, it is possible to explore the differing trends observed above, although it must be acknowledged that small numbers sometimes render robust comparisons difficult to achieve.

Across the sub-district infants were born into a variety of domestic situations. Where the families could be observed in the census of either 1881 or 1891, the majority (2,106) were conjugal family units headed by both a father and a mother. Occasionally one parent was absent on census night, but a further 121 families were headed by a mother, either unmarried or a widow. Because of the disparity in life chances experienced by illegitimate infants, these families will not be analysed here.[9]

The fathers of the majority of infants, 64 per cent, worked within the shoe industry, but 14 per cent were employed in agriculture and 22 per cent in a range of other occupations required to support a rapidly expanding population, such as builders, shopkeepers and teachers. The number of children observed being born into a family during the 1880s varied from 1 to as many as 8 infants. Of course this did not necessarily reflect the final size of the family, simply the stage in each family's child-bearing history at which they were 'captured' in the sources. Some of the families were 'newly created', whilst others were more established, having had children born before the 1881 census.[10]

---

8 Neonatal rates appear to have declined in the most rural parishes, but the numbers involved were too small to allow this to be confirmed.

9 Families headed by a single father were included in the general analysis because it is likely that in the case of the death or absence of the biological mother, a surrogate mother would have been found to care for the infant. In addition it was not possible to identify when mothers had died or left.

10 'Newly created' families were those for whom there was evidence that the man was unmarried in 1881 and that marriage occurred during the study period. Marriage information was not available to the study. It is recognised that there may have been other newly created

Of the 1,412 conjugal family units found still resident within the sub-district in 1891, 308 (22 per cent) had experienced infant death during the 1880s. However, the percentage of families in Rushden experiencing infant death was higher at 24 per cent, than in the other urban towns (20 per cent) and in the rural parishes (17 per cent). The number of infant deaths within a family also varied; 232 families experienced 1 death, 64 families 2 deaths, 6 families 3 deaths and another 6 families experienced 4 infant deaths during the decade. The majority (59 per cent) of families in which more than one infant died were found in Rushden, potentially contributing to the higher rate of infant mortality in the town.

In Higham Ferrers sub-district, in-migration clearly contributed to the growth in population experienced across the 1880s. However, the migration flow was not entirely one way. Of the 2,106 conjugal families, including those that were 'newly created', who had at least one infant born in the sub-district between 1880 and 1891, 991 families (47 per cent) had been present in the 1881 CEBs and remained in the sub-district in 1891. A further 147 families (7 per cent) were present in 1881, but left before they could be enumerated in the following census. However, the 1891 census enumerated 423 families (20 per cent) who had arrived after 1881, had children and remained to be enumerated in 1891. Out of the 2,106 families registering a birth in the sub-district in the 1880s, therefore, 545 or 26 per cent were not present on census night in either 1881 or 1891.

The occupation could be identified for the fathers of infants born into 1,412 conjugal family units during the 1880s and observed in the 1891 Higham Ferrers sub-district CEBs. The sub-district was at the heart of the shoemaking industry in Northamptonshire and shoemaking was the major form of employment in the area with 903 (64 per cent) of the families headed by a shoemaker. A further 192 (14 per cent) were engaged in agriculture. The other main occupations involved men working in building, trade or shop-keeping, together with professional groups such as doctors, dentists and clergy. Ironstone workers were found in Irthlingborough, where there were several quarries, and rail workers in Wymington – the site of a large rail development during the 1880s. For the purpose of this study this group were known collectively as 'other occupations' because individually they were too small for analysis. The impact of the type of occupation on infant mortality was therefore explored using three categories; shoemakers, agricultural workers and 'other occupations'.[11]

Shoemaking was not evenly distributed across the sub-district. In Rushden, shoemakers accounted for 78 per cent of all heads of conjugal families, whereas in Higham and Irthlingborough less than 64 per cent of families were headed by shoemakers and in the rural parish villages only 31 per cent of families had a shoemaker as their head. The variation in population size meant that 56 per cent of

families who were not identified in the CEBs because the information about the father was not available.

11 It had been hoped to consider infant mortality by social class but determination of social status from the description of occupation in the CEBs proved problematic as, for example, the term 'shoemaker' was used as a generic term to describe those who worked on their own, or were employed within a factory setting as managers or labourers.

families in the sub-district who had a shoemaker at their head lived in Rushden, while 35 per cent lived in the small urban parishes and 9 per cent in the rural parishes.

Amongst the families having infants in the sub-district in the 1880s there was little variation in the experience of infant mortality among the occupation groups. Twenty-two per cent of shoemaking families, 24 per cent of agricultural families and 21 per cent of families whose head worked in other occupations, lost at least one infant before its first birthday. However, the number of families experiencing more than one death varied slightly, so that shoemakers had an IMR of 113 deaths per 1,000 live births, while the agricultural workers' rate was 106 per 1,000 and that of the group in other occupations 103. While this was not a statistically significant difference, when neonatal mortality is considered infants of shoemakers were seen to be at a significantly higher risk. Of every 1,000 infants born alive to shoemakers, 41 would not survive the first month of life. This compared to 34 infants born to agricultural workers and 26 for those born to fathers in other trades. If the infants of shoemakers survived their first four weeks, their chance of survival – 76 per 1,000 – was in the same range as the 75 for infants of agricultural labourers and the 79 for infants born to fathers in other occupations. It would thus appear that the infants of shoemakers were particularly susceptible to neonatal mortality.

If the infant mortality experience of the infants of shoemakers is compared by place of birth an interesting paradox is revealed. Although in general rural infants had an overall rate of mortality lower than that of their urban counterparts, Figure 6.5 indicates that the IMR for small-town-born children was only 90 per 1,000 live births, while that for those born in the rural parishes was higher at 123. The higher rates of rural mortality are seen in both neonatal and post-neonatal rates, although the relatively small numbers of births to shoemakers in the rural parishes mean that the differences are not statistically significant. However, the difference in overall IMR between Rushden (126 per 1,000 live births) and the two small urban areas (90 per 1,000) was statistically significant.[12]

The distribution of infant deaths among the families of agricultural labourers within the sub-district was very different from that of shoemakers. In particular, infants born to agricultural labourers living in Rushden experienced a significantly higher overall IMR compared with their peers in either the smaller towns or the rural parishes.[13] This was driven by a particularly high P-NMR of 149 deaths per 1,000 live births, much higher than the rates experienced by the infants of shoemakers living in the town, and by the infants of agricultural labourers residing in Higham, Irthlingborough or the rural parishes (Figure 6.6). The reputation of agricultural workers for having low IMRs would appear to rest with the high proportion of their number who live in small settlements, rather than in anything intrinsically healthy about the occupation. In his study of Preston, Morgan (2003) hypothesised that horses were the main cause of infant diarrhoea because of the large number of flies the horse manure attracted. If urban based men with agricultural occupations kept cattle and pigs close to residential buildings, then they, their families, and possibly their neighbours may also have suffered from the scourge of flies (Hall and Harding 1985). It is interesting,

12 Significant at the 95 per cent level.

13 Significant at the 99 per cent level.

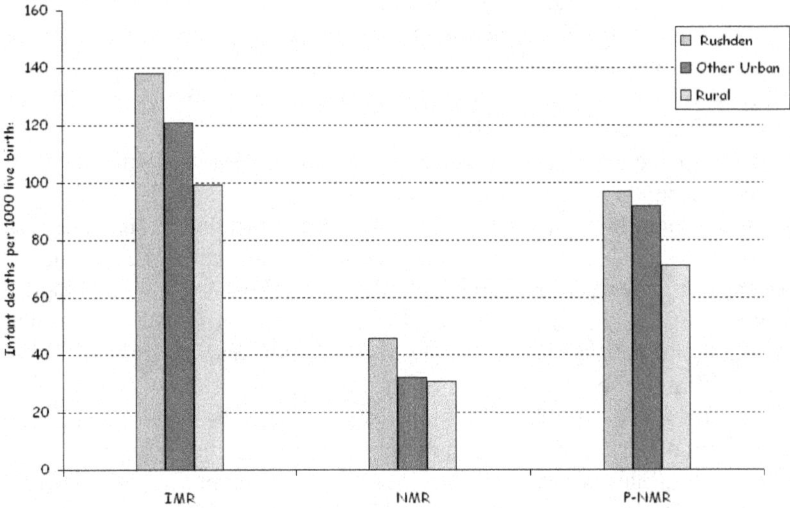

Figure 6.5 Comparisons of IMR, P-NMR and NMR for infants of shoemakers by place of birth, 1880-1889

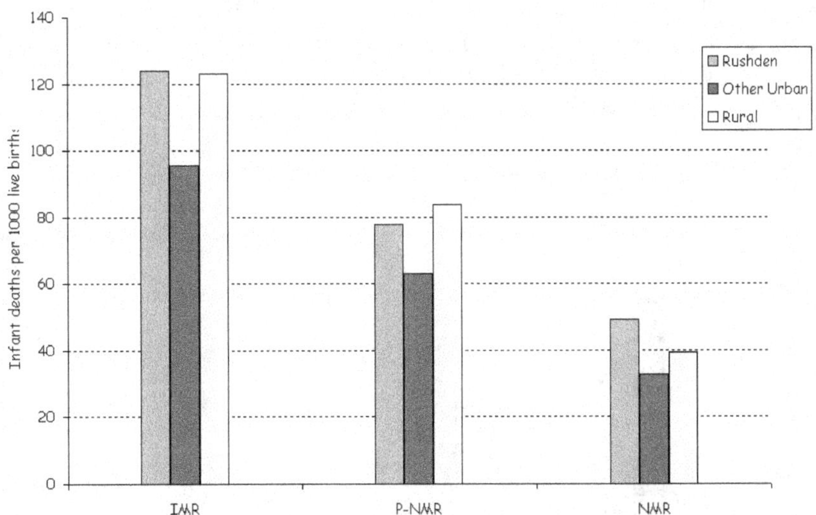

Figure 6.6 Comparisons of IMR, P-NMR and NMR of infants of agricultural labourers by place of birth, 1880-1889

therefore, to speculate on why the older infants of agricultural labourers in Higham and Irthlingborough did not suffer to an extent similar to those in Rushden.

The number of agricultural labourers' infants dying in the neonatal period was very small, just 15 altogether, rendering meaningful comparison with the children of shoemakers across the three types of settlement far from robust. However, as comparison of Figures 6.5 and 6.6 shows, neonatal rates amongst shoemakers' children do appear to be notably higher.

For infants born to fathers in other occupational groups (Figure 6.7) the penalty for the post-neonatal infant of being born in Rushden was again clearly identifiable with the difference in mortality between them and those infants born in rural parishes being highly statistically significant.[14] For infants born in Higham and Irthlingborough the situation was again seen to be healthier than for those in Rushden and the difference statistically significant.[15] Once again the difference between Rushden, the smaller towns and the rural parishes was driven by levels of post-neonatal mortality, there being no significant difference between levels of neonatal mortality.

Thus, cohort IMRs differed markedly across the sub-district, and among occupations, not always in ways which might have been predicted. Infants born to shoemakers ran an elevated risk of mortality in the first month of life, wherever they were living. In Rushden the rates of infant death were comparatively low for the infants of shoemakers who survived the first month, but particularly high for the infants of agricultural workers suggesting detrimental exogenous influences were present within those families working on the land or with animals. Surprisingly, the mortality rates for infants born to shoemakers in the rural parishes were seen to be unusually high, comparing unfavourably both with those of other infants in rural areas as well as those born in Higham and Irthlingborough. Indeed the rural shoemakers' children ran virtually as great a risk of death as those of Rushden's

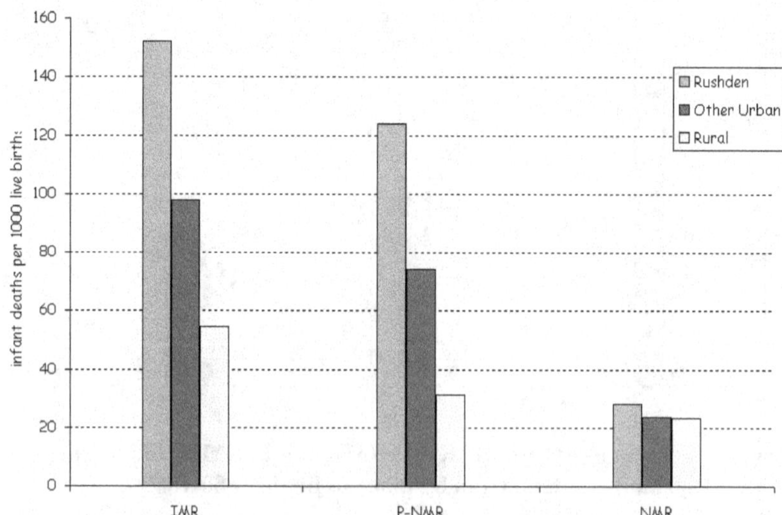

Figure 6.7 Comparisons of IMR, P-NMR and NMR of infants of men working in occupations other than shoemaking and agriculture by place of birth, 1880-1889

14 Significant at the 99 per cent level.

15 Significant at the 95 per cent level.

shoemakers. Despite small numbers these results were statistically significant and must therefore be taken as indicative that certain, as yet unidentified, features of the lives of shoemakers in the rural parishes lessened their infants' chances of survival.

Maternal Occupation

Newman devoted a whole chapter of *Infant Mortality: a Social Problem* to 'The occupation of women' (Newman, 1906: chapter V), focussing mainly on the effect of married women's work away from home on exacerbating rates of infant mortality, although frequently illustrating that the relationship was far from straightforward. There is little doubt that the success of the shoe industry in Northamptonshire was founded on the employment of women and children, as they were cheaper to employ (Greenall, 2000). However, of the mothers listed in the vaccination registers, only 16 per cent were listed in the 1891 CEB as having an occupation and only 8 per cent were listed as being employed in shoemaking. It is likely that this is an unrealistic reflection of female employment in the area, because although wages were low for women, employment was readily available during the 1880s as shoemaking underwent rapid expansion. It may be that as many women with children worked from home as 'outworkers', with *ad hoc* working arrangements, they were not seen, or did not see themselves, as employed for the purpose of the census. According to Higgs (1987), the guidance given to enumerators was unclear with regard to married women's paid work within the home, especially part-time or casual employment such as that of the shoe worker. Higgs citing Branca indicates that:

> … the duties of a housewife may have been seen as the proper occupation of a married women, whatever her actual experience of work in the factory (Higgs, 1987: 64).

If in some situations work outside the home was not listed in the census, then it is even less likely that those who worked at home, dividing their labour between caring for the household and shoemaking activities, would be listed as being employed. The combination of roles undertaken by women shoemakers was described in 1886 by one reporter in Northampton who wrote:

> … the feminine portion of the community vary the operations of putting on quarters and seaming linings with excursions to the kitchen to see how the dinner is cooking or superintending the mysteries of the family wash (*Boot and Shoe Journal*, 1886: 14).

Where women with children were listed in the census as having an occupation there was no guarantee that they had been employed, or employed in the same occupation, during their pregnancy, at the time of birth or during the infant's first days and months (Garrett, 1998).

Garrett and Reid (1995) argue that the quality of care a child received while its mother worked was also important. As the majority of the shoe-working mothers worked within the home, it would seem likely that childcare remained under the watchful eye of the mother rather than being the sole charge of unskilled older siblings or others. Working within the home also ensured that mothers who were able to do so could breastfeed their infants. The work was also not thought to be as hazardous for the pregnant woman as that seen in the heavy manufacturing industry or other occupations where the work involved long hours of standing, lifting heavy loads or working in unsanitary conditions (Newman 1906; chapter V). Therefore if the key to death in early life lay in women's work, the situation in the Higham Ferrers sub-district would seem relatively favourable and yet, as we have seen, the NMR remained high, particularly amongst shoemakers.[16] Wherever they were born within the sub-district the infants of shoemakers, with an NMR of 41 per 1,000 live births, fared worse than the infants of non-shoemakers who lost only 29 infants in the first month for every 1,000 babies born. Similarly illegitimate infants born to mothers working in the shoe trade fared much worse during the first month with an NMR of 175 per 1,000 births compared to those whose mothers were employed in other settings with an NMR of 21, although it should be noted that the numbers involved were very small.

Shoemaking does not have a reputation as a 'dangerous trade', such as that carried by certain parts of the pottery industry (Holdsworth,1997), but Millward and Bell (2001) indicate that whilst the IMR for Northampton town, a centre of shoemaking, between 1907 and 1910 was relatively low at 109 per 1,000 live births, the NMR was higher than average at 42 per 1,000. Similarly Graham (1994) found that there were strong links between shoemaking, women's employment and levels of premature birth in his study of Northampton between 1911 and 1931. Clarke and Mason (1985) recorded abnormally high perinatal mortality rates – deaths occurring after 28 weeks of gestation and in the first week of life – amongst women in the Leicestershire shoemaking industry in the years 1976-82 and a similar situation was also found in Montreal, Canada (McDonald and McDonald, 1986). Both Clarke and Mason and McDonald and McDonald concluded that there may have been a foeto-toxic agent in either the glue used in the closing process or in the leather itself, which resulted in foetal or early neonatal death. There are no sources available to give an insight into rates of miscarriage or stillbirth in nineteenth-century Higham Ferrers, but, since shoemaking is the common factor in these studies and those of early twentieth-century Northampton, it is possible that such foeto-toxic agents were present in shoemakers' homes in the sub-district during the 1880s, providing a plausible, but as yet unsubstantiated, explanation for the relatively high rates of neonatal mortality observed.

Conclusion

This chapter has employed a relatively unknown source, the vaccination birth and infant death registers (although see also Chapter 8), to consider infant mortality in the shoemaking district of Higham Ferrers. Although not as detailed as the civil registers of births and deaths the documents, generated by the administration system designed to ensure that every child had the chance to be vaccinated against smallpox,

16 The NMRs of shoemakers' infants remained significantly higher even when allowance was made for multiple births, maternal age and short birth intervals.

do enable an exploration of infant mortality which can carry the researcher into the villages, streets, workshops and individual homes of the sub-district. Further, because the linking of data from the birth registers to those covering deaths permits exact age at death to be calculated, a greater understanding of the distribution of mortality across the first year of life, and how that may vary both socially and spatially has been achieved.

Despite remarkable population growth experienced during the 1880s, the registration sub-district of Higham Ferrers appears to have enjoyed relatively stable rates of infant mortality across the decade. However, when the town of Rushden was considered separately from the smaller urban settlements and rural areas in the registration-district, it could be seen that Rushden had indeed incurred an 'urban penalty', as the growth of infrastructure struggled to keep pace with the influx of inhabitants. In contrast the neighbouring rural parishes saw marked improvements in infant survival chances, despite few improvements in housing, drainage or sewerage. The growth of Rushden over the decade rested on the expansion of the shoe industry, which is believed to have relied heavily on the labour of women and children at this period. It might be expected that shoe-workers would be subject to an urban penalty, but in fact it was the children of shoe-makers living in the rural parishes who ran very high risks of death compared to their neighbours. Shoemakers in Rushden appear to have enjoyed levels of overall infant mortality which were lower than men in other occupations, particularly those of agricultural labourers whose children suffered high rates of post-neonatal mortality. No matter whether they lived in a town, a smaller urban area or in the countryside, shoemakers' children seem to have been particularly prone to dying within the first month of life, suggesting that something in the life of shoemakers was pernicious to women, and hence to their offspring.

Newman, in considering the role of 'The occupation of women' in infant mortality (Newman, 1906: Chapter V), concentrated primarily on occupations and towns where a high proportion of mothers were removed from the home and into factories or workshops. He concluded that there was no doubt that there was a close relationship between such work and high infant mortality as a result of:

> … (a) accidents from machinery, materials, and other external agents; (b) injury or poisoning from toxic substances, or injury from excessive dust, fumes, vapour, or extremes of temperature … (c) injury through fatigue and strain, long hours, insufficient periods of rest for food …or the carrying of heavy weights … (d) injury derived from defective sanitary conditions … or an insufficiency or unsuitability of sanitary conveniences; and (e) too short a period of rest at the time of childbirth (Newman, 1906: 132).

As the vaccination records of Higham Ferrers have revealed, even women who worked from home in rural surroundings did not necessarily escape the dangers created by the demands of industry, and urban living could pose a greater threat to the survival of some infants than to others, a theme to which we return in subsequent chapters.

Chapter 7

Urban-rural Differences in Infant Mortality: A View from the Death Registers of Skye and Kilmarnock[1]

Eilidh Garrett

> The diseases from which infants die are various. But for obvious reasons they are less diverse than those causing death in adults. The influences surrounding the life of an infant are less numerous and on the whole less complex. Its environment is not only smaller in compass and sphere but more uniform throughout than that of the adult. It is less exposed to weather, to cold, to a changing diet, to the dangers and exigencies of occupation, and so forth. On this account it is possible to delimit the main causes of death in infancy with some exactitude (Newman, 1906: 44-45).

Introduction

On page 35 of *Infant Mortality: a Social Problem* Newman wrote:

> … the distribution of (the infant) death rate in England and Scotland leads us to believe that there may be, broadly, a line of cleavage between urban and rural conditions, a study of which may throw some light on to the causes of infant mortality (Newman, 1906 [hereafter '1906')).

He had considered the distribution of infant mortality across the counties of England and Wales and had demonstrated that 'agricultural' counties enjoyed much lower infant mortality rates (IMRs) than 'industrial' ones, particularly those where mining or textiles formed a major part of the economy (1906: 23). In Scotland he was able to take his argument a step further, as the *Annual Reports* of the Local Government Board for Scotland, 1899-1904, returned two IMRs for each Scottish county; one for the 'landward', or rural, portion and one for the 'burghal' or 'urban' portion (1906: 30). The urban areas were much more densely populated, and generally had higher rates of infant mortality than the rural areas surrounding them.[2] Newman concludes that:

---

1 This chapter draws on work undertaken as part of an ESRC-funded project: *Determining the Demography of Victorian Scotland through Record Linkage* (award no. RES-00023-0128). A debt of thanks is also owed to the General Register Office for Scotland, Edinburgh for special permission to transcribe the civil registers of the study communities.

2 It is not entirely clear, from the text accompanying the table Newman presents on page 30, whether he is comparing the figures for each county with that of its urban component,

> there is a relation existing between conditions arising out of urban life which are unfavourable to the life of infancy and there are conditions existing in rural districts which are on the whole favourable to infancy (1906: 38).

To further prove his point Newman turns to the Registrar-General's *Report* for 1891 where IMRs for the 'rural' counties of Hertfordshire, Wiltshire and Dorset were compared with those of Preston, Blackburn and Leicester, three towns chosen for their poor rates of infant survival and representative of mining and textile districts where such rates were commonplace. For good measure the figures for five 'mining and manufacturing counties', which were 'neither purely industrial and yet contain(ed) a considerable intermixture of rural elements', were also tabulated (1906: 38). He chose Staffordshire, Leicestershire, Lancashire, the West Riding of Yorkshire and Durham. The table Newman presents very clearly demonstrates that the towns were indeed more lethal for infants, and that this effect increased as the infants grew older (1906: 39). Out of 100,000 infants born in the rural areas some 5,000 died within the first month of life, while of the same number of births in the five counties 8,000 died and in the three towns just over 9,000 perished. In the following eleven months of life a further 5,000 rural infants would die, yet in the five counties this figure would be 9,000 and in Preston, Blackburn and Leicester 13,000.

Newman also derived certain 'broad facts' concerning infant mortality from the Registrar General's tables; four points of similarity and three points of difference between rural and urban situations. The similarities he listed as:

> a. Infant mortality is highest in the first day of life in both the rural and the urban groups.
>
> b. Mortality, although remaining high, then declines throughout the remaining days of the first week.
>
> c. The mortality is at its maximum in the first week, falls enormously in the second week, remains at much the same level through the third week, and falls again in the fourth.
>
> d. Passing from weeks to months the mortality falls enormously in the second month… then it continues very slowly to fall during the remainder of the year (1906:40).

He noted the dissimilarities to be:

> a. The aggregate mortality in the towns is twice as high as it was in the three rural counties.
>
> b. The town rate is throughout the twelve months higher than the rural rate except on the first day of the first week.

---

or whether he is comparing the rural portion of the county with the urban. Exceptions to the general rule that urban areas would be less healthy for infants than rural areas are provided by Orkney, Shetland and Kinross (1906: 30).

c. The town rates are most in excess of the rural rates in the later months of the year (1906: 41).

Finally he considered what was killing the infants. In both the 'rural counties' and 'urban centres' the main causes of death in the first month of life were 'premature birth', 'congenital defects', 'atrophy' and 'immaturity'. From the second month respiratory diseases and diarrhoea began to take their toll along with whooping cough; other infectious diseases would carry off infants later in their first year. Diarrhoea, in its epidemic form, was seven times more lethal in the three towns than it was in the rural counties, although 'all diseases of infancy' were 'heavier in the towns than [in] the counties'. In Newman's view, therefore, high infant mortality was a 'problem of town life'(1906: 42). Rural life, as represented by the agricultural counties was, he opined, 'favourable to infancy, probably owing to social conditions and domestic habits' (1906: 42).

Newman was not the first to commentate on the unequal survival chances of infants. He cites the comments of both Farr (1885) and Ogle (1885) who had addressed the issue two decades previously.[3] More recent researchers such as Williams and Galley (1995), Woods and Shelton (1997), Woods (2000) and Garrett *et al.* (2001) have also pursued the same topic. Until very recently however comparisons within England and Wales, where access to the registers of vital events is somewhat problematic for academics, have largely relied on aggregate statistics for counties, registration districts or registration sub-districts published by the Registrar General. Individual level data indicating the industrial, urban and domestic conditions directly impinging on individual children and their families are very scarce. Comparative studies, using such data to examine conditions in more than one setting, are even rarer.[4] Thus over the last century it has not proven possible to directly test Newman's assertions that it was the 'social conditions and domestic habits' prevailing in the countryside which were protecting rural infants.[5]

The Study Communities, Skye and Kilmarnock

The present chapter turns to Scottish data, considering material collected for a study of the demography of Scotland in the late nineteenth century.[6] The 1861, 1871, 1881, 1891 and 1901 census enumerators records for the crofting communities on the Isle

3 Although Newman cites Farr's work published in 1885, Farr was publishing on the topic of 'healthy' versus 'unhealthy' districts decades earlier. See, for example, Farr (1859, 1864).

4 Williams' work on Sheffield (e.g. Williams, 1992) and Reid's work on early twentieth century health visitor records in Derbyshire are welcome exceptions (e.g. Reid, 2001; Chapter 10 in this volume), as, in combination, are the studies using vaccination registration data being undertaken by postgraduates at the Open University; see Drake, 2003, 2005; Clark, 2003; Davies, 2003; James, 2003; Smith, 2003; and Chapters 6 and 8 in this volume.

5 Kemmer (1997) used civil register material to consider infant mortality, but her comparison was between two major Scottish cities; Aberdeen and Dundee.

6 See footnote 1 above.

of Skye, Inverness-shire and the industrial centre of Kilmarnock in Ayrshire were transcribed and the civil registers of births, deaths and marriages encompassing the intervening four decades for the registration districts covering the two communities were also computerized, making it possible to investigate infant mortality at the level of individual children and their families. Figure 7.1 shows the location of the two study communities and Table 7.1 lists for each of them the numbers of demographic events registered in each decade alongside the population returned at each census.

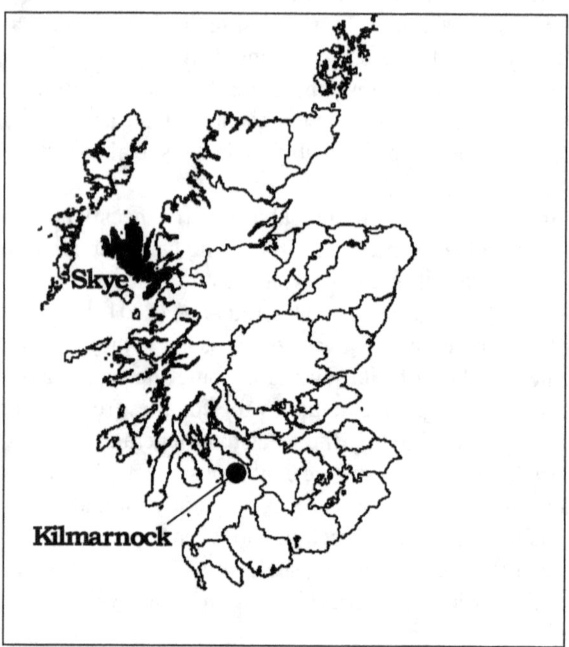

Figure 7.1 The location of Skye and Kilmarnock

The Isle of Skye extends for approximately 560 square miles, giving a population density in 1881 of about 32 persons per square mile. Kilmarnock registration district, in contrast, covered roughly 15 square miles but contained just over 1,720 persons per square mile in 1881. The 'rural' nature of Skye and the 'urban' nature of Kilmarnock are further substantiated by the occupational profile offered by the male heads-of-household enumerated in each location in 1881 (Table 7.2).

Virtually half of all men heading households on Skye returned their occupation as 'crofter' and a further 26 per cent were working in other jobs on the land or at sea. In Kilmarnock, even although the registration district included Kilmarnock's 'landward' parts as well as the 'burgh', only three per cent of male heads worked in agriculture. A quarter worked with metals or machines or in mining, while 15 per cent worked making textiles, carpets or items of dress, particularly bonnets, for which the town was renowned. More than one man in ten gave his occupation as

Table 7.1 The population returned at each census, 1861-1901, for the Isle of Skye and town of Kilmarnock, and the number of births and infant deaths occurring in the intervening decades in each community, along with IMRs

SKYE

Census year[a]	N of population	Decade[a]	N of births	N of infant deaths	IMR[b]
1861	19,604	**1860s**	4,911	520	106
1871	18,099	**1870s**	4,909	497	101
1881	17,683	**1880s**	4,384	452	103
1891	16,372	**1890s**	3,525	322	91
1901	14,606				
		1860s-1890s	17,729	1,791	101

KILMARNOCK

Census year[a]	N of population	Decade[a]	N of births	N of infant deaths	IMR[b]
1861	23,556	**1860s**	9,664	1,258	130
1871	24,072	**1870s**	9,863	1,369	139
1881	25,864	**1880s**	9,078	1,178	130
1891	27,968	**1890s**	9,553	1,235	129
1901	33,142				
		1860s-1890s	39,158	5,040	132

Notes: a) The 1860s were taken to run from 1 Jan 1861-31 Dec 1870. The other decades were defined in the same manner
b) IMR calculated as the number of infant deaths per 1,000 live births
Source: Census enumerators' books and civil registers

'labourer'.[7] If we take Newman's rule of thumb of an IMR of approximately 100 infant deaths per 1,000 births, as indicative of 'rural' conditions, and an IMR of 170 as symptomatic of districts where 'mining and manufacturing' were carried out, then according to the figures in Table 7.1 Skye's IMRs were well in line with its rural setting. Infants in Kilmarnock appear to have enjoyed better survival chances than might have been expected given the town's industrial profile.

7 For fuller descriptions of the two communities at the end of the nineteenth century see Mackay (1992), Nicolson (1994) and Hunter (2000).

Table 7.2 The distribution of male heads of household by occupational sector: Kilmarnock and Skye, 1881

KILMARNOCK Occupation	% of all male heads of household	SKYE Occupation	% of all male heads of household
Agriculture	3	Crofting	49
		Other agriculture	14
		Fishing and seafaring	12
White Collar	5	White Collar	4
Retail	7	Retail	3
Metals	9		
Machines and Machinists	8		
Mining	8		
Carpet making	4		
Textiles	5		
Bonnet making, dress, shoes	6		
House building and maintenance	8		
Railway	3		
Labourers (undefined)	12		
All other occupations	22	All other occupations	18
Total	100	Total	100

Source: Census enummerators' books

Scottish Civil Registers as a Source

Skye and Kilmarnock were chosen for study as representative of 'urban' and 'rural' communities of approximately equal size in 1861. Before the study commenced it was not known what details the demographic records would reveal concerning health, health services and registration practices within the two communities, therefore in the discussion which follows it should be remembered that Skye and Kilmarnock were *not* chosen for their respective positions on a spectrum of record keeping.

In the nineteenth century the civil records of Scotland were, as they are today, administered separately from those of England. Registration was compulsory, but there was no charge (Sinclair, 2000: 36). Deaths had to be registered within eight days of the death occurring, a time limit which still stands.[8] Ideally, a completed death certificate should contain a considerable amount of information.[9] The person informing the registrar of the death was asked to supply the date, time and address of death as well as the name, sex, marital status and age of the deceased. If the deceased was, or had been, married the name of their spouse was noted. If the spouse was male

8 Births had to be registered within 21 days. Thanks to Christopher Beaton of the Records Enterprise Section, GROS for checking the historical regulations concerning registration.

9 The information collected did undergo some changes over time – see Sinclair (2000).

then his rank or occupation was also given, if female then her maiden name was to be supplied. There was a separate column where the name and occupation of the deceased's father was entered along with the maiden name of their mother and her occupation, although in the nineteenth century the latter was usually only entered if she was unmarried. If either parent was dead this was noted beside their name. No matter what their age, if the deceased was illegitimate this fact was entered, even if both parents' names appeared on the certificate. The length of the final illness and the cause of death were recorded, with note being taken of whether the latter had been certified by a medically qualified person, usually the doctor attending the deceased in their final days or hours. In a very few cases the cause of death had to be verified by *post mortem* examination or, if foul play was suspected, the death was referred to the Procurator Fiscal, the Scottish equivalent of a coroner.

When the laws governing the Scottish registration system were laid down in the 1850s, cognizance was taken of the fact that many of the country's inhabitants lived in remote, far flung communities where a doctor could not always be reached in time. In such cases the person reporting the death was allowed to state that the deceased had had 'no medical attendant' during the course of their last illness, and the informant was then permitted to state the cause of death. Williams (1996) notes that similar practices were permitted in England, where civil registration had begun in 1837 and that in 1858 some 17 per cent of death certificates did not carry a medically certified cause of death. By the 1890s only 3 per cent of English deaths went uncertified.[10]

Having given all the required information the informant supplied their own name and their relationship to the deceased. They had to state whether they had been present at the death and, if not, give their address at the time. Finally the date and place of registration were entered, and the certificate signed by the registrar.

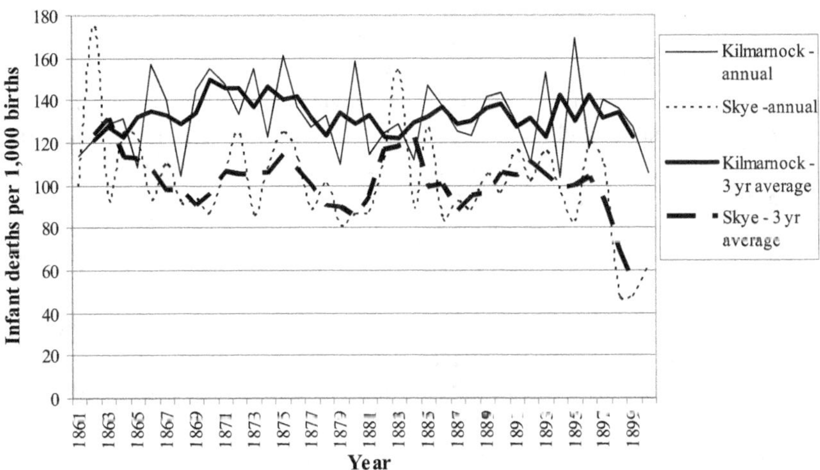

Figure 7.2 Annual IMR and 3 year moving average IMR: Kilmarnock and Skye, 1861-1900

10 See also Vögele (1998: 24).

In combination with the number of births occurring each year, culled from the registers of births, the numbers of infant deaths in the death registers allow the path of IMRs over the last four decades of the nineteenth century to be calculated for both Skye and Kilmarnock (Figure 7.2). What was apparently a combination of influenza, whooping cough and diphtheria on Skye in 1862 meant that mortality on the island was, on average, as high as it was in urban Kilmarnock at the beginning of the study period. Thereafter however the IMRs on Skye fluctuated around 100 deaths per 1,000 births, until the last few years of the century, when they declined dramatically. Only another severe outbreak of 'whooping cough' in 1883 again brought the IMR in Skye above that of Kilmarnock, which lay, on average, between 120 and 140 infant deaths per 1,000 births throughout the four decades observed. Only in the early-1870s and the mid-1890s did Kilmarnock's the three-year moving average IMR edge over 150 per 1,000. The early-1870s rise appears to have been due to a combination of bad years for both diarrhoea and bronchitis, while diarrhoea contributed to, but was not the sole reason for, the rise in the 1890s.

The wealth of detail available on the death certificates of Skye and Kilmarnock can be used to go beyond basic questions concerning the medical conditions which killed infants to examine other aspects surrounding their deaths. Consideration of the records suggests that the medical care of infants, and the processes, if not the system, under which their deaths were registered differed markedly between the two communities being studied. This has several implications for our understanding of the factors underlying the differences in infant mortality between the island and the town, and also for our understanding of mortality records more generally.

Interpreting Cause of Death

Many previous authors, including Hardy (1994), Williams (1996), Woods and Shelton (1997) and Vögele (1998), have drawn attention to the problems of disease classification in historical sources, and the use of published, aggregate data has been hampered by the changing classification schemes used by the reporting authorities. Hopes that recourse to original, individual level certificates would enable a clearer picture of the path of mortality, and the history of disease, to be drawn have proved ill-founded. The cause of death appearing on a death certificate is the result not simply of medical expertise, easily classified under a particular heading by the registrar. Which cause of death was entered on the certificate depended firstly on whether or not the deceased received medical attention in the course of their last illness, secondly on the qualifications and circumstances of the person, professional or not, reporting the death, and thirdly on the regulations and process governing registration of the fact that a death has occurred.

Where someone had not had a medical attendant in their last illness this was sometimes entered as 'no medical attendant' on the death certificate, indicating that a doctor could not confirm the nature of the ailment. However the term 'not medically certified' is also frequently found on certificates. In such cases it is not possible to tell whether someone had not been seen by a medically qualified person prior to death, or whether they had been seen, but that it had not proven possible, for whatever reason, to acquire that doctor's certification of the cause of death. Perhaps the doctor lived at a distance and neither party felt willing or able to make an additional journey; perhaps the doctor was away attending another patient and the funeral arrangements could not be postponed; perhaps the family were unable, or unwilling, to meet with further expense. For whatever reason, from Figure 7.3, it can be seen that infants on Skye were much less likely to have the cause of their death certified than were infants in Kilmarnock where, by 1900, such certification was practically universal.

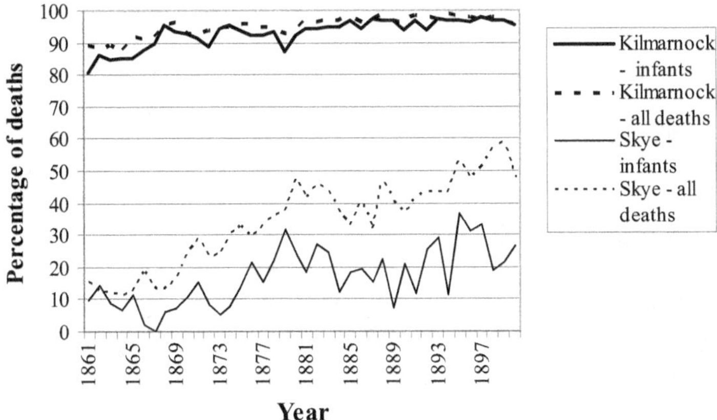

Figure 7.3 The annual percentage of infant deaths and all deaths which were medically certified: Skye and Kilmarnock, 1861-1890

Of the 5,040 infant deaths registered in the civil registers of Kilmarnock between 1861 and 1900 4,691, or 93 per cent, had a medically certified cause of death. In 261 (75 per cent) of the remaining 349 cases, the reason for non-certification was given as 'no medical attendant'.[11] Among infants who succumbed before they were four weeks – or 28 days – old, 11 per cent were not attended by a doctor shortly before their death, while among older infants only 3 per cent passed away without medical attention. In all, some five per cent of fatally ill infants in Kilmarnock in the last four decades of the nineteenth century did not received medical help. Was this through choice, poverty or circumstance?

If a mother in labour was not attended by a doctor there may have been no time to summon one when things went wrong. Nine per cent of all infants who had their death medically certified in Kilmarnock survived for less than 24 hours after birth, but amongst those whose deaths were not medically certified, 30 per cent were less than a day old. Further, while 12.7 per cent of certified infant deaths were ascribed to 'premature birth', 29.5 per cent of deaths of 'uncertified' children were ascribed to this cause.[12]

11 In 98 cases it was not clear whether the infant's cause of death had been medically certified, a post mortem carried out, or the death referred to the Procurator Fiscal.

12 Cause of death classification is the author's own, derived from the causes entered on the death certificates.

Table 7.3 **The percentage distribution of selected causes of death amongst neonates, post-neonates and all infants, where causes of death are medically certified and uncertified: Kilmarnock, 1861-1900**

	Neonates				Post-neonates				All infants			
	Uncertified		Certified		Uncertified		Certified		Uncertified		Certified	
Standard cause of death[a]	N	%	N	%	N	%	N	%	N	%	N	%
Bowels	28	16.8	57	4.2	17	18.3	126	3.8	45	17.2	183	3.9
Diarrhoea/cholera	0	0.0	46	3.4	3	3.2	534	16.0	3	1.1	580	12.4
Bronchitis	2	1.2	71	5.2	11	11.8	794	23.8	13	5.0	865	18.4
(Broncho-)pneumonia	0	0.0	15	1.1	0	0.0	169	5.1	0	0.0	184	3.9
Consumption[b]	0	0.0	6	0.4	7	7.5	122	3.7	7	2.7	128	2.7
Croup	0	0.0	2	0.1	2	2.2	36	1.1	2	0.8	38	0.8
Debility	50	29.9	298	21.9	15	16.1	116	3.5	65	24.9	414	8.8
Premature birth	71	42.5	540	39.7	6	6.5	54	1.6	77	29.5	594	12.7
'Unknown'/not given/blank	7	4.2	4	0.3	3	3.2	3	0.1	10	3.8	7	0.1
All other causes	9	5.4	321	23.6	29	31.2	1,377	41.3	39	14.9	1,698	36.2
All deaths	167	100.0	1,360	100.0	93	100.0	3,331	100.0	261	100.0	4,691	100.0
% of deaths certified	**89.1**				**97.3**				**94.7**			

Notes: a) Standard causes – for discussion see text. b) 'Consumption' includes deaths registered as being from 'consumption', 'phthisis' and 'tuberculosis', where the seat of the disease was unspecified, or said to be 'pulmonary' in nature

Source: Civil Registers of Deaths, Kilmarnock, GROS

Altogether almost two thirds (64 per cent) of non-certified infant deaths in Kilmarnock occurred to babies who were less than a month old, whereas only 29 per cent of certified infant deaths were in this age group. Thus in order to compare the causes of death between the 'certified' and the 'uncertified' groups we must allow for age. Table 7.3 tabulates cause of death among neonates (those aged less than 28 days), post-neonates (those aged 28 days or over, but less than 366 days) and all infants. The causes are 'standardised' rather than 'classified' because they do not attempt to categorise the deaths into any particular classification scheme, but rather to standardize the information given on the death certificates, to allow for mis-spellings, the use of synonyms such as 'pertussis' and 'whooping- (or 'hooping-' or 'chin-') cough', and for the entry of multiple causes of death. The cause of death used in the table was usually the first one listed, although there were a few exceptions, for example where the second cause was more specific than the first as in the entry 'fever, measles'. The table shows that approximately 40 per cent of both 'certified' and 'uncertified' neonatal deaths were the result of 'premature birth'. The much larger proportion of neonates whose death was not certified meant, however, that when 'premature birth' was considered as a cause of death amongst all infants, the percentage of deaths due to this cause was far higher amongst 'uncertified' that 'certified' infant deaths.

'Debility' is also over-represented amongst all 'uncertified' infants, being principally an ailment of neonates in Kilmarnock, although lay people were rather more apt than doctors to use this term for older infants. Twenty-five per cent 'uncertified' infants reputedly succumbed to this cause of death, whereas only nine per cent of 'certified' infants did so.

The fact that the terms used to describe the cause of death were those of lay people in some cases and of doctors in other cases must also be considered. Table 7.3 suggests that in Kilmarnock doctors were rather less likely to say that they did not know the cause of death, than were lay people. Only seven (0.1 per cent) of the 4,691 medically certified death certificates had their cause of death entries left blank, or entered as 'unknown'. Among the 'uncertified' infant deaths ten, or almost four per cent, were of 'unknown' cause. Table 7.3 shows, however, that doctors were much more likely to report diarrhoea as a cause of death, than were non-medical informants who were more likely to report 'bowels' as the killer complaint. In the vast majority of these cases the actual words used on the certificate were 'inflammation of the bowels'. The symptoms of the two conditions may have been very similar, but it is possible that 'diarrhoea', with its complicated spelling, was seen as a 'medical' term by lay people, who preferred to use the vernacular when reporting the death verbally to the registrar. When a doctor provided a diagnosis he did so in writing, and this would have been taken, or sent, to the registrar (Sinclair, 2000: 46). It is of interest that 16 per cent of post-neonatal deaths certified by doctors were from diarrhoea, whereas only 3.4 per cent of certified neonatal deaths were. However, deaths certified as being from bowels made up a roughly similar proportion in each age group, therefore doctors may have been of the opinion that two separate conditions were involved. Terms such as 'suffocation', 'syphilis' and 'tetanus' were also given by doctors as causes of death, but lay people either avoided their use, or either did not, or chose not to, recognize their symptoms. Again 'bronchitis', 'pneumonia' or 'broncho-pneumonia' were far more likely to be given as diagnoses by doctors than non-qualified informants, who were rather more prone to use 'consumption', 'phthisis' or 'tuberculosis', particularly amongst post-neonates, but it is not clear whether the second group of terms was being used as synonyms for the first or not.

When we turn to Skye, the picture becomes more complex. First of all, the proportion of deaths certified by a medically qualified person was much lower than that in Kilmarnock, as Figure 7.3 showed. In the 1860s fewer than 20 per cent of infant deaths had a medically certified cause of death, and by the late 1890s the figure had scarcely risen beyond 30 per cent. In Kilmarnock infants were slightly disadvantaged relative to the rest of the population but that disadvantage had disappeared by the end of the 1890s. Figure 7.3 indicates that on Skye the proportion of infants being seen by a doctor before death in the 1860s was little different than that amongst the population as a whole, but over the later decades of the century the rate of certification improved less rapidly among the youngest children than it did for adults and older children.

The difference in the proportion of infants who had their deaths medically certified suggests that the population of Skye had poorer access to medical aid either in terms of proximity or of ability to pay any fees the medical man saw fit to charge.[13] The journey to fetch a doctor must have been a long and time-consuming one for many inhabitants of Skye: doctors were few and the population far flung and thinly spread. We do not know how far a parent might have had to travel to solicit the help of a doctor, nor how long a wait they might have had before the doctor managed the journey to where their child lay sick. The doctor then had to be paid both for his attendance and for any medicine prescribed.

The medical men had to meet the cost of traveling to the patients out of their fee, and if the population were generally healthy, and reluctant or unable to pay for their treatment when they needed it, there was little incentive for medical men to become 'Highland Doctors' (McRae, 2005). Dr McTavish, a surgeon at Kilarrow, writing in the mid-nineteenth century, illustrates why doctors may have been less than enthusiastic to attend a call:

> … the discomfort a medical man endures by being obliged to ride or drive eight or ten miles in severe weather and then forced to remain for many hours in a miserable hovel without warmth and almost invariably without food, is unknown to those who have not experienced it (quoted in McRae, 2005).

Dr Livingston, working near Ballachulish on the mainland, displays a similar frustration with his lot:

> … after all that fatigue and expense I am sorry to say that not more than three out of ten (patients) can pay anything. So in a pecuniary point of view, I would be much better stopping at home had it not been for the suffering of humanity (quoted in McRae, 2005).

---

13 The workings of the Poor Law, in providing medical aid for the needy, have not yet been studied in the two communities.

Table 7.4 **The percentage distribution of selected causes of death amongst neonates, post-neonates and all infants, where causes of death are medically certified and uncertified: Skye, 1861-1900**

	Neonates				Post-neonates				All infants			
	Uncertified		Certified		Uncertified		Certified		Uncertified		Certified	
Standard cause of death[a]	N	%	N	%	N	%	N	%	N	%	N	%
Bowels	2	0.2	3	3.0	2	0.3	7	3.7	4	0.3	10	3.5
Diarrhoea/cholera	0	0.0	1	1.0	2	0.3	1	0.5	2	0.1	2	0.7
Bronchitis	5	0.6	6	6.1	26	4.2	50	26.5	31	2.1	56	19.4
(Broncho-)pneumonia	0	0.0	2	2.0	0	0.0	13	6.9	0	0.0	15	5.2
Consumption[b]	0	0.0	1	1.0	6	1.0	4	2.1	6	0.4	5	1.7
Pleurisy[c]	109	12.5	0	0.0	67	10.9	1	0.5	176	11.8	1	0.3
Cold[c]	7	0.8	0	0.0	23	3.7	0	0.0	30	2.0	0	0.0
Croup	18	2.1	1	1.0	23	3.7	4	2.1	41	2.8	5	1.7
Influenza[c]	7	0.8	0	0.0	29	4.7	2	1.1	36	2.4	2	0.7
Debility	4	0.5	10	10.1	2	0.3	4	2.1	6	0.4	14	4.9
Premature birth	38	4.3	17	17.2	1	0.2	1	0.5	39	2.6	18	6.3
'Unknown'/not given/blank	561	64.3	1	1.0	244	39.7	1	0.5	805	54.2	2	0.7
All other causes	146	16.7	58	58.6	241	39.2	107	56.6	387	26.0	165	57.3
All deaths	872	100.0	99	100.0	614	100.0	189	100.0	1,486	100.0	288	100.0
% of deaths certified			**10.2**				**23.5**				**16.2**	

Notes: a) and b) See Table 7.3, c) Neither 'pleurisy', 'cold' nor 'inluenza' featured amongst the causes of death registered in Kilmarnock so none of these causes appear in Table 7.3. Source: Civil Registers of Deaths, Skye, GROS

Between 1861 and 1900 only 16 per cent of Skye infant death certificates included a medically certified cause of death. Amongst neonates only one in every ten deaths was medically certified, and amongst post-neonates slightly fewer than one in every four. Comparison of Table 7.4 with Table 7.3 shows that a second difference in certification between that of Skye and that of Kilmarnock was that there was a far greater use of the term 'unknown' on Skye. When a doctor was in attendance a cause was almost invariably given, but on a total of 59 per cent of neonatal death certificates the cause of death was registered as 'unknown'; for post-neonatal certificates the figure was 30 per cent. This makes it very difficult to draw comparisons between the Skye data and that from Kilmarnock, and also when comparing the 'medically certified' and non-'medically certified' deaths on Skye. Although comparison of medically certified causes of death in both communities is possible, until more is known about medical services on the island, the suspicion remains that infant deaths attended by a doctor on Skye were in some way 'selected'. Did parents feel doctors could treat bronchitis but that they had little to offer when a child came down with 'pleurisy', for example?

What is evident from Table 7.4 is that on Skye too, doctors and lay people were apt to use different terms. Doctors reported 8 per cent of the neonatal and 33 per cent of the post-neonatal deaths they attended as being due to 'bronchitis' or '(broncho-)pneumonia', while the lay people favoured 'cold', 'influenza' 'croup' and, in particular, 'pleurisy'; terms scarcely used by the medical men. While indicating that respiratory diseases were a major risk for infants on Skye, scarcely a surprise perhaps given the prevailing weather and the poor quality of their housing, the various causes of death reported give rise to several questions. Can the terms used by the lay people be considered synonymous with the medical term 'bronchitis'? When the causes of death were tabulated for publication by the General Register Office was recognition given to the qualifications of the person certifying the death, and did that affect into which category the cause of death was put? Was the lay diagnosis of 'pleurisy' counted in with the medical 'bronchitis', or was it placed in with 'diseases of the lung' or even in 'other causes'? While such questions matter little in Kilmarnock, as the proportion of lay certified deaths was so small, they take on much greater significance when considering the rural population of Skye, and may affect our ability to accurately compare the two settings. Without further work on death certificates from other rural communities it is not possible to say whether or not Skye is an exceptional case. However, in any comparative study variations in the proportion of lay versus medically reported causes of death could potentially affect the relative mortality profiles presented by cause for the different communities, and compromise comparisons made.

Neither doctors nor laymen reported deaths from diarrhoea as a major cause of death on Skye, although doctors did certify a rather higher percentage of 'stomach' and 'bowel' related complaints. However, if the general population did not see diarrhoea as a cause of death, but only as a symptom of another condition they had been unable to diagnose, then they may have chosen not to give diarrhoea as the cause of death. With over half the causes of deaths being unknown this possibility cannot be dismissed. Premature births are another case in point. While doctors on Skye diagnosed this as the cause of death in 17 per cent of the deaths they certified, only four per cent of the lay informants did so. It seems unlikely that this major cause of death, one of the main killers of infants in Kilmarnock, did not exist on Skye, and yet it was not apparently acknowledged by the population. Perhaps women on the island were unsure of the date on which their babies were due, and therefore could not say that the baby had been born prematurely, whereas the doctors could recognize clinical signs that the child had arrived too early. Or perhaps the mothers just assumed that new babies often died and did not view their early arrival as a major factor in their death. 'Weakness' was a word used by informants, particularly of neonatal deaths, but, as in Kilmarnock, 'debility' was predominantly used by doctors.

Table 7.5 illustrates the problem of comparing rates of particular causes of death amongst infants in our 'rural' and 'urban' settings. The table lists IMRs for selected causes of death and indicates the percentage of each cause which was medically certified. The fact that certain causes of death on Skye were much more likely to be medically certified than others, coupled with the high proportion of deaths of 'unknown' cause, makes interpretation of the table virtually impossible and suggests that it is unlikely that the underlying risks to infant survival on Skye relative to those of Kilmarnock will ever be fully understood.

Woods and Shelton (1997) have drawn attention to the problems caused by regional variations in the proportion of deaths ascribed to 'other causes' in the Registrar General's returns for England and Wales. If a relatively high proportion of the population of a registration district, or even a county, were rural, had restricted access to medical assistance and therefore failed to return a certified cause of death then many 'unknown' causes of death may have found their way into the 'other causes' category. Williams, writing of the mid-nineteenth century, reports that:

> ... it was precisely where medical men were thinnest on the ground – rural, isolated districts – that medical certification tended to be most deficient (Williams, 1996: 61).

Things improved over the second half of the century but the regional pattern,

> ... remained remarkably stable, with rural areas continuing to experience the highest rates. In North Wales, 11 percent of deaths were still uncertified in 1881 and even in the 1890s 8 percent were uncertified, compared to only 1 percent of deaths in London at both dates (Williams, 1996: 61).

It should be noted that Williams was discussing *all* deaths which went uncertified, She emphasizes that, as on Skye, deaths of infants were much less likely to be certified than those of adults or older children.

Table 7.5 Infant mortality rates, by selected cause of death, showing percentage of the deaths from each cause which were medically certified: Skye and Kilmarnock, 1860s-1890s

	Skye		Kilmarnock	
Standard cause of death[a]	IMR per 10,000 births	% of deaths medically certified	IMR per 10,000 births	% of deaths medically certified
Unknown/not given/blank	455	2.5	4	41.2
Bowels	8	71.4	60	80.3
Diarrhoea	2	50.0	153	99.5
Bronchitis/broncho-pneumonia/pneumonia	57	70.0	278	98.8
Consumption[b]	6	45.4	35	94.8
Pleurisy	100	0.6	0	-
Cold	17	0.0	0	-
Croup	26	10.9	10	95
Influenza	21	5.3	0	0.0
Debility	11	70.0	125	86.4
Premature birth	32	31.6	176	88.5
All other causes	311	29.9	455	97.8
All infant deaths	1,001	16.2	130	94.7
N of births	17,729		38,158	

Notes: a) and b) see Table 7.3. Source: Civil Registers of Births and Deaths, Skye and Kilmarnock, GROS

Death Registration on Skye

A further point concerning regional differences in England and Wales is brought to mind when considering the Skye data. It is noticeable, when working with the Skye civil registers, that all vital events were recorded in English – even names were Anglicised in these official records – yet this was a second language for the majority of Skye's inhabitants, and a foreign tongue for some.[14] The 1901 census enumerators' books disclose that at that date 79 per cent of the island's population aged 16 or over spoke both Gaelic and English, and a further 20 per cent spoke only Gaelic.[15] We must therefore imagine that in many instances causes of death were being reported, or symptoms described, to the registrar in Gaelic to then be translated into English. How accurate were the registrars in their translation? If they could not manage a suitable translation was the cause of death simply entered as 'unknown'? This raises the question of whether a similar situation arose in Welsh-speaking parts of Wales. How did the population and the registrars cope there? Woods and Shelton (1997) demonstrate that, while other age groups in Wales had high proportions of 'other causes' of death reported, amongst infants the proportion was 'surprisingly low' in both North and South Wales (Woods and Shelton, 1997: 43, Maps 4-6: 44). 'Diseases of the brain' were however 'especially prominent' as a cause of death amongst infants and young children in Wales, suggesting that this may have been the cause of death favoured when diagnosis, translation difficulties or a lack of knowledge prevented a more accurate assessment (Woods and Shelton, 1997: 43, Map 10: 63, Map 19: 91).

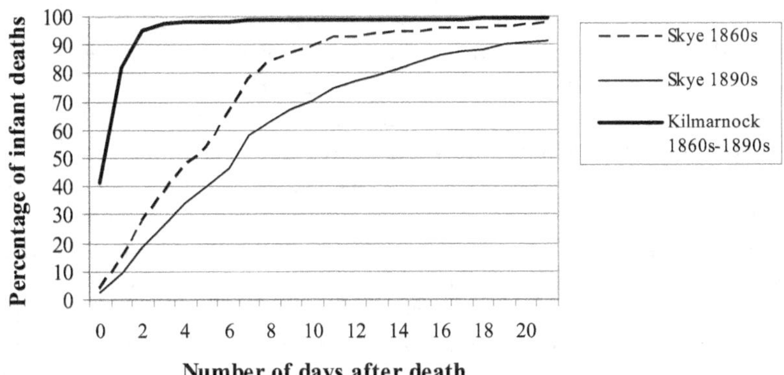

Figure 7.4 The cumulative percentage of infant deaths registered by number of days elapsed after death: Skye 1860s and 1890s and Kilmarnock, 1860s-1890s

14 For a history of the use of the English language in the Highlands of Scotland, see Withers (1988).

15 The 1891 and 1901 census questions on language were almost universally answered on Skye. Only 48 (0.5 per cent) persons stated that they were monoglot English speakers. In Kilmarnock the column containing the language question was seldom filled in. Researchers must either assume that the person enumerated spoke only English, or treat the question as 'unanswered'.

The remoteness of life for the population of Skye, and no doubt other rural populations in the nineteenth century, relative to their urban cousins can be gleaned from another item on the death certificates: the date on which the death was certified. Subtracting this from the date on which the death was reported to have occurred, gives an indication of how relatively far flung and remote the population was.[16] Figure 7.4 compares the time it took to register the deaths of infants in Kilmarnock 1861-1900, with the time which elapsed between an infant dying and this event being registered on Skye in the 1860s and the 1890s. The date of registration was not transcribed for all 40 years on Skye and only those cases were both a date of death *and* the date of registration are known were considered when compiling Figure 7.4. Two points in the Figure stand out. First is the fact that over 95 per cent of infant deaths in Kilmarnock were registered within two days of the death occurring; more than 40 per cent were registered on the day death occurred. On Skye fewer than five per cent of dead infants had their demise registered on the day it occurred. It took until the fourth day after death before half of the Skye deaths were registered in the 1860s, and a week for half the deaths to be registered in the 1890s. By the eleventh day after death had occurred 90 per cent of all infant deaths on the island had been registered in the 1860s, but it was almost three weeks before the same proportion had been registered in the 1890s. The law, it should be remembered, stipulated that deaths should be registered within eight days. Why the reporting of deaths on Skye should have become slower in the 1890s is not clear. Birth registration also took longer on average in the 1890s than the 1860s. In the latter decade 50 per cent of registered births were reported to the registrar within 14 days, in the 1890s it took until 20 days after birth before 50 per cent of births were registered.[17] It is possible that, as the reporting of civil events was a predominantly male task in both communities, the downturn in the island's economy, reflected in the accelerated decline in population over the 1880s and 1890s apparent in Table 7.1, coupled with increasing levels of temporary migration to the mainland and beyond, meant that it took longer to find someone able to travel to the registrar.[18] It would seem likely, given Figure 7.4, that many infants were buried before their death was registered, despite a legal requirement for a death certificate to be obtained before an interment could proceed (Flinn *et al.*, 1977). Graveyards on Skye, and elsewhere in Scotland's North-West, are often sited apart from churches, in locations where the soil was deep enough to receive coffins, and even the smallest communities would have their own. The journey to the graveyard would have been far shorter than the journey to the Registrar's for many of Skye's inhabitants and the possibility has to be considered

---

16 Data on the names of the registrars and where registration took place (both noted on the civil registration certificates) were not systematically collected for the births and deaths of Skye and Kilmarnock. Had they been it would be possible to work out the exact distances the informants had had to travel to fulfil the legal requirement to register these events. It is hoped the information will be entered in the future, to give further insight into this question.

17 By law births had to be registered within 21 days, so it is probable that a child dying in the first few days of life would have its birth registered more promptly than if it had survived.

18 The death registers of both Skye and Kilmarnock show that in both communities over 80 per cent of infant deaths were reported to the registrar by male informants; the vast majority being closely related to the deceased.

that some infant deaths may not have been registered, particularly those occurring soon after a journey had been made to register the birth. Identifying the number of such cases with certainty is, however, far from easy.

Death Registration in Kilmarnock – The Role of Doctors

The death registers of Kilmarnock allow us to consider another potential source of confusion when considering cause of infant death: differential diagnosis between doctors. That this factor may have affected the reporting of deaths has been suspected by several authors, Williams (1996) and Kemmer (1997) among them, but without a considerable body of death certificates, the scale of the problem has been little researched.[19] So far the current study has only considered data from Kilmarnock for the period 1891-1900, so findings can only be suggestive, but they are indicative that doctors from different graduating cohorts and backgrounds may have certified causes of death differently.

Between 1891 and 1900 the names of 80 different doctors appear on the certificates relating to deaths registered in Kilmarnock. Many of the doctors were only resident in the town for a few weeks or days, possibly acting as *locums* for doctors who were absent or ill, and others may not have dealt with infant patients. In all, 42 doctors signed at least one infant death certificate in the course of the decade. Twenty-four of the doctors signed the certificates of fewer than five infants, while nine signed 50 or more. The latter doctors are considered in Tablc 7.6. It can be seen that out of the nine, six share surnames, suggesting a family connection, and where this is the case, initials are used to distinguish individuals. Dr Rankin certified 175 infant deaths over the decade; 15 per cent of the total 1,200, while Dr D. Lawrie certified 147 (12 per cent) and Dr J. Robertson 125 (10 per cent). Drs Prentice and J. Lawrie had not had time to contribute such large numbers to the total as they were only certifying deaths in the town from 1896 onwards, presumably having arrived in that year. Dr Prentice had previously worked in the Victoria Infirmary, Glasgow and Dr Lawrie at the Royal Hospital for Sick Children in Edinburgh (*London and Provincial Medical Directory*, 1901). Although at least six of the nine doctors listed in Table 7.6 were connected in some capacity to Kilmarnock Infirmary, few of their tiny patients died in hospital. Kilmarnock 'Fever Hospital and Infirmary' had been set up by 'voluntary efforts' in 1868, but it was not until 1891 that a 'Children's Block' was opened (Mackay, 1992: 186). Up to that date only five infants were recorded as dying in the Infirmary, and over the 1890s only fourteen more infant patients died as in-patients. The doctors must therefore have attended the vast majority of dying infants in their own homes. The columns headed by letters in Table 7.6 indicate the proportion of deaths from a particular cause which were certified by each of the nine doctors. Thus 32 infant deaths over the 1890s in Kilmarnock were certified as being due to conditions of the 'Bowels' (Column A, Table 7.6). Eighteen of these, 56 per cent of all 'bowels' deaths, were certified by Dr Rankin and a further six (19 per cent), by

19 Kemmer (1997: 14) notes 'differences in diagnostics' between 'medical attendants' in Dundee and Aberdeen.

Table 7.6 The percentage of selected causes of death certified by doctors certifying over 50 infant deaths in Kilmarnock, 1891-1900

					Cause of Death									
Doctor	A	B	C	D	E	F	G	H	J	K	L	M	**% of all infant deaths certified by doctor**	**N of all infant deaths certified by doctor**
Rankin	56	23	10	18	22	18	2	4	3	78	0	11	15	175
Lawrie, D.	9	15	11	16	16	7	17	14	11	0	21	14	12	147
Robertson J.	0	8	6	11	6	13	31	11	15	0	10	9	10	125
MacLeod, W.	0	6	16	13	1	10	7	9	3	0	12	16	8	96
Prentice	0	6	14	3	6	2	0	8	15	0	12	4	7	87
MacLeod, D.	19	9	1	13	0	16	2	11	1	3	2	9	7	86
Robertson, R.	0	2	13	11	7	1	10	6	14	0	12	5	6	76
MacDonald, D.	6	2	5	3	13	3	7	4	7	0	8	7	6	66
Lawrie, J.	3	5	9	0	0	1	2	6	15	0	0	4	5	58
% of cause certified by:														
9 doctors listed	94	75	85	87	70	72	79	73	84	81	75	77	76	916
All other doctors	6	25	15	13	30	28	21	27	16	19	25	23	24	284
N of all infant deaths from cause	32	128	93	38	90	137	42	160	110	37	52	57		1,200

Notes: A: Bowels, B: Bronchitis, C: (Broncho-) pneumonia, D: Convulsions, E: Debility, F: Diarrhoea/cholera, G: Meningitis, H: Premature birth, J: Stomach, K: Teething, L: 'Consumption'– (pulmonary) tuberculosis, (pulmonary) consumption and (pulmonary) phthisis, M: Whooping cough

Source: Kilmarnock death certificates 1890-1900.

Dr D. MacLeod. Thus, in all, these two doctors certified 75 per cent of all 'bowels' deaths. It is therefore of interest that both of these doctors, when identified in the 1891 census, are found to have been relatively elderly: Dr Rankin was reported to be 59 and Dr MacLeod, 69. The doctors often provide a note of their qualifications alongside their signature, but the *London and Provincial Medical Directory* furnishes further details, such as their *alma mater* and the date of their graduation. According to the *Medical Directory* both Dr D. MacLeod and Dr Rankin qualified in the 1850s, and perhaps this led to rather 'old-fashioned' diagnoses. The same two doctors were the only two in the list of nine, to certify 'teething' (Column K, Table 7.6) as a cause of death in the last decade of the nineteenth century, Dr Rankin certifying 78 per cent of all the deaths from this cause. While he certified 15 per cent of all infant deaths, Dr Rankin certified 23 per cent of such deaths from 'bronchitis' and 22 per cent of those from 'debility', but only 2 per cent of deaths from 'meningitis', 4 per cent of those from 'premature birth' and 3 per cent of those from 'stomach' related causes. Dr J. Robertson, on the other hand certified over 30 per cent of all infant deaths from 'meningitis' occurring in the town, although he certified only 10 per cent of all infant deaths. Dr Robertson had qualified in 1889 and was only 27 years old in 1891.

However, as doctors attached to the Infirmary would have had particular specialisms and interests, depending on whether they were surgeons, like Dr D. MacLeod, or physicians, such as Dr J. Robertson, the cases they treated may have been selected from a particular disease profile, and this could result in their apparently excessive contribution to the certification of deaths from particular causes. Considerably more work to uncover the biographies, expertise and roles within the health care system operating in Kilmarnock of the town's medical personnel will be necessary before a full understanding of their influence on the cause of death statistics published for the town can be achieved.

Seasonality of Death

Use of cause of death information has been demonstrated to be problematic in both our rural and our urban communities. However, their mortality profiles can be compared in other ways which cast light from different perspectives on the insults to infant health on rural Skye and urban Kilmarnock.

Figure 7.5 shows the seasonality profile of infant deaths in both Kilmarnock and Skye for each of the decades between 1861 and 1900. The figure has been compiled, using a 'crop year', i.e. beginning the annual cycle in October, rather than the more conventional January. The figures do not reflect absolute levels of mortality, but the relative distribution of deaths across the year. If deaths in each decade were spread across the year in the proportion that each month contributed days to the year, then each monthly point would have an index of 100. When the number of deaths in a particular month is smaller than such a distribution would lead one to expect then the index for that month is less than 100, if larger then the index is greater than 100. In broad outline, and allowing for fluctuations over the decades, the seasonal profiles of Skye and Kilmarnock are not dis-similar, despite very different seasonal patterns

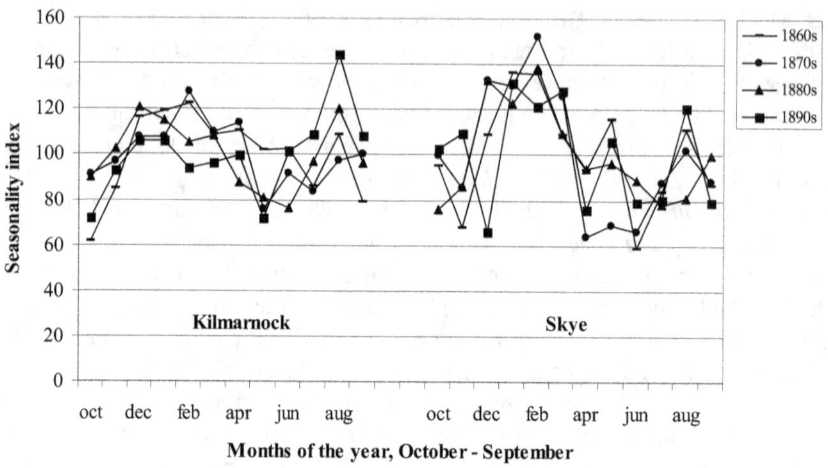

Figure 7.5 The seasonality of infant mortality: Kilmarnock and Skye, by decade 1860s-1890s

in births.[20] Both have excess deaths in the winter months between December and March, and a marked peak in August with relatively few deaths between April and July and September and November although May on Skye saw an upward 'blip' in two of the four decades. However, the winter peak is much more pronounced on Skye, while the August peak is more pronounced in Kilmarnock, although this is due to a particularly high peak in the 1890s. As outlined by Newman (1906, Chapter VI) and others (Morgan 2002, Hall and Drake in this volume, for example), summer peaks in mortality were commonly associated with urban areas and outbreaks of epidemic diarrhoea. However, the August peak is also quite discernable on Skye, especially in the 1890s, and this raises questions about the general lack of diarrhoea deaths recorded in Skye's death registers, or whether diseases other than diarrhoea were contributing to the August peak both on Skye, and possibly elsewhere. The concentration of deaths in the winter months shown in Figure 7.5 does, however, appear to confirm the preponderance of respiratory related diseases in the mortality profiles outlined in Tables 7.3 and 7.4 above.

---

20 Births on Skye were highly seasonal in the second half of the nineteenth century, with a dearth of births in March, April and May, presumably as a result of the absence of menfolk during the previous summer. A distinct peak of births occurred in December and January, presumably conceived in the spring before the men went to sea or seasonal migration to the mainland began. Births in Kilmarnock were much more evenly spread across the year.

Age Specific Mortality Patterns within the First Year

Housing, as has already been suggested, has to take part of the blame for the preponderance of respiratory diseases amongst infants on Skye. Writing in 1890/91 Dr Grant, County Medical Officer of Inverness, describes a typical 'black house' in which many of the crofting families on Skye and the neighbouring areas lived:

> The house is divided into three apartments with one entrance. The first apartment is used for housing the cattle, the manure and house refuse being allowed to accumulate here for months and in some cases until required for agricultural purposes. The second apartment used as (the) kitchen, is divided from the first by a rudely constructed partition, a few feet high. The fire is on the middle of the floor and a small hole in the roof serves to let out the smoke. As a rule there is no window, but a small pane of glass is fixed in one corner, where the roof joins the wall. Beyond the kitchen is the third apartment, used as a bedroom with two or more beds according to the number of inmates. This apartment is also used as a store room (Ferguson, 1958: 109-110).

The houses partly earned their 'black' soubriquet from the soot deposits which gathered round the inner walls from the peat fire which, coupled with the lack of a chimney, meant that the inhabitants, particularly when shut in by winter weather, would have lived in a persistently smoky, if fragrant, atmosphere which must have aggravated any respiratory condition.[21]

Given the apparent proximity of livestock and their effluvia to the living quarters of their owners, it is perhaps remarkable that more stomach and bowel complaints did not appear in the mortality record of Skye, as might have been expected were similar living conditions to be found in an urban area. Perhaps infants, being exposed

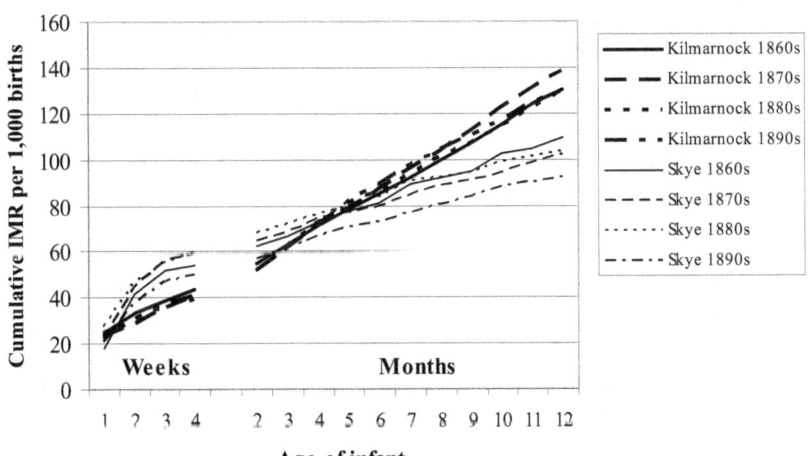

Figure 7.6 Cumulative IMR in the first four weeks and following 11 months of life: Kilmarnock and Skye, 1860s-1890s

21 In fact the classification of 'black' and 'white' crofters houses was used by the 1884 'Crofting Commission' to differentiate between houses built solely by their occupants (the black) and those built with help from the local landed proprietor (the white) (Ferguson, 1958: 108).

to such hazards from very early on, gained a measure of immunity to pathogens which their urban cousins, being far less exposed to such accumulations of 'filth' in the early weeks of their life could not hope for.

To examine this possibility, the age profiles of infant deaths in both Skye and Kilmarnock were examined. For each of the four decades Figure 7.6 graphs cumulative IMRs for the first four weeks of life, and then for each subsequent month over the first year. The right hand part of the figure indicates that, in common with other rural-urban comparisons, (Galley and Woods, 1999; Reid, 2002), the town of Kilmarnock, had considerably higher IMRs than those of Skye, particularly in the six oldest months. For both communities the mortality rates climb steadily from the second month onwards, although those for Kilmarnock rise more steeply, indicating infants were at greater risk of post-neonatal mortality. On Skye post-neonatal mortality rates (P-NMRs) appear to have fallen in the 1890s, particularly amongst those surviving to the tenth month or beyond. The age-specific IMRs on which Figure 7.6 is based show that the P-NMR for this age group fell from 14 deaths per 1,000 births in the 1860s to just under 8 deaths in the 1890s. The same figures show that in Kilmarnock the oldest infants – those in their tenth to twelfth months – also saw a marked decline in mortality from the 1870s onwards. Amongst those in the second and third months of life in the town mortality rose over the 1860s, 1870s and 1880s but fell in the 1890s, whereas amongst those in the fourth to sixth months of life infant mortality rose to peak in the 1890s; those aged seven to nine months experienced a notable decline in mortality between the 1870s and 1880s and only a minor rise in the 1890s. It would appear, therefore, that the infants contributing most to the marked August peak in infant mortality seen in Kilmarnock in the 1890s were those in the fourth to sixth months of life. This age group was the most likely to be being introduced to solid foods.

It is unlikely that infants were being weaned entirely at this age. It should be noted that Dr Stark, Newman's principal informant on Scotland (Newman 1906: 229-231), was of the opinion that no infant was fed 'spoon-meat' 'until it is nine months old or has cut its front teeth' as the Scots were 'aware that spoon-meat given before that period is extremely apt to bring on convulsions' (1906: 229).[22] Stark cited the 1860 death rate of 35 deaths per 1,000 births from convulsions in England against the Scottish rate of 6 per 1,000 as evidence of the wisdom of the Scottish method of infant management (1906: 231). Dr Ross, Officer of Health for Barvas parish, Isle of Lewis, which lay across the Minch from Skye, reported to Newman that breastfeeding would be continued for at least eighteen months, or 'even up to two or three years' (1906: 231), but it is likely that many infants would have been having soft or pureed items introduced into their diets, or be picking up dirt from the floor by the time they were six months old.

22 *The Concise Scots Dictionary* (Robinson, 1985) defines 'spune-meat' as 'soft or liquid food eaten with a spoon'.

Neonatal Mortality

Perhaps the most unexpected aspect of Figure 7.6 however, is the fact that in the left hand portion of the graph the points for the third and, particularly, the second week cumulative IMRs for Skye were much higher than those recorded for Kilmarnock. Generally, rates within the first week of life were somewhat lower in the rural community, although a dramatic increase from the 1860s meant that by the 1880s first week IMRs on Skye rose above the more consistent rates of Kilmarnock, before declining markedly in the 1890s. Whether the rise over the first three decades was 'real', or an artifact of the registration process remains a question for future examination. The very high rates of mortality in the second week of life and high rates in the third week on Skye meant that while neonatal mortality in Kilmarnock lay at around 40 deaths per 1,000 births, neonatal mortality rates (NMR) on the island reached 60 deaths per 1,000 births in both the 1870s and 1880s, and lay above 50 deaths in the 1860s and 1890s. In contrast post-neonatal mortality in Kilmarnock hovered around 90 deaths per 1,000 births across the four decades, roughly double the rate for Skye. With more than half of all infant deaths occurring in the first month of life, the age-specific infant mortality distribution on Skye was therefore highly unusual for a nineteenth century community

Neonatal deaths have been long considered to represent 'endogenous' deaths; deaths which are largely due to causes related to the health of the mother, condition of the foetus and complications of delivery which medical skill could do little to alleviate. By the early nineteenth century in England, Wrigley *et al.* (1997: Table 6.4, 226) found a NMR of just under 50 deaths per 1,000 births, a figure which had declined steeply over the previous century and a half. Galley and Woods (1999: 50-55) suggest that a NMR of 60 deaths per 1,000 births would usually be associated with a total IMR of around 200 deaths per 1,000 births. Skye's IMR was only half of this, emphasizing the island's exceptionally high rates of neonatal mortality. Such rates suggest that either reporting of deaths later in the first year of life was very erratic, or that the female population were exceptionally unhealthy, producing sickly offspring, or that the process of birth was particularly dangerous in this otherwise relatively risk free rural environment. The low death rates of older infants do not seem out of place when compared with rates among older children or the adult population which were also highly favourable when compared to their urban counterparts. Thus it also seems unlikely that maternal health on Skye was so inferior to that of the urban women of Kilmarnock as to engender the island's elevated NMRs.

In previous work it was suggested that high levels of death in the second week of life could signify the presence of the tetanus bacillus on Skye (Garrett and Davies, 2003) and it was therefore with great interest that Newman is noted quoting Dr Ross who reported that 'twenty years ago', so in the mid-1880s, 'many infants were lost every year in their first week of life from *trismus neonatorum*, brought about ... by mismanagement after birth', which 'caused gastro-intestinal disorder and convulsions' (1906: 231). Even writing at the beginning of the twentieth century, Dr Ross was keen to blame the custom of feeding new born babies castor oil, melted butter or sugared water for causing the *trismus*, or lock-jaw, in infants but it seems likely that in fact the cause was infection with the tetanus bacillus. Still a feared killer in the Developing World today this bacillus lives harmlessly in the intestines of animals, but can form spores which are often found in soil (Wingate and Wingate, 1988: 467-468). Steel, describing the terrible toll which infant tetanus wreaked among the population of the remote Hebridean island of St. Kilda in the nineteenth century, where the disease was called the 'sickness of eight days' because of the age at which the sufferer was most likely to die, suggests that the treatment of the umbilicus at birth was the most likely route by which an infant became infected (Steel, 1994). Once the cord had been cut, birth attendants, drawn from the local female population and seldom with any training, would dress the stump with a rag. This rag would often be smeared with something resembling an ointment; Steel specifically mentions salt butter being used in Barvas, the parish for which Dr Ross was responsible, while on St. Kilda fulmar oil was reputed to have been used. Whether the rags used were far from clean and carried the tetanus spores from the beaten earth floors, or whether the spores were present either in the unguent smeared on the open wound or on the instruments used to cut the umbilical cord is unlikely to ever be fully ascertained. As Steel remarks of St. Kilda: 'no one knows for sure whether such practices ever took place …' as 'no outsider, not even the minister or the schoolmaster, knew what went on when a baby was born'; the same would appear to have been true of other remote rural communities (Steel, 1994: 153).

The birth and death registers of the two communities studied here allow the enquiry concerning the very high rates of second week deaths on Skye to be pursued a little further, although the island's registers contain no references to 'tetanus', '*trismus*' or 'lock jaw'. In 54 per cent of the 357 deaths of infants in their second week of life occurring between 1861 and 1900 on Skye the cause of death was 'unknown'. A further 17 per cent were ascribed to 'pleurisy', 9 per cent to 'birth, weakness or debility', 3 per cent to 'whooping cough' and 1 per cent to 'bowels, diarrhoea or stomach disorders'. Although more than twice the number of births occurred in Kilmarnock than on Skye, only 264 infants died in their second week of life. Almost half, 47 per cent, were alleged to die from 'birth, weakness or debility' while 16 per cent died from complaints to do with the 'bowels, diarrhoea or stomach'. Three per cent succumbed to 'tetanus'. In fewer than 1 per cent of deaths, as seen above, was the cause reported to be 'unknown'. It is likely that medical men in Kilmarnock were aware of tetanus and its symptoms as Dr Donald Macleod, mentioned above as a major certifier of infant deaths, had published a paper on that very topic in 1859 (*London and Provincial Medical Directory*, 1901).

Birth conditions or practices surrounding the birth seem to be a more likely culprit, and this argument is strengthened when we consider the geographic pattern of second week mortality on Skye. Figure 7.7 indicates that IMRs for Kilmarnock and Skye were not very different in the fourth week of life. As discussed in regard to Figure 4.6 survival chances of Skye infants in their first week of life deteriorated until by the 1880s they were higher than those of Kilmarnock However across the four decades IMRs in the second week of life were roughly three times higher than those in Kilmarnock, and the Skye rates for the third week of life also consistently the higher by a considerable margin

The seven registration districts covering Skye are co-terminus with the seven parishes of Bracadale, Duirinish, Kilmuir, Portree (which includes the island of

Raasay), Sleat, Snizort and Strath. Figure 7.7 shows the IMRs for each of the first four weeks of life, for each of the registration districts. From the figure it can be seen that in the 1860s and 1870s each of the districts had rates of second week mortality which were higher than those for Kilmarnock, however Kilmuir and Snizort, on

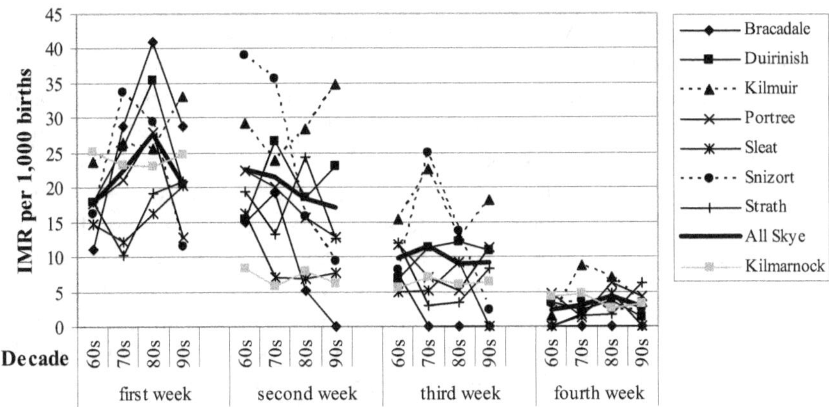

Figure 7.7 Infant mortality rates in the first four weeks of life: Skye, its parishes and Kilmarnock, 1860s-1890s

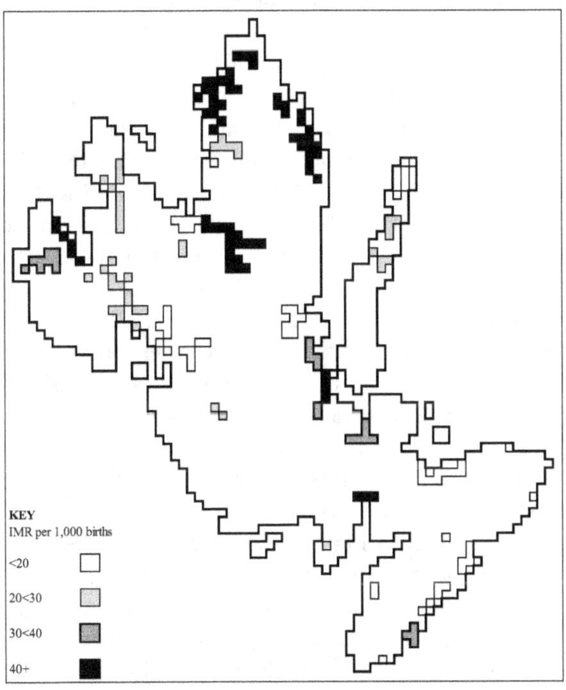

Figure 7.8 Concentration of second and third week deaths, by 'birth clusters': Skye 1861-1900, showing IMR(7<21 days) per 1,000 live births

Skye's north-eastern peninsula, had rates which were well above the overall Skye figures for these decades. From the 1870s onward the parish of Snizort saw marked improvement in its second and third week IMRs, but conditions in Kilmuir do not seem to have improved as second week survival rates for this parish deteriorated over time.

Many of the settlements on Skye were very small and saw relatively few births over the forty years between 1861 and 1900, therefore if a finer spatial analysis is to be undertaken IMRs have to be calculated for 'clusters' of settlements. In Figure 7.8 historic Ordnance Survey maps have been used to locate each settlement on the island within a kilometre grid square. The number of births occurring 1861-1900 within each square kilometre has then been summed. Contiguous squares, touching along a side or at a corner, where births occurred have then had their births added together. Each group of squares forms a 'birth cluster'. Only clusters where at least 100 children were born over the 40 year are shown on Figure 7.8 which shows the IMR for the second and third week of life (IMR(7<21days)) for each cluster.[23] This exercise is not an exact one, and the small numbers of deaths may affect the rates calculated, but Figure 7.8 suggests that certain communities were particularly susceptible to infant deaths in the second and third weeks of life, although it should be remembered that any IMR(7<21days) of over 20 lies above that found in urban Kilmarnock. Out of the 36 clusters shown in Figure 7.8 eight had an IMR(7<21days) of over 40 per 1,000 births. Five of these lay on the north-eastern peninsula, four in the more northerly parish of Kilmuir and one in the parish of Snizort. The settlements in the Snizort cluster lay primarily on the western shore of Little Loch Snizort, a long sea-loch, but settlements on the eastern shore were also included. One could envisage a midwife, 'handy woman' or birth attendant being summoned by boat from across the loch were a woman to go into labour on the opposite shore, rather than the much longer journey by land being made. All of the clusters in Kilmuir parish had an IMR(7<21days) of over 40 per 1,000 births. Two of the four clusters lay on the parish's east coast, two on the west coast. The more southerly of these lay immediately adjacent to a Snizort cluster with an IMR(7<21days) of only 20 to 30 deaths per 1,000 births, and very close to that lay a small, but densely populated cluster with an IMR(7<21days) of less than 20. The other three clusters with particularly elevated levels of IMR(7<21days) lay in the parishes of Duirinish, Portree and Strath. The parish of Bracadale was very sparsely populated, and so had few birth clusters, but none of them had a second and third week IMR of more than 30. The southern peninsula of Sleat was remarkable for the number of its clusters where the IMR(7<21days) lay at less than 20 deaths per 1,000 births.

The patterns of IMR(7<21days) shown in Figure 7.8 suggest that either there were practices common to very local areas which were putting infants at greater risk than elsewhere on the island, or that certain individuals, possibly local birth attendants, may have been responsible for introducing pathogens either via the instruments they used, or through the way they treated the umbilical stump, which was placing the newborns in their care at greater risk. The latter explanation would

23 The second and third week deaths were used in combination to give a greater number of deaths.

certainly fit the case of Snizort: the replacement of just one individual who had been attending women in the 'elevated mortality' settlements, could have resulted in the dramatic decline in second and third week deaths observed between the 1870s and 1880s. In Kilmuir it is likely that more than one individual was involved, given the number of clusters and their relative locations, and this may have helped maintain the high rates of mortality seen in Figure 7.7.

Can an accusatory finger be pointed even more precisely? It would seem not. Each of the Skye censuses from 1861-1901 lists one or two women, usually elderly widows, as 'midwives', but none is observed in more than one census. No 'handy women' or 'birth attendants' are listed, however, and if women in the remote communities depended on such help during labour, only the elevated rates of mortality now suggest that certain of the women called upon may have exacted a heavy price for their presence. Figure 7.8 serves as a reminder, however, that despite housing conditions being of a very low standard across Skye, high mortality amongst neonates was by no means inevitable, and rates could vary markedly within a very short distance.

Conclusion

This chapter has attempted to compare mortality experience in two Scottish communities, one rural and one urban, using individual level data extracted from the civil registers of births and deaths. On the evidence uncovered at this level, there would seem to have been large variations in the standard of the information collected. This appears to have been largely related to the proportion of deaths which were not attended, or at least certified, by a doctor in remote rural areas, although the qualifications and interests of the doctors in the towns may also have affected the patterns of cause of death recorded in urban areas. Comparison of the factors underlying mortality patterns observed at the aggregate level should thus be considered to be resting on shaky foundations, until more is known about the reporting mechanisms and medical services available in any two places being compared. It has further been shown that the registration system did not work as efficiently in the rural area studied as it did in the towns. Legislation regarding the time limit under which events should be registered appears to have held little sway in an area where journeys could be long and difficult. In a community where death rates in the first few days and weeks of life were much higher than later on in the first year, it was perhaps tempting to delay registering a birth until one was sure that one would not have to make a second journey shortly afterwards to register a death. Distance, and possibly the need to wait for a person to become available to make the necessary journey, would seem to underlie the delays seen in the registering of deaths.

A further aim of the study was to assess whether the four 'similarities' and three 'dis-similarities' which Newman observed in urban and rural rates derived from the Registrar General's returns could be detected in Skye and Kilmarnock. Let us consider these in turn. For the first similarity; infant mortality did indeed seem to be highest in the first day of life both on both Skye and in Kilmarnock. However, as we have shown, mortality on Skye appears to have followed an unusual path. Rather than declining through the remaining days of the first week, as those in Kilmarnock did and as Newman's second 'similarity' would have predicted, the rates began to swing upwards again, particularly after the fifth day. On Skye Newman's third similarity was also proven false: mortality did not fall 'enormously' from the first to the second week, although it followed the expected trajectory in Kilmarnock. Nor was it clear that Newman's fourth similarity held true. Mortality rates in neither Skye nor Kilmarnock showed a tendency to slowly fall across the rest of the first year of life. As Figure 7.6 shows they stayed remarkably stable.

Turning to Newman's 'dis-similarities', overall infant mortality in Kilmarnock was certainly higher than that on Skye, but not twice as high. After the first month of life the rates in Kilmarnock did run at virtually double those of Skye, although without a discernable increase in this ratio as age increased as Newman would have expected. However in the first month of life infants on Skye ran higher risks in both the second and third week; rates were not higher throughout the twelve months, as Newman had claimed. The aggregate statistics and the systems designed to gather them did not, therefore, provide infallible rules and indeed may have masked many of the finer comparative points influencing mortality in both urban and rural areas. However we are left to wonder whether Skye and Kilmarnock are truly representative of their wider constituencies – the rural and urban areas of Scotland first and of Great Britain second? Until further local level, comparative, in-depth studies are undertaken, we cannot be sure.

Chapter 8

Diarrhoea: The Central Issue?[1]

Eric Hall and Michael Drake

Readers of George Newman's *Infant Mortality: a Social Problem* can be in no doubt that for him, infantile diarrhoea was a central issue. Not only did he devote an entire chapter to it, but diarrhoeal statistics appeared in 24 tables and underpinned 5 charts, whilst the disease itself was referred to on 89 of his 356 pages. But even these statistics underplay the centrality of the disease for Newman's analysis. Whether he was discussing the various diseases that so often proved fatal for infants, or the social, domestic or occupational determinants of them, deaths from diarrhoea were so uppermost in his perception of infant mortality that he felt bound to devote a great deal of attention to them. True, diarrhoea was not the most common cause of death, as he recognised. That distinction fell to the so-called wasting diseases: congenital defects, injury at birth, want of breast milk, atrophy, debility and marasmus. But unlike diarrhoea, the wasting diseases did not show the often significant differences across time, space, social class and status; that is, whether or not a child was born outside or inside marriage. It was these differences that indicated, not only that diarrhoea itself was, in essence, a social problem, but, more importantly, that something could be done about it.

Today infant diarrhoea is still a major killer of young children. Although infant mortality ascribable to diarrhoea has been all but banished from Western society, across huge swathes of the developing world it continues to claim its victims. Estimates of the number of children who die each year vary, but around two millions is a widely reported figure. Given the number of births annually is around 130 millions, this would suggest a death rate of around 15 per 1,000 live births. Such a figure is not far from that prevailing in many English towns at the time Newman was writing (BPP, 1913, xxxii: 1). It is not only this that Newman would have found familiar. So too are the measures that are promoted at the present time, to counter the disease:

- Improving access to clean water and safe sanitation
- Promoting hygiene education
- Exclusive breast-feeding
- Improved weaning practices
- Immunizing all children; especially against measles
- Using latrines
- Keeping food and water clean

1 The work reported here was supported by the Wellcome Trust and the Open University.

- Washing hands with soap (the baby's as well) before touching food
- Sanitary disposal of stools. [2]

From this list only immunisation against measles would not have been familiar to Newman.

Much of the empirical data in Newman's study was drawn from local enquiries, conducted, for the most part, by Medical Officers of Health (MOHs), including himself. He drew, for instance, on data from Bedfordshire, where – although he doesn't admit to it in *Infant Mortality: a Social Problem* – he was part-time MOH, whilst at the same time being MOH in Finsbury (Currie, 1998: 18). The study presented in this chapter is in keeping with this, in that it discusses the incidence of diarrhoea in a local area. Ipswich was a town that grew rapidly in the nineteenth century, by 1900 over half its working population was in industrial occupations. It remained throughout, the major port of East Anglia and continued to serve as the market centre for its mainly agricultural hinterland (*Harrods Directory,* 1873: 214; Census Reports, 1871-1931).

This chapter examines diarrhoea during the period 1876-1930, a time span with Newman's contribution at its centre, his book being published in 1906. Thus, with hindsight, it is possible to see Newman's work in a wider temporal context. Sources will be drawn upon that were unavailable to Newman, or at least were not apparently used by him, namely the civil registers of births and deaths. The former were obtained via the Vaccination Registers for the town. These registers, which contained copies of the civil registers of births, were kept as part of the government's attempt to make a reality of compulsory vaccination against smallpox (Drake, 2005). They have been little used, but may be found in Record Offices across the country. The deaths, including causes of death, are taken from a copy of the civil register of deaths obtained apparently by Ipswich's MOH, presumably to help him in his work.

The main purpose of this chapter is to examine the role of diarrhoea in relation to the infant mortality rate (IMR). What contribution did it make to that rate? How did that contribution change? The timing of diarrhoeal deaths across the year will also be considered. Newman noted, as did many contemporaries, that there was a marked rise in such deaths during the summer months. But what happened to them when diarrhoea was on the wane? Finally the sources available allow for the possibility of examining diarrhoea in relation to social class over time. How did this vary?

There were 10,975 infant deaths recorded in the Ipswich death register, between 1876 and 1930, and of these 1,621 (14.8 per cent) were diagnosed as being caused by diarrhoea/enteritis. This number hides considerable variations within the period. For example for the 5 years 1896 to 1900 inclusive, 399 of the 1,432 infant deaths (27.9 per cent) were from diarrhoea/enteritis, the peak years being 1898 (103 cases or 34.3 per cent of the total deaths in the year) and 1899 (126 cases or 38.2 per cent of deaths). These figures are in marked contrast with those at the end of the period under consideration. Thus for the years 1916-20, 1921-25 and 1926-30 the corresponding totals and percentages were, respectively, 29 cases (4.8 per cent of all deaths), 34 cases (7.4 per cent) and 15 cases (4.1 per cent). It is possible that the

2 These recommendations may be found at: http://rhydrate.org/diarrhoea

Registrar General's guidance, as to what should be placed in the diarrhoea-enteritis category, brought about changes in diagnosis and categorisation that make direct comparisons difficult over the period generally. In fact the Registrar General voiced his exasperation with the way the term was used on a number of occasions. For instance in 1900 he noted that, because of the confusion over what should and what should not be termed 'diarrhoea', 'the statistics with reference to a large and fatal group of epidemic diseases, which are of exceptional importance to medical officers of health, are in danger of becoming vitiated to such an extent as to be worthless for preventive purposes' (Registrar General, 1900: 28). He also noted that there was 'a widespread objection on the part of medical practitioners to the employment of the term "diarrhoea" in certifying the cause of death, probably because the term is generally held by the public to imply a mild disease insufficient in itself to cause death.' (Registrar General, 1900: 28). Whether or not this had a class dimension is impossible to say, although if it did, it would have important implications for what follows. In Ipswich also, as in other places, the MOH's personal opinion of what he felt was the appropriate classification may have been influenced by a desire to assist his case for environmental improvements.

From Figure 8.1, it is clear that diarrhoeal deaths were at their height, for the period under consideration here, towards the end of the nineteenth century. Had there been no deaths from diarrhoea then the IMR could have been as much as 40 per 1,000 lower. However, it is also apparent that, in Ipswich, at least, the IMR, net of diarrhoea, would not have fallen before the first decade of the twentieth century. What is, perhaps, more interesting is that, if we ignore the sharp rise in diarrhoeal deaths in the 1890s, they and those ascribed to other diseases and conditions followed the same trajectory throughout the period. This trajectory was one of stability from the 1870s to the 1900s, when a secular decline began.

Seasonality of Diarrhoea/Enteritis Deaths

An analysis of the 1,621 deaths ascribed to diarrhoea/enteritis in the years 1876-1930, revealed that although such deaths occurred in every month of the year, by far the highest proportion were in July, August and September. These three months accounted for 1,279 deaths or 78.9 per cent of the total. The link between summer and deaths from diarrhoea was well known. From the outset, the first MOH for Ipswich, Dr Elliston (1873-1906), frequently referred to diarrhoea in his annual reports. For example, in his very first report he noted that:

> ... this disease is always fatally prevalent in Ipswich every autumn. Fifty out of the 54 deaths this year are children under five years of age, they are unable at this tender age to withstand the unfavourable influences by which they are surrounded in the shape of foul open cesspools and improper food; these deaths nearly all occur among the poor in the crowded neighbourhoods, where the drainage is faulty (Ipswich MOH Report, 1874: 9).

Although Elliston refers to 'autumn' as being the season for the highest fatalities, an analysis of the 39 infant deaths in 1873 shows that 19 were in August, 18 in September and 2 in October. It seems likely that Elliston's definition was slightly

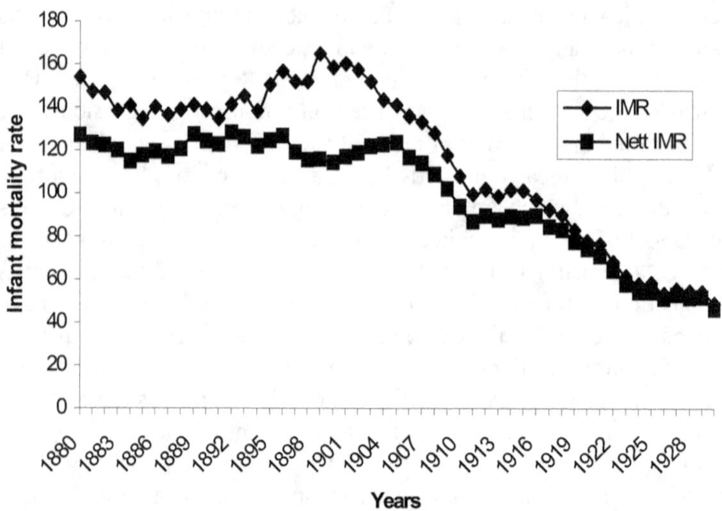

Figure 8.1 Five year moving average of the IMR and net of diarrhoea/enteritis deaths: Ipswich, 1876-1930

different to the commoner interpretation of June, July and August as summer, but it was perhaps the case that fatalities of older children took place slightly later than those of infants.

It is possible to correlate temperature and diarrhoea more closely by drawing on records of the former from the archives of the Meteorological Office. For the entire period 1876-1930 the correlation coefficient was 0.60. But when the period was split into two halves – 1876-1906 and 1907-1930 – the coefficients were 0.71 and 0.59 respectively. In other words, there was a stronger correlation between the two sets of data in the first period than in the second. This would suggest that measures to reduce the effect of hot, dry weather on the incidence of diarrhoea had had some effect.

One can go further than this by relating the hottest summers with the death rates ascribable to diarrhoea. Table 8.1 shows the IMR for diarrhoea/enteritis in the 11 years when deaths from the disease were at their highest, in descending order. Comparison is then made with the temperatures for the two three month periods when deaths from the disease were also at their highest. There is clearly a relationship between the hottest periods and the highest incidence of deaths from diarrhoea and enteritis. Of the 22 three monthly periods, 9 occurred in the hottest periods recorded nationally over the 55 years. It is of interest that 1911, which had the highest June-August temperature in the series – indeed it was the highest since records began in 1660 – had the lowest IMR of the series, attributable to diarrhoea/enteritis. This again suggests, given that 1911 was a peak year for diarrhoea nationally, that the local initiatives introduced a few years earlier to counteract the disease, were having an effect, a point noted by the MOH in Ipswich, Dr. Pringle, at the time.

For diarrhoea the years of least prevalence were 1879; 1881; 1885; 1894; 1888; 1902; 1907; 1909 and 1910. All had diarrhoea/enteritis IMRs of less than 10. Were

Table 8.1 The relationship between temperature (degrees centigrade) and diarrhoea/enteritis deaths in the 11 hottest years in Ipswich between 1876 and 1930

Year	Diarrhoea/ enteritis IMR	Average June, July, August temperature (°C)	Ranking of Column 3 for years 1876-1930	Average September, October, November temperature (°C)	Ranking of Column 5 for years 1876-1930
1899	70	16.93	2	10.17	11
1898	56	15.13	23	11.23	1
1895	56	15.27	20	10.00	17
1880	48	15.23	21	9.03	40
1878	42	15.97	7	9.00	41
1884	41	15.87	9	9.73	24
1893	38	16.47	3	9.33	34
1906	36	15.60	14	10.70	3
1904	33	15.17	22	9.13	38
1900	32	15.83	10	10.23	8
1911	29	16.97	1	9.77	22

Source: Death returns, Ipswich 1872-1930; Meteorological Office, Bracknell

these comparatively low figures reflected in the temperature data? With regard to June, July and August temperatures 1879 ranked fifty-third of the 55 years (53/55); 1881 (39/55); 1885 (38/55); 1894 (41/55); 1888 (54/55); 1902 (47/55); 1907 (55/55); 1909 (51/55) and 1910 (35/55). These nine years were amongst those with the coolest average summer temperatures, over the period reviewed. For September, October and November the years were ranked 1879 (47/55); 1881 (28/55); 1885 (48/55); 1894 (30/55); 1888 (36/55); 1902 (26/55); 1907 (19/55); 1909 (39/55) and 1910 (43/55). Although this is not such a significant finding it is worth noting that these years were also among the coolest years between 1876 and 1930.

A similar correlation exercise was carried out between deaths from bronchitis and pneumonia and the coldest winter months of December, January and February. No significant correlation was found.

Diarrhoea: Temporal and Spatial Dimensions

Taken together the vaccination and death registers make it possible to analyse cause of death both over time and across space. We can do this in a way that goes beyond the published statistics. What follows then is an examination of the incidence of deaths from diarrhoea by month in the two sub-registration districts of Ipswich – the Eastern and the Western – over the period 1876-1930. Because monthly totals of infant deaths from diarrhoea are small at this spatial level, the period has been divided into three sections: 1876-90; 1891-1910 and 1911-30. The first of these was a period when diarrhoea was quite rife in some years; in the second the 1890s saw high mortality from diarrhoea and the 1900s the start of a rapid fall; whilst from 1911 to 1930, the IMR fell very rapidly, with deaths from diarrhoea, by the end of the period, of negligible importance.

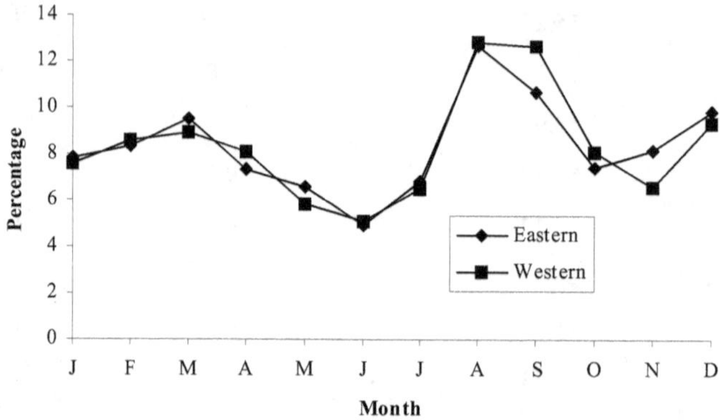

Figure 8.2 Monthly deaths from diarrhoea as a percentage of annual deaths from the disease in the Eastern and Western sub-registration districts of Ipswich, 1876-1890

Figure 8.2 shows that in the years 1876-1890, the incidence of deaths from diarrhoea, in both sub-districts, followed the same course for the first eight months of the year. In September , a month in which diarrhoea occurred more frequently (see above), and October, there were more deaths in the Western sub-district. The fit was not so close for the years 1891-1910 (Figure 8.3), but again, the Western sub-district had a higher percentage of deaths attributable to diarrhoea than the Eastern in September as well as in August.

There were only a small numbers of deaths from diarrhoea in the years 1911-1930 (Figure 8.4), but despite this the two sub-districts followed much the same track, although again the Western sub-district had a higher percentage of its annual deaths from diarrhoea in August. Note too that although there was still a summer peak, a greater percentage of deaths now occurred in the earlier part of the year. It is also worth pointing out that until the 1910s, there was a substantial number of deaths from diarrhoea throughout the year, so that even without a summer peak many children would have died from this disease.

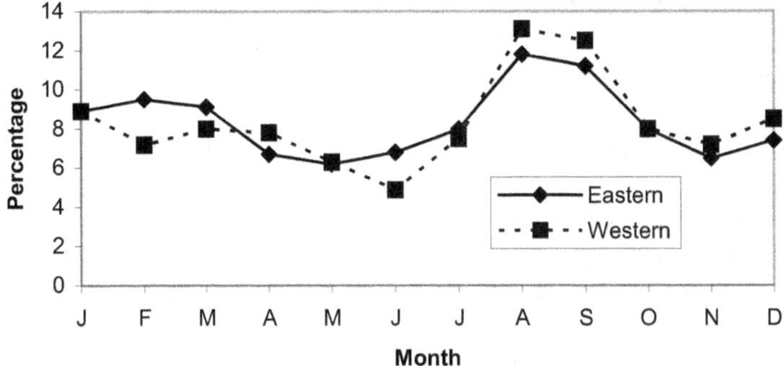

Figure 8.3 Monthly deaths from diarrhoea as a percentage of annual deaths from the disease in the Eastern and Western sub-registration districts of Ipswich, 1891-1910

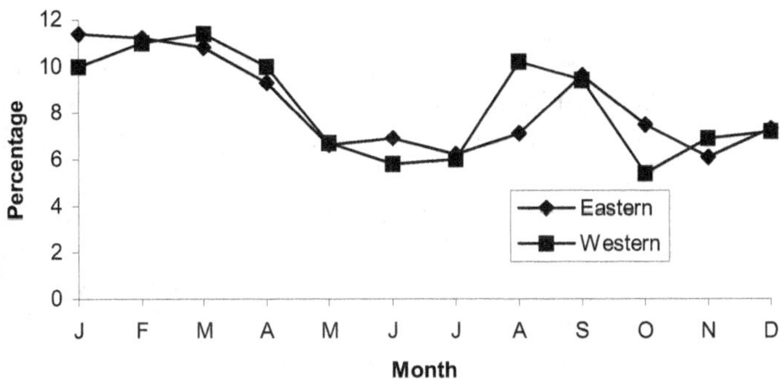

Figure 8.4 Monthly deaths from diarrhoea as a percentage of annual deaths from the disease in the Eastern and Western sub-registration districts of Ipswich, 1911-1930

That the Western Sub-district had a higher percentage of its annual deaths from diarrhoea in the summer months than did the Eastern could be, in part at least, accounted for by the different environments of the two areas. For instance, newer housing, on higher ground, was built in the Eastern sub-district and although it was bounded by the river, to the south, the actual buildings adjacent to the river were, in the main, factories or other premises of an industrial nature. By contrast the Western Sub-district was noted for its poor housing in the Stoke area of the Borough, with its crowded streets adjacent to the Docks and river, an area too in which sewers were installed later than in the east. The conclusion must be, however, that the trajectory of deaths from diarrhoea was the same across the borough, at least so far as its constituent sub-registration districts were concerned. This applied across the whole of the period 1876-1930.

Whilst deaths from diarrhoea/enteritis varied widely from year to year, this was also the case with some other causes of death. Take, for example deaths put down to prematurity, marasmus and debility, which, one might think were unlikely to do so. Deaths from this grouping numbered 657 or 48.5 per cent of all deaths in the five years 1901-1905 as compared to 507 or 35.4 per cent of all deaths in the previous five years. This explains most of the increase in the net IMR level in the latter years. The number of deaths from Bronchitis and Pneumonia showed only small differences over the two periods.

Between 1896-1900 and 1901-1905, the IMR in Ipswich fell from 160 to 142. This was almost entirely due to a fall in deaths from diarrhoea/enteritis, for, as we have seen premature deaths and those attributed to marasmus rose considerably. To what extent this was due to improvements in the local environment or to a return to wetter and cooler summers is open to debate. The number of privy middens fell from 8,000 in 1894 (one for every 7 inhabitants) to 125 in 1906 and a comprehensive sewage system, was begun in 1897 and finally completed in 1904 (Ipswich MOH Report, 1908: 38).

Diarrhoea: A Class Phenomenon?

Here the investigation of the incidence of diarrhoea in Ipswich is carried further by investigating possible links between it, social class and income. Property values are used here as a surrogate for income. In an attempt to overcome the problem of small numbers, deaths in Classes I and II have been combined and the data for all classes has been aggregated into five year periods. Classes III and IV are not considered in view of the anomalies present in occupational descriptions such as 'baker', where it is impossible to tell, from the entries in the Vaccination and Death Registers, whether an individual is a worker or an employer (Blumin, 1968; Mills and Mills, 1989). The mortality experience of the combined Class I and II families is thus compared with that of Class V; all occupations containing the description labourer, other than that of agricultural labourer) and, those dubbed Class IX: illegitimate infants who were separately identified in both the Vaccination and the Death registers. The number IX has been chosen so as to avoid confusion with the Classes I to VIII in the 1911 Census. The argument for including Class IX is that throughout the period studied here, illegitimate infants and their mothers belonged to what, today, we would call an 'underclass'. Much attention was paid to illegitimacy, partly reflecting a deep moral outrage and partly because it had financial implications for the community. It is possible too that as the IMR of children born outside marriage was commonly double that of those born within it, especially in the major cities, it was felt that lessons could be learned from the experience of illegitimate infants that might be relevant to that of legitimate ones (Drake, 2005).

Table 8.2 indicates the numbers of births occurring in selected time periods in each of the classes chosen for analysis, together with the number of total births in Ipswich as a whole, by way of context.

Table 8.3 shows the IMR in quinquennial periods from 1876-1930 in the classes represented in Table 8.2 and in Ipswich as a whole, indexed against the IMR for 1876-80. The percentages are of total births in Ipswich.

Table 8.2 **Recorded births in Ipswich for selected social classes, 1876-1930: number born to each class and the percentage of all births in Ipswich**

	Births											
	1876-80		1881-90		1891-1900		1901-10		1911-20		1921-30	
Class	N	%	N	%	N	%	N	%	N	%	N	%
I & II	369	4.5	607	3.6	611	3.2	610	3.5	578	3.5	605	3.9
V	1,378	16.9	3,071	18.2	3,755	22.8	4,288	18.9	3,145	18.9	3,739	24.3
IX	405	5.0	931	5.5	1,025	5.2	979	5.4	890	5.4	471	3.1
Total	8,158	100.0	16,906	100.0	17,740	100.0	18,772	100.0	16,603	100.0	15,387	100.0

Source: Vaccination registers Ipswich, 1876-1930

Table 8.3 Changes in the IMR by social class in 5 year periods, Ipswich 1876-1930, indexed against the IMR per 1,000 live births for 1876-1880

	Class			
	I & II	V	IX	Ipswich
IMR 1876-1880	125	136	247	154
1876-1880	100	100	100	100
1881-1885	64	104	81	88
1886-1890	70	109	106	92
1891-1895	81	127	113	97
1896-1900	112	117	106	103
1901-1905	82	114	91	92
1906-1910	60	104	72	70
1911-1915	39	88	88	66
1916-1920	59	53	58	50
1921-1925	14	43	44	38
1926-1930	16	36	38	31

Note: The row labelled IMR shows the infant mortality rate for 1876-80

Source: Vaccination registers and death returns, Ipswich, 1876-1930

Table 8.3 shows that the IMR for Classes I & II fell to 16 per cent of its 1876-80 level by 1926-30 having risen above it by 12 per cent in the five years 1896-1900, after which a dramatic decline began, save for the 1916-20 period. The IMR of Class V peaked in the early 1890s, when it was 27 per cent above its 1876-80 level. But it too then began to decline, although not so precipitously as that of Classes I & II. By 1926-30 the IMR of Class V was virtually a third of its 1876-80 level. Illegitimate mortality followed much the same pattern as that of Class V, but as it had started from a much higher figure (its IMR in 1876-80 being 247 as against 136 for Class V) its fall to 1926-30 was greater than that of any other class. The overall experience in the Borough was, then, that the IMR peaked between 1896 and 1900, at slightly higher than the base period of 1876-80, but then fell gradually until after the First World War when the rate accelerated to finish at under a third of the 1876-80 level.

As noted above, the number of deaths for Classes I & II was, in some periods, very small. Thus between 1921 and 1930 there were only 11 deaths from 605 births. The number of illegitimate births too fell dramatically towards the end of the period, from an average of over 90 births per year to under 50. This may indicate a change in the moral climate or better access to birth control. Notwithstanding any reservations, it is clear that the decline in the IMR was a general phenomenon although more pronounced amongst the upper classes and for illegitimate infants. As regards the latter, the initial high level of the IMR gave the greatest scope for improvement in absolute terms.

The Association between Class and Disease

Dr Elliston, like his peers, often talked of the 'filth diseases', especially diarrhoea/enteritis. Did the infants of the upper classes in Ipswich fare better than those of the lower classes, so far as these diseases were concerned? There were a total of 11,621 infant deaths recorded as having taken place within Ipswich between 1873 and 1930. Of these deaths 2.5 per cent were of infants from Classes I & II combined and 22.4 per cent from Class V. Illegitimate infants (Class IX) made up 9.3 per cent of all deaths. From 1876-1930, the IMRs for these classes were 80.2, 124.0 and 212.3 respectively.

Some 1,621 infants died from diarrhoea in Ipswich, between 1876 and 1930 (14.8 per cent of all infant deaths). This overall proportion was remarkably close to the proportion succumbing to the disease in each of classes, namely 14.7 per cent in classes I & II combined; 13.9 per cent in class V and 16.2 per cent in class VI. Although this shows that around one in seven of all infant deaths in each class died from diarrhoea/enteritis, it did not mean that the IMR attributable to the disease was the same. In fact the IMR attributable to diarrhoea/enteritis was 12.1 in Classes I and II; 16.8 in Class V and 34.4 in Class IX.

Further investigation revealed that the fathers from Classes I & II whose children died from 'filth' diseases included a doctor, several teachers, the family of a newspaper proprietor who was at one time a member of the Public Health Committee, a stockbroker, and two of the leading manufacturers in the Borough – Ransome, of the main engineering firm and Churchman, a tobacco manufacturer. The addresses at which deaths took place are consistent, in the main, with the social class category adopted, not only in regard to location but also in terms of their value, as recorded in the Valuation Lists. Having noted this it seems that although the more affluent families were less likely to lose infants to diarrhoea than the labouring class, the difference was not very great, whilst the trajectory of deaths from the disease was pretty much the same for all classes across the years (see Figure 8.5). This may have been due to poor drains and inadequate scavenging affecting all classes, if, as is believed, the water supply was of good quality,. Perhaps there was a private hygiene/contact/cleanliness origin for the spread of the disease. Although this might seem less likely in the case of the higher social classes, it must be borne in mind that their infants were more likely to be cared for by servants, or non-family members, who may not have applied at all times the standards of hygiene their employers might have hoped for.

Tables 8.4, 8.5 and 8.6 provide a breakdown of all infant deaths by disease, analysed for the five years 1876-1880 and then in decades for the rest of the period studied. This helps put deaths from diarrhoea into the context of other diseases affecting the social classes under consideration.

It must be acknowledged, yet again, that because the number of infants born to families in Classes I & II was small, the number who died was generally very small too. Any conclusions drawn must then be indicative only, giving rise to hypotheses that could be tested on larger samples in other locations. Even allowing for the fact, however, that mortality was very low, or non-existent with respect to some causes, in Classes I & II, it is noticeable that improvements overall dated from, at the latest, the

Table 8.4 Number of infant deaths and IMRs per 1,000 live births by cause of death in Classes I & II: Ipswich, 1876-1930

	1876-1880		1881-1890		1891-1900		1901-1910		1911-1920		1921-1930	
Cause	**N**	**IMR**	**N**	**IMR**	**N**	**IMR**	**N**	**IMR**	**N**	**IMR**	**N**	**IMR**
Bronchitis	4	10.8	8	13.2	4	6.5	4	6.6	1	1.7	1	1.7
Premature	9	24.4	2	3.3	8	13.1	8	13.1	5	8.7	1	1.7
Convulsions	5	13.6	5	8.2	6	9.8	5	8.2	2	3.5		
Marasmus	2	5.4	2	3.3	7	11.5	4	6.6				
Debility	7	19.0	6	9.9	11	18.0	10	16.4	6	10.4	1	1.7
Diarrhoea/enteritis	9	24.4	5	8.2	17	27.8	6	9.8	4	6.9		
Pneumonia			4	6.6	4	6.5	5	8.2	3	5.2		
Pertussis/whooping cough	4	10.8	6	9.9	1	1.6	2	3.3	1	1.7		
Other	6	16.3	13	21.4	14	22.9	11	18.0	14	24.2	8	13.2
Total	46	124.7	51	84.0	72	117.7	55	90.2	36	62.3	11	18.3

Source: Death Returns, Ipswich, 1876-1930

Table 8.5 **Infant deaths and IMRs by cause of death in Class V: Ipswich, 1876-1930**

	1876-1880		1881-1890		1891-1900		1901-1910		1911-1920		1921-1930	
Cause	**N**	**IMR**	**N**	**IMR**	**N**	**IMR**	**N**	**IMR**	**N**	**IMR**	**N**	**IMR**
Bronchitis	19	13.8	46	15.0	74	19.7	68	15.9	35	11.1	39	10.4
Premature	18	13.1	42	13.7	64	17.0	88	20.5	55	17.5	28	7.5
Convulsions	22	16.0	35	11.4	50	13.3	43	10.0	18	5.7	4	1.1
Marasmus	14	10.2	42	13.7	70	18.6	71	16.6	16	5.1	12	3.2
Debility	35	25.4	77	25.1	86	22.9	114	26.6	57	18.1	24	6.4
Diarrhoea/enteritis	18	13.1	50	16.3	134	35.7	91	21.2	26	8.3	12	3.2
Pneumonia	10	7.3	20	6.5	38	10.1	29	6.8	28	8.9	5	1.3
Pertussis/whooping cough	6	4.4	23	7.5	19	5.1	20	4.7	5	1.6	1	0.3
Other	45	32.7	111	36.1	85	22.6	111	25.9	74	23.5	74	19.8
Total	187	136.0	446	145.3	620	165.0	635	148.2	314	99.8	199	53.2

Source: Death Returns, Ipswich

Table 8.6 **Number of infant deaths and IMRs per 1,000 live births in Class IX by cause of death: Ipswich, 1876-1930**

	1876-1880		1881-1890		1891-1900		1901-1910		1911-1920		1921-1930	
Cause	**N**	**IMR**	**N**	**IMR**	**N**	**IMR**	**N**	**IMR**	**N**	**IMR**	**N**	**IMR**
Bronchitis	8	19.8	14	15.0	26	25.4	16	16.3	14	15.7	4	8.5
Premature	7	17.3	18	19.3	28	27.3	31	31.7	36	40.4	7	14.9
Convulsions	10	24.7	19	20.4	16	15.6	5	5.1	5	5.6	1	2.1
Marasmus	10	24.7	34	36.5	40	39.0	29	29.6	8	9.0	3	6.4
Debility	21	51.9	47	50.5	40	39.0	32	32.7	21	23.6	11	23.4
Diarrhoea/enteritis	17	42.0	29	31.1	65	63.4	35	35.8	12	13.5	4	8.5
Pneumonia	4	9.9	5	5.4	17	16.6	7	7.2	17	19.1	1	2.1
Pertussis/whooping cough			1	1.1	1	1.0	3	3.1	4	4.5	1	2.1
Other	23	56.8	48	51.6	43	42.0	41	41.9	43	48.3	16	34.0
Total	100	247.1	215	230.9	276	269.3	199	203.4	160	179.7	48	102.0

Source: Death Returns, Ipswich, 1876-1930

second decade of the twentieth century. It appears evident that bronchitis was much less of a problem in the higher social classes than it was amongst labourers' families. Similarly convulsions and marasmus were also less evident although debility was a problem in Classes I & II for most of the period studied. Diarrhoea/enteritis declined as a cause of fatality in all classes after 1900 but still accounted for over 8 per cent of deaths of illegitimate infants in the last 20 years studied, and almost the same proportion of labourers' infants, whereas in Classes I & II no such deaths occurred in these two decades. A caveat must be entered here. The reporting of cause of death was far from being an exact science. Diagnosis varied from doctor to doctor and, where no medical practitioner was involved, from parent to parent. This was especially the case with diarrhoea (see above).[3]

In spite of the small numbers involved, it is apparent that the number of infant deaths peaked at almost the same time for each group studied. From that peak in the 1890s, the decline was greatest for Classes I & II (almost 85 per cent between the 1890s and the 1920s). Comparable percentages for Class V and Class IX infants were 67.9 and 82.6.

Figure 8.5 shows in graphical form, the IMR attributable to diarrhoea by class in Ipswich between 1876-80 and 1926-30.

It can be seen, first, that throughout the period, the highest IMR occurred amongst infants born outside marriage and, second, that the profile of the diseases' impact, over time, was all but identical for each class. As already noted, deaths from diarrhoea/enteritis fell considerably from a peak in the late 1890s. After 1911 there were comparatively few such deaths. Whatever brought about this decline, which

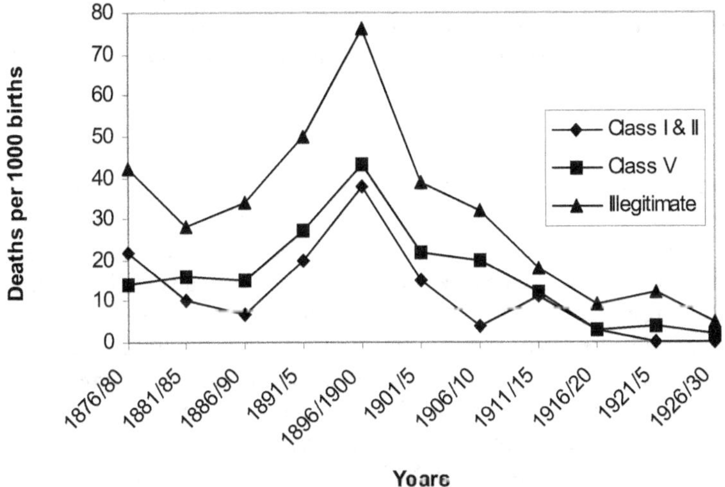

Figure 8.5 The IMR resulting from diarrhoea/enteritis, by class: Ipswich, 1876-1930

3 For a further discussion of this issue, see Garrett in this volume.

saw the number of deaths fall to virtually zero, it was experienced by all classes although, clearly, it took longer for classes V and IX to approach the very low levels experienced by Classes I & II after 1911.

As the causes of diarrhoea attracted much comment, both locally and nationally, it was decided to use the National Archives Inland Revenue Valuation Lists (Valuation Lists), prepared in 1910-13 as part of the Finance Act of 1909, to gain some indication of the income levels of the three groups, although.these lists provide a snapshot at a specific time and may not be wholly applicable for years before or after the valuation was made.

Table 8.7 Percentage of deaths from diarrhoea/enteritis by property value band[a] in years of high and low numbers of such deaths: Ipswich, 1890-1911

	Values - £ sterling							
	0-50	51 -100	101 -150	151 -200	201 -250	251 -300	301 -350	351+
Low Years[b]	13.0	31.5	31.5	12.3	4.3	3.1	0.6	3.7
High Years[c]	13.0	28.5	29.5	14.5	4.6	3.4	0.8	5.7
Ipswich valuations	10.8	22.3	22.0	19.9	8.5	5.6	2.5	8.4

Notes: a) Property bands are as defined in 1910-13. b) Low years with under 50 infant deaths from diarrhoea/enteritis were: 1890, 1891, 1892, 1894, 1902, 1903, 1905 and 1907-10. c) High years with 50 or more infant deaths from diarrhoea/enteritis were: 1893, 1895, 1896, 1897, 1898, 1899, 1900, 1901, 1904, 1906, 1911

The bottom row of Table 8.7 shows the percentage distribution of property, by valuation in Ipswich in the years 1910-13. Rows two and three show the percentage number of deaths from diarrhoea/enteritis, by the valuation of the properties in which they occurred, in years when diarrhoea/enteritis claimed a relatively low number of victims ('Low' row two) and a relatively high number ('High' row three). Had the incidence of deaths been the same in each property band, then we would have expected, for instance, that 10.8 per cent would have occurred in properties with a valuation of £0-50; 22.3 per cent in properties valued at £51-100 and so on. In fact, in both the years with a high incidence of diarrhoea and those with a low incidence, there were more deaths than expected in the lower property bands and fewer in the higher. This would suggest that children living in the higher valued properties had a smaller chance of dying from diarrhoea/enteritis than those living in the lower valued ones. To what extent this was because of advantages associated with the properties themselves (such as location, quality of construction, running water, or WCs) or from those associated with the higher incomes enjoyed by persons in such properties, one cannot say.

Discussion

Ipswich grew rapidly throughout the period 1870-1930, mostly by natural increase. It had a mixed economy which grew over the period, with its role as East Anglia's major port and commercial activities for its largely rural hinterland being supplemented by a number of manufacturing industries including iron founding, agricultural machinery, ship-building, brewing and tobacco.

At the start of the period, Ipswich was in a poor state as regards sanitation and scavenging, although neither inadequate water supplies nor overcrowding seem to have been problems. The propensity of its inhabitants to store their excrement beside their homes for commercial purposes – being surrounded by agricultural land meant that demand for manure was high – was an object of comment (Ranger, 1856: 5-6). So too was its child mortality rate, it being comparable with that of the major industrial centres of the North and Midlands (Glyde, 1850: 41).

The completeness and richness of returns of births and deaths for the town has permitted a detailed analysis of all 99,893 births and 11,939 infant deaths recorded from 1872-1930, along class and spatial lines, as well as by cause of death and the timing of change.

The IMR fell from the 1870s (it was then 148 deaths under one year per 1,000 births) to the 1880s, rose in the 1890s and then began a secular decline in the 1900s, falling, by the 1920s, to 53 deaths under one year per 1,000 births. This pattern was common to both the eastern and western parts of the town, although the levels were somewhat, if not consistently, different. The onset of the secular decline, when viewed in decadal terms, is close to that of England and Wales as a whole. However, when one goes behind the aggregates, it emerges that the decline began at various times for the different social classes. Furthermore when deaths from diarrhoea are stripped out, far from the start of the secular decline being pushed back a decade or so, as suggested for the kingdom as a whole (Woods, Watterson and Woodward, 1989: 129) it remained unchanged until around the turn of the century.

Deaths from diarrhoea were, as others have found, closely linked to changes in temperature. However, although the summer months saw the highest incidence of the disease until the 1910s, the other months accounted for the majority of deaths. From 1911-30, deaths from diarrhoea fell sharply and a winter peak replaced that of the summer. Diarrhoea was, then, endemic, a factor of some importance when causation is being considered.

The decline of deaths from diarrhoea had a significant impact on the age distribution of infant deaths. Until 1896-1900, fewer than 20 per cent of all deaths in the first year of life in Ipswich occurred at seven days or under, thereafter the percentage rose steadily, reaching 22.3 in 1906-10, 24.9 in 1916-20 and 38.3 in 1926-30. A similar pattern occurred with deaths under one month: the percentages being, respectively, under 30, 39.2, 42.8 and 56.2. The corollary of this is that deaths at ages over one month fell from 74 per cent in the years 1896-1900 to 44 per cent in the years 1926-30. The fall in mortality in the higher age groups was particularly dramatic between 1896-1900 and 1906-10. Thus, for example, the death rate of infants aged 3-6 months per 1,000 births fell from 37.1 in the former period to 17.8

in the latter. This might suggest that improvements in weaning, brought about by the ministrations of the recently appointed Health Visitor, could have played a part.

IMRs by cause of death and class over a long period, are the most interesting quantitative results of the unique set of data that forms the foundation of this study; a set that has allowed us to push our analysis back to the 1870s; something that is very difficult to do without individual level data. Although social classification is not without its problems, long use has given it a central role in much social scientific enquiry (Blumin, 1968; Mills and Mills, 1989) . The overall IMRs had a not dissimilar profile for the classes examined here over the years 1870-1930, although the beginning of the secular decline began around 1900 for Classes I & II and after 1905 for Class V (all labourers other than agricultural labourers) and Class IX (unmarried mothers, here described as an 'underclass'). However, in absolute terms there were some sizeable differences, the IMR of Classes I & II falling by 106 deaths (a fall of 85 per cent between 1876-80 and 1921-30)) under one year per 1,000 births; that of Class V by 83 (61 per cent); and that of Class IX by 145 deaths (59 per cent) per 1,000 births over the period as a whole (Tables 8.4-8.6).

When individual proximate causes of death are considered, the contribution of diarrhoea/enteritis to total deaths amongst the upper classes had a similar profile to those of the lower classes, although at lower levels throughout. More interesting, perhaps, is that the IMR attributable to the disease was almost as high amongst the infants of the upper classes as amongst those of labouring class, especially during the epidemics of the late 1890s when it was about 30-40 deaths per 1,000 births in both classes. At this time the IMR attributable to diarrhoea was 75 per 1,000 amongst the infants of Class IX (Figure 8.5).

An analysis of deaths from diarrhoea/enteritis, according to the rateable values of the properties in which they occurred (here taken as a surrogate for family income) showed a higher percentage than expected in the lower valued properties and a lower than expected one in the more highly rated ones (Table 8.7). This measure is, perhaps, a better indicator of social class differences than the traditional one based upon occupations.

What brought about the decline in the IMR from around 1900? What role did human intervention play? It is natural to look, in the first instance at least, at the work of the two MOHs who between them covered the entire period considered here: G.S. Elliston (who held the post from 1873-1906) and A. Pringle (1906-1934). As we have seen, there was little change in the IMR during Elliston's time in office, whilst Pringle oversaw a rapid and virtually continuous decline in it from the day he arrived in Ipswich. Although there is a danger of falling into the *post hoc ergo propter hoc*[4] fallacy, an examination of these two men's work does appear to provide some clues as to why the IMR should behave as it did in Ipswich in the years 1870-1930.

Elliston was a native of Ipswich; Pringle was not. Elliston was Ipswich's first MOH – there were only two candidates for the post, the other man also being a native of Ipswich – and the length of time it took to make his appointment suggests a reluctance on the part of the town's elite, which controlled the council, to accept the need for it (Ipswich Borough Council Minutes, 1873:146). That his appointment was

4 'After this, therefore because of this'.

to be only part-time and at a relatively low salary is further evidence of this somewhat half-hearted approach. He had no special qualifications for the job. Pringle, on the other hand, responded to an advertisement which called for qualifications in public health and was appointed expeditiously to a full-time post at a reasonable salary (Ipswich Public Health Committee Minutes, 1905). He had experience of public health from his previous posts in industrial Lancashire (*The Medical Directory, 1906*). Elliston's main concerns were sanitation, particularly sewers, middens, refuse removal from a town which, when he took up office, had a higher rate of 'infantile mortality' (defined by him as deaths under age five) than many of the largest 18 largest towns in the country. Pringle also enjoyed a much more fruitful relationship with the council than had Elliston: the chair of his public health committee was particularly supportive and personally provided finance for one of Pringle's initiatives (*Suffolk Chronicle and Mercury*, 1963: 14) .

Pringle was much more involved with infant mortality *per se* than was Elliston, recognising that the mortality of other age groups had fallen in Ipswich, during the latter's tenure. He inaugurated a scheme for visits by a trained female sanitary inspector to a majority of the newly born and by the time he left office there were four health visitors carrying out this role, despite a fall in the number of births. He also started mother and baby clinics which rapidly proved popular, and improved the quality of midwives who came to attend a majority of births. Pringle presided over a decline in infant mortality that, when he arrived, was virtually at the level Elliston found it (146 deaths under one year per live 1,000 live births as against 150) and when he left it was, at 36, the lowest ever recorded for a town the size of Ipswich.

It is rather tempting to suggest that had Elliston done what Pringle did, then infant mortality would have fallen sooner in Ipswich. This would, however, be to ignore a number of factors. First, the 'media' profile of infant mortality was higher in Pringle's time although it had been growing. Symptomatic of this was that in the year Pringle arrived in office, 1906, the first National Conference on Infant Mortality took place. The conference came shortly after a report of the Inter-Departmental Committee on Physical Deterioration had expressed serious concerns about the health of infants. Similar worries can be traced back to the late nineteenth century, but the intensity of concern was more muted during Elliston's tenure of office (see Chapter 2 in this volume).

Second, Pringle operated in an environment where central government initiatives aimed at reducing infant mortality were growing. Elliston, and indeed his local authority, operated much more in isolation. Thus, even when a Local Government Board Inspector produced a pretty damning report on the sanitary condition of the town, it received no publicity (Wheaton, 1894-95; *Ipswich Public Health Committee Minutes*, 1894: 144). It is true that a comprehensive sewerage system was finally inaugurated in 1897 and completed by 1904, but it was symptomatic of the reluctance of the authorities to act, that the council's sewerage committee had spent two years discussing the matter.

A third factor that needs to be considered is that much of the success of Pringle's initiatives could be said to have been predicated on the work done, albeit slowly, under Elliston's watch. Perhaps the most telling statistic here, as noted above, is the virtual elimination of privy middens. Pringle. however, would not have this, noting pointedly in his Annual Report for 1909, that 'general sanitary improvements' in the borough had 'exerted no influence upon the infantile mortality rate up to 1906'. He admitted that the sharp fall after that date might have been due to 'accidental circumstances' but noted that 'it was passing strange that the accidental circumstances should continue in existence for a period of three consecutive years'. If such they were, as has been shown, they continued for much longer than that (Pringle 1910: 39). It should, perhaps be concluded, that whilst the sanitary improvements – the abolition of middens, the rapid disposal of refuse, the creation of a comprehensive sewerage system with the concomitant introduction of water closets, and the demolition of insanitary dwellings – were a necessary condition for the fall in the rate of infant mortality, they were not a sufficient condition. For that the individual level initiatives inaugurated under Pringle, were needed. These led, in his words, to 'the more intelligent management of infancy by an instructed community of mothers' (Pringle, 1927: 30); a statement with which George Newman would have readily concurred.

Chapter 9

Infant Mortality, a Spatial Problem: Notting Dale Special Area in George Newman's London[1]

Graham Mooney and Andrea Tanner

Introduction

Public health was amidst a transformation at the beginning of the twentieth century. The policies of environmental clean up which had so dominated late nineteenth century sanitation were giving way to different forms of intervention that focused on the behaviour of individuals, their interactions with each other and with the environment. Sociologically-based critiques have identified infant mortality, and the policies aimed at reducing it, as one of the signifiers of this transformation (Armstrong, 1986, 2002; Wainwright, 2003). Indeed, Armstrong has argued that the very subtitle of George Newman's 1906 book was indicative of how 'the conceptual space of infant mortality' had shifted 'from one traversed by sanitary axes to primarily social dimensions' (Armstrong, 2002: 33). Demographic analyses of infant mortality by legitimacy and class contributed to the demarcation of a 'separated space of the independent [infant] body' that contrasted with 'the physical space of nature, geography and meteorology' that had gone before (Armstrong, 2002: 33). This chapter engages with these issues by considering the emergence of infant mortality as a focus of intervention in Kensington, West London, at the end of the nineteenth century.[2] It disputes Armstrong's environmental/societal disconnect of the conceptual

1 The work for this paper was undertaken as part of wider project at the Centre for Metropolitan History, Institute for Historical Research, 'Mortality in the Metropolis, 1860-1920', funded by the Wellcome Trust. We gratefully acknowledge the Trust's financial support. We would like to thank staff at Kensington Public Library for their help and co-operation and Anna Shepherd for research assistance. Bill Luckin, Derek Keene and audiences at a seminar in Cambridge and a conference in Liverpool provided comments on previous versions.

2 The Kensington registration district was divided into two sub-districts: Kensington Town, the area north of the High Street (comprising 1,586 acres); and Brompton to the south (694 acres), which took in such areas as Earl's Court, South Kensington and Brompton, its southernmost boundary being the Fulham Road. After Kensington became a metropolitan borough in 1901, the registration district absorbed the Kensal district, which had previously belonged to the Chelsea registration district. There were two parliamentary divisions, namely North Kensington and South Kensington, whose boundaries were the same as those of the registration sub-districts, apart from the Kensal area, which remained in Chelsea only for parliamentary elections. In local politics, Kensington Borough had nine wards, each served by local councillors. In North Kensington, the wards were Golborne, St. Charles, Norland

space of infanthood. While not disagreeing that public health interventions were becoming more concerned with individuals and their behaviours, we suggest that, in Kensington at least, the social formulation of the infant mortality problem was directly related to its environmental one. In other words, environment and society in the early twentieth century continued to be linked in a way that sought to mesh deprived living conditions with a corrupt moral environment that begat social problems. We focus on infant mortality in an impoverished section of north Kensington, Notting Dale, and the crèche system developed there as a philanthropic response to the social 'problem' of working mothers. Crucial here is the notion that despite their charitable origins and associations with working class 'improvement', crèches (day nurseries) were attacked as places that contributed to moral disintegration by allowing mothers of infants and young children to continue work outside the home. Although regarded in some quarters as a potential 'stream of knowledge upon all matters relating to child life' (Toogood, 1910: 86), the daily crèche regime did not involve the close monitoring of mothers and babies that has been associated with state-sponsored domestic health visiting and infant welfare centres. Newman himself regretted that crèches were 'but a palliative method of prevention, and are not to be compared for a moment in value with education and training of the mother' that could be found in these latter places (Newman, 1906: 277).

Victorian Kensington was widely regarded as a healthy place to live. The prosperity of many of its inhabitants – Charles Booth described Brompton in the south of the area as 'the richest district in London' (Booth, 1903: 17) – and the female bias of its population predictably inclined the district towards a favourable position in the capital's mortality tables.[3] This picture, however, is incomplete (Reeder, 1968). While it is true that Kensington lacked the usual prerequisites for the high age- and cause-specific adult death rates associated with many of the poorer districts in London, as the 1890s progressed its infant mortality record grew steadily worse and failed to achieve parity with the capital as a whole until the First World War. The five streets in Notting Dale responsible for Kensington's poor track record of infant mortality wore a succession of identities in the nineteenth century, first as 'the Potteries', then as 'the Avernus' and finally 'the Notting Dale Special Area' (NDSA). Pressure brought to bear by local citizens and the media directed a reluctant Vestry's attention towards Notting Dale. Defining and analysing 'unhealthy' sections of a city for the spatial focus of intervention was a notable strategy of public health in these years, serving to legitimate and perpetuate existing economic, political and moral structures (Niemi, 2000). In the case of high rates of infant mortality in Kensington,

---

and Pembridge.The Notting Dale Special Area was in Norland ward. South Kensington's wards were Holland, Earl's Court, Queens Gate, Brompton and Redcliffe. The wards were the basic geographical unit for the borough and its predecessor, the Vestry. The Medical Officer of Health's (MOH) mortality statistics were usually presented for the Vestry/Borough as a whole and sub-divided into Kensington Town and South Kensington (Brompton, although the official registration sub-district, was occasionally abandoned as the designation after 1901). The MOH also produced birth and mortality statistics by ward in his post-1901 Annual Reports.

3 By 1901 females constituted 64 per cent of Kensington's population over the age of 15 years. For London as a whole, the corresponding figure was 54 per cent.

married women working as laundresses in NDSA were singled out as posing a threat to the life chances of their offspring. While Kensington's public health department emphasised the value of maternal education and childcare schemes, such municipal provision proved politically untenable as a burden on the rates (Marks, 1996). Sensitive to the needs of the district's wealthier inhabitants, public health officials did not advocate restricting women's re-entry into the local service economy after giving birth, but instead tacitly approved the charitable crèche system that fulfilled a need the local state was largely unwilling to meet.

Politics and the Environment in Notting Dale

For the parish Vestrymen, residents and estate agents alike, the image of Kensington that they wished to promote throughout London and the country as a whole was of the 'village of palaces' (Erlington, 1922: 3). Politically, it was like many other London parishes, in that the Vestry was in the hands of tradesmen of the middling sort and a coterie of slum landlords until 1870. The fiercely contested election that year for control of the Board of Guardians signalled a change in the political make-up of Kensington local government (Kensington Board of Guardians Minutes, 29 April 1870). Henceforth, a group of urban gentry took a more prominent part in local politics. In 1879, the Kensington Ratepayers' Association confirmed its growing influence within the parish by fielding all the successful candidates for election to both the Vestry and the Board of Guardians (*West London Observer*, 19 and 20 April 1879, 28 May 1879). The members of this association were well-educated, politically sophisticated gentlemen committed to keeping a lid on the rates. Nevertheless, by the 1890s the local authority regarded itself as one of the more progressive in London in matters of public health and welfare; their MOH, Dr Thomas Orme Dudfield, was famous for his assiduous and imaginative interpretation of the sanitary legislation, particularly with regard to housing defects (Tanner, 1998); the rates-funded public baths and washhouses had opened in 1888; a public library followed in 1891; Avondale Park was created for the poor of Notting Dale in 1892; and Kensington was the first local authority to employ women inspectors of workshops (McFeely, 1988: 22).

The sub-district of Kensington Town – approximately half a mile north west of Holland Park, and a few hundred yards north of the grand Royal Crescent off the Uxbridge Road – was home to Kensington's working class. Here were to be found the cabmen, labourers and street sellers who had to live within easy distance of the livings to be gained from their wealthy neighbours. When the parish was being developed in the 1850s and 1860s, this northernmost area included the Potteries, a network of narrow streets filled with derelict cottages. The Potteries was not only socially separate from the rest of Kensington, but physically isolated, for the only way out of it was a narrow alley called 'Cut Throat Lane' (Curle, 1967: 95). At this time, brick making was the principal summer occupation of the men (and hence the name Potteries), laundrywork the well-paid, year-round work of the women, and pig-keeping the lucrative (and sanitarily contentious) sideline of almost everyone

(Malcolmson, 1975: 34-7; Valentine, 2005). In January 1893, a local newspaper compared the area with the worst sinks of the East End:

> … whole streets steeped to the lips in iniquity, veritable hotbeds of everything vile and abandoned, and hundreds, if not thousands, of people living under conditions of life that are a foul disgrace to our civilisation … (*Daily News*, 24 January 1893)

This anonymous article, provocatively entitled, 'A West End Avernus', was supported by letters of confirmation from local clergymen, and prompted a furious correspondence in both the Kensington and the Westminster press (Kensington Vestry Minutes, 14 March 1893; *Kensington News*, 25 February, 11 and 25 March 1893).[4] The Vestry's Works and Sanitary Committee swiftly refuted the newspaper claims that the area was as bad as it had been painted (*Kensington News*, 31 January 1893). Yet in the following month, copies of a resolution of a meeting of influential local inhabitants called for 'serious attention' to the sanitary and immoral condition of the district (*Kensington News*, 25 February 1893). Shamed into action by this adverse publicity, the Vestry hastily performed a survey of the district and conceded that there was indeed evidence of overcrowding and uncleanliness, and that some dustbins were in need of repair (Kensington Vestry Minutes, 14 March 1893). The recommendations that a temporary additional sanitary inspector undertake more frequent house-to-house inspection were approved and swiftly acted upon (Kensington Vestry Minutes, 26 April 1893). In March 1896, however, the MOH's monthly report highlighted the continued unsatisfactory conditions and a second survey was carried out the following June by a special committee of the Vestry (Kensington Vestry Minutes, 11 March 1896). Headed by the chairman of the Vestry – a clergyman who was shortly to be appointed Bishop of Peterborough – the 12-man committee included a barrister, a member of the London County Council (LCC) and the Vestryman who represented the five streets under examination (Kensington Vestry Minutes, 11 March 1896). It visited the area twice; collated oral evidence from three clergymen, the MOH, the Poor Law Relieving Officers, the police, and one member of the Vestry; and received six written reports, including three from the school board (Kensington Vestry Minutes, 15 July 1896).

Taking advantage of the once-only quinquennial census of London in 1896,[5] the committee could not fail to form a clear impression of the area under scrutiny, a small locality that was soon to be re-christened Notting Dale Special Area by Vestryman Bamber (Kensington Vestry Minutes, 4 November 1896). The five streets – Bangor Street, Crescent Street, Kenley Street, St. Clement's Road and St. Katherine's Road – which approximated that of its previous incarnation as the Potteries in the 1850s and 1860s, had appallingly high crude death rates, with all but Kenley Street's being double that of Kensington as a whole, which in 1896 was 16.8 per 1,000 (Table 9.1).

---

4 In Roman mythology, Avernus was a crater near Cumae in Campania which was believed to be the entrance to the underworld. It was also a name for the underworld itself. By the late nineteenth century, it had come to be an acceptable euphemism for Hell.

5 This census was conducted in London to meet criticism that equalization of the annually-assessed rates across the capital would be unreliable if based on a decennial census of population (Davis, 1988: 184).

Table 9.1 Overcrowding and crude death rate (CDR, per 1,000 population) in Notting Dale Special Area, 1896

	Population	Houses	Persons/ House	Deaths[a]	CDR
Crescent St.	709	39	18.2	33	46.5
Bangor St.	874	39	22.4	45	51.4
St. Katherine's Rd.[b]	1,326	110	12.1	56	42.2
Kenley St.[c]	404	52	7.8	12	29.7
St. Clement's Rd.	543	42	12.9	28	51.6
Total	3,856	282	13.7	174	45.1

Notes: a) Deaths refer to 1895. b) The data here are for all of St. Katherine's Road although only the worst section of the street, with a housing density of 16.2 people per house and a CDR of 44.5, was incorporated into NDSA. c) Previously William Street

Source: Dudfield (1896)

The previously expressed views of the Vestry that these high rates were the fault of the inhabitants seemed to be rubberstamped by the committee:

> The people inhabiting the area are largely made up of loafers, cab runners, beggars, tramps, thieves and prostitutes, and undoubtedly a deplorable state of things exists in the district, the ordinary laws of decency being disregarded or set at defiance (Kensington Vestry Minutes, 15 July 1896).

Overcrowding was the culprit of human immorality, argued C.E.T. Roberts, the Vicar of St Clement's in Notting Dale. Despite having performed missionary work all over the east end, Roberts had never seen 'anything so inhumanely foul as this Kensington slum' (*The Times*, 7 January 1899). The question was not simply one of the five streets, however. The influence of overcrowding in NDSA went beyond its four thousand or so inhabitants, since '...there is a large outer ring of another 4,000 permeated and infected by this inner ring, the immorality of which, owing to overcrowding, is almost as bad as that in the 'Avernus' itself' (*The Times*, 7 January 1899). Even eighteen months later, the committee's chairman, Mr Wheeler, informed readers of *The Times* that the '... solution of the problem of rich and poor, of wealth and poverty, of crime, vice, and misery ...' lay in '... more stringent laws, to better forms of local government, and above all to the co-operation in that government of the best of our citizens' (*The Times*, 26 December 1898).

In the light of this comment and Roberts's remarks about the insidiousness of overcrowding, it is perhaps not surprising that the Vestry's response to the report was to use the existing tools of government to remedy environmental defects. Housing conditions swiftly became the focus of sanitary effort. Common lodging houses were initially targeted for special consideration, but although the NDSA (including all of St. Katherine's Road) had no less than 216 registered common lodging houses with a total of 1,525 rooms for anyone able to pay 4d. or 6d. per night, these were already subject to a relatively high degree of monitoring and regulation under the Local Government Act of 1883 (West London Observer, 10 January 1885; Rose, 1988: 56-63). In contrast, it was in the remaining one hundred or so houses let as furnished lodgings that 154 of the NDSA's 189 deaths in 1896 occurred. The letting of furnished rooms was a profitable and exploitative exercise: a keeper could net 30s a week in rent, the value and quality of the furniture provided being negligible. The committee requested clarification of LCC's definition of 'common lodging house' in the hope of applying it to some of these houses let as lodgings, but the Council refused to sanction any legal extension of the term to cover furnished rooms. The Vestry subsequently embarked on a programme of registering all multi-occupancy dwellings in the parish, a temporary sanitary inspector was appointed for a year for the five streets, and the number of house inspections increased dramatically. Kensington was the first London Vestry to register multi-occupancy dwellings and in 1896, 255 of the 275 houses in the NDSA were inspected a total of 398 times, and the same 255 houses were re-inspected on 1,157 occasions (Dudfield, 1896: 151).

Further attention was given to streets and yards, the state of which – 'saturated with human ordure, and filth of every description' – struck the committee as being dangerous to local children who chose to play in them rather than the nearby Avondale Park (Kensington Vestry Minutes, 15 July 1896). All property owners in NDSA were ordered to pave their yards and the Vestry itself paid more than £2,750 to asphalt the streets (Kensington Vestry Minutes, 23 September 1896 and 9 March 1898).[6] A new system of dust collection was instigated and a plan of sanitary improvements – including additional lavatories and better access to clean water for the inhabitants – had largely been completed within a year of the investigation (Kensington Vestry Minutes, 2 June 1897). Increased inspection of the area (undertaken without any additional sanitary staff) resulted in greater numbers of notices to remedy being issued (covering 251 out of 257 houses in the five streets) and many prosecutions for overcrowding (Dudfield, 1897: 126-30).[7] At the same time, the Vestry petitioned the LCC to acquire some of the worst housing in the NDSA, so that they may be let as lodging houses under the provisions of the Housing of the Working Classes Act of 1890 (Kensington Vestry Minutes, 28 July 1897; *The Times*, 7 January 1899). These official approaches, and requests from influential residents and local clergy, met with

6 Wheeler reported a figure of 'something like £4,000' in *The Times*, 26 December 1898.

7 Prosecutions are recorded in Kensington Vestry Minutes, 29 June 1898, 19 October 1898, 16 November 1898, 14 December 1898 and 25 January 1899.

refusal, the LCC informing the Vestry that it had sufficient powers to regulate the area under existing legislation.[8]

This episode typified the Vestry's fractious relationship with the LCC, which had formed in 1889. Kensington displayed mounting hostility throughout the 1890s to what it saw as interference in its own affairs. In 1895 it had tried to contract out of the LCC totally by incorporating the Vestry under the Municipal Corporation Acts (Robson, 1948: 93). As more duties were imposed on the Vestry under the Public Health (London) Act of 1891[9] and measures to equalise the rates began to cost the parish large sums, the mood of the elected representatives of Kensington had swung against progressive policies (Davis, 1988: 184). Animosity reached something of a zenith when the LCC investigated the sanitary condition of Kensington in 1899 (*The Times*, 25 December 1899). Throughout the two-year run-up to the enquiry, both the Local Government Board and the LCC continued to refer to the NDSA as the 'Avernus district', much to the annoyance of the Vestry members (Kensington Vestry Minutes, 7 October 1896, 27 January and 10 February 1897, 14 June 1899). The LCC investigation downplayed the immorality and bad habits of the inhabitants, and rather than emphasize the Vestry's environmentally-based response to the 1896 report on NDSA, it pointed to Vestry parsimony for allowing Notting Dale to become such a blight on the reputation of Kensington. The LCC's assistant MOH, Dr William Hamer claimed that had the Vestry listened to Dudfield's pleas to increase sanitary staff, '... the West End Avernus would never have attained its notoriety ...' (Hamer, 1899: 21).

Infant Mortality and Working Women in Kensington

Hamer's explosive report specifically mentioned the high rate of infant deaths in the parish, though that was not something the Vestry could possibly have been surprised about. From the mid-1890s Dudfield's monthly reports to the Vestry frequently alluded to the problem of infant mortality. Some elected representatives of the Kensington ratepayers rejected that the Vestry was responsible for meeting the needs of babies born into the small world of NDSA, at least not before the poor of the area had reformed themselves first. The number of child deaths was very high, wrote Vestrymen Mr Bamber and Dr Maguire, and people of the area weere having more children than they 'ought to have':

8 A few years later, the new borough council purchased all the properties in Kenley Street, in addition to several more in the adjoining streets. The majority of the houses were renovated, but those on the south side of Kenley Street were rebuilt, and subsequently let to the previous occupants, as well as those residents of Notting Dale thought likely to make good tenants. The new properties, which were opened in 1906, had displaced 350 occupants, and re-housed a total of 438 (Kensington Public Library, Record of the Work Carried out by the Council of the Royal Borough of Kensington Under Part III of the Housing of the Working Classes Act 1890, 1906: 7).

9 54 & 55 Vict. C. 76. 1891, An Act to Consolidate and amend the Law Relating to Public Health in London.

> The Vestry could not help that fact; the matter had nothing to do with the Vestry; philanthropy might do something in the matter, but the Vestry could not ... what they must get before they could alter the existing condition of things was a better class of mother' (*Kensington News*, 10 April 1897).

The committee appointed to consider Hamer's report continued to allege that parental neglect, '... either wilful or in ignorance ...', was the main cause of infant mortality, and that lack of knowledge of what constituted food suitable for infants was largely to blame. It held parents responsible for high infant death rates, taking their offspring into the streets in all weathers to beg and sell flowers and then allowing them to sleep in saturated clothing (Kensington Vestry Minutes, 31 January 1900).

As with the mapping of 'unhealthy' areas in other British cities, the demarcation of NDSA underlined the distinctions between healthy, morally-upstanding citizens and their slovenly, unhealthy neighbours (Niemi, 2000). Unlike other places, though, the vivid empirical comparisons in Kensington were not based on ward-level information that carried with it a political, that is electoral, reality (Marks, 1996: 96-8, 268-73). The data for NDSA were fashioned by singling out streets and parts of streets that were historically associated with environmental deprivation in their previous guises, the Potteries and the 'Avernus', reinforcing the established image of the segregation of health status in the district. While it was thus politically expedient for Kensington's elected representatives to point the finger at the overcrowding, habits and morals of 4,000 or so inhabitants, the formal naming as NDSA helped dissipate pejorative associations of the negative moniker 'Avernus'. It also signified that the Vestry was devoting resources to this 'special' place. This method of spatial targeting was not uncommon in other cities (Niemi, 2000), and its promise of economic efficiency was attractive to urban politicians. But in Kensington there was a drawback: pencilling a boundary line around the problematic area paradoxically opened it up to an intense scientific analysis of infant mortality that raised the spectre of ongoing demands on Kensington's ratepayers. Dudfield used the NDSA to conduct a sustained inquiry into the reasons for Kensington's consistently bad infant mortality. In 1896, the Infant Mortality Rate (IMR) of NDSA was at a shocking 432 per 1,000 live births (*Kensington News*, 27 June 1896). Despite the sanitary actions outlined above, such high levels were maintained, and the rate even exceeded 500 per 1,000 live births in 1899 (Table 9.2). For NDSA to have an infant death rate as high as anything in the East End was an unacceptable black mark on Kensington's reputation. To emphasize the impact NDSA had on Kensington's overall IMR, Table 9.2 removes the births and infant deaths occurring in NDSA from the overall totals for Kensington. As a result, Kensington's IMR drops very close to, if not always below, the IMR for London as a whole.

Kensington Vestry did virtually nothing to address directly the problem of infant mortality. Its effort to commemorate Queen Victoria's Diamond Jubilee in 1897 by launching a scheme for NDSA that included the provision of a crèche failed to garner sufficient support among the philanthropically inclined in the parish of the Queen's birth (*Kensington News*, 24 July 1897; Prochaska, 1995: 133). In stark contrast to the mushrooming of crèches facilities in the first decade of the twentieth century (see below), the appeal raised only £2,000, which was less than one-third the sum

Table 9.2 Infant mortality rates (IMR per 1,000 live births) in London and Kensington, 1896-1906

	London IMR	Kensington Borough IMR		Town sub-district IMR		Brompton sub-district IMR	Notting Dale Special Area		
							Births	Deaths	IMR
1896	162	172	*(163)*	181	*(171)*	124	118	51	432
1897	159	162	*(152)*	166	*(154)*	140	130	56	431
1898	167	176	*(168)*	188	*(179)*	110	117	49	419
1899	167	174	*(163)*	184	*(171)*	122	120	61	508
1900	160	174	*(164)*	187	*(176)*	104	113	54	478
1901	149	157	*(149)*	161	*(152)*	132	112	46	411
1902	140	143	*(135)*	147	*(138)*	119	98	41	418
1903	131	139	*(131)*	147	*(139)*	89	136	46	338
1904	146	141	*(137)*	148	*(144)*	98	113	31	274
1905	131	139	*(132)*	145	*(137)*	107	119	41	345
1906	133	132	*(126)*	142	*(136)*	74	104	32	308

Notes: Figures in parentheses give the IMRs of Kensington Borough and Town sub-district after subtracting the births and infant deaths occurring in Notting Dale Special Area.
See Note 2 in the text for a description of the areas included in the table

Sources: Kensington MOH Annual Reports, 1896-1906

required (Kensington Jubilee Commemoration Scheme Minute Book, 19 May and 1 June 1897; *The Times*, 21 December 1899). The lesson for the Vestry was a harsh one: the Kensington public clearly wished a more impressive and central memorial to Victoria, and were unwilling to support a meagre building in a part of the parish that was both infamous and obscure.

It was not until the MOH appealed to the Vestry on the grounds of 'national efficiency' that action was taken. Dudfield persuaded the Public Health Committee to sanction a survey into the circumstances of the infant deaths in Kensington in 1901, carried out under his supervision, and using the borough's female workshop and factory inspectorate for 'the delicate duty of interviewing bereaved mothers' (Dudfield, 1902: 62). In 1901, a total of 581 babies died in Kensington, 506 of these to mothers living in Kensington Town. Norland ward, in which the NDSA stood, accounted for 129 of the dead infants. Previous analyses of infant deaths in the MOH reports had used only the occupational status of the father, derived from the babies' death certificates (Dudfield, 1897: 25-6). Now, the inspectors interviewed 257 of the bereaved mothers, and managed to elicit the maternal employment status and the feeding method used in all but two of these cases. Most of the mothers looked after their own babies throughout their short lives, although only 35 per cent of them breast-fed without recourse to any artificial foods, a very low figure for London at that time (cf. Fildes, 1992, 1998). The majority of the artificially fed infants who died were given cow's milk mixed with either barley or limewater, or proprietary brands of farinaceous food. Other diets included proprietary food without cow's milk, and milk with thickeners such as biscuits, arrowroot, flour or bread. Condensed milk, sold door-to-door by grocers and oilmen, was used to feed a further 29 infants. The preferred method of feeding was by bottle and rubber teat, and the average child was given milk thickened by sopped bread or flour at the age of two months. One baby, whose mother's milk had failed after 10 days, was weaned on 'nursery biscuits and milk, gravy and potatoes, a drop of tea and anything going'; another was spotted in a pram at the age of six months sucking on a fish's head. Fifteen of the babies had been given whisky or brandy, up to half a teaspoonful in each bottle (Dudfield, 1902: 66).

The majority of dead infants in Kensington came from families headed by labourers and artisans (Dudfield, 1902: 65). Almost 50 per cent of the 146 occupied mothers who lost children were employed in the laundry trade, while 22 per cent were classed as domestic servants and 14 per cent charwomen. Some of the responsibility for the many premature and perinatal infant deaths in Kensington was blamed on the mothers

> ... continuing at work when they should be resting at home ... The law forbids return to work at workshops and factories, until four weeks after confinement, but there is reason to believe that too often poor women resume their place at the washing-tub or ironing-board at a much earlier period (Dudfield, 1902: 65; Brooks, 1986: 154).

Whilst the report emphasised that there were no grounds for thinking that mothers were deliberately neglectful or unkind, their ignorance led to 'neglect of the children in the critical early days of their brief existence' (Dudfield, 1902: 66).

The 1901 survey was a prescient exercise for its time and anticipates information that only became widely available after the notification of births was introduced in 1907 (Mooney, 1994; Newsholme, 1906; Wainwright, 2003). The findings of the survey helped to pave the route towards more in-depth enquiries in Kensington in two significant ways. First, the MOH was now satisfied that much of the borough's infant mortality was preventable and that the 'moral duty' to conserve life in its early stages must be primarily borne by the State through sanitation, instruction in the laws of health, crèches and the provision of a municipal milk supply (Dudfield, 1902: 67-8). Second, some simple, though vital, connections had been made that linked employment in general, and laundrywork in particular, with feeding methods and premature death in Kensington.

Laundrywork and Infant Feeding

Charring and laundrywork were dominated by married and widowed women (Malcolmson, 1981: 445). In 1901, 29 per cent of Kensington's occupied married and widowed women were employed in the laundry and washing service, compared to 16 per cent in London as a whole (Census Report, 1901: Tables 32 and 35). Laundrywork was especially important in North Kensington, where women either supplemented the family income, or, more frequently, acted as the chief breadwinner. Around the turn of the century NDSA alone had 211 laundries, proving itself worthy of the nickname 'Laundryland' (Malcolmson, 1970: 63, 1975: 45). Most women worked between four and four and a half days a week (Monday and Tuesday being traditionally quiet), and were paid hourly for washing and piecework rates for ironing. The London season meant that there was very little work in August and September. In many ways it was the ideal occupation for married women, in that it utilised domestic skills, required little or no capital, and could be increased or decreased according to the financial circumstances of the family.

The work was physically demanding, and the hours were long, sometimes stretching to between 70 and 80 hours in a busy week, especially for women working in small hand laundries or at home (Squire, 1924: 22). Rates of pay, considered very high at between two shillings and half-a-crown a day for washers, and between three shillings and three and sixpence for ironers in the 1860s, had not increased by the Edwardian era (Malcolmson, 1986: 13). The wages were certainly better than in domestic service and the sweated clothing industry, but hardly enough to keep a family in comfort. Constant standing in high temperatures and a damp atmosphere meant that many women suffered from rheumatism and ulcerated legs, and others arguably weakened their constitutions by taking on extra work in pregnancy, in order to put something by for the new baby (Malcolmson, 1986: 17; Ross, 1986: 78).

Laundrywork was already subjected to a high degree of state intervention through the Factory and Workshop Acts (1891) and the Public Health (London) Act (1891), which gave sanitary authorities responsibility for the regulation of workshops. Although the Vestry complained about the additional workload this legislation entailed, in November 1893 Kensington had been the first local authority in the country to appoint women workshop inspectors, Lucy Deane and Rose Squire

Table 9.3 Legitimate births surviving to one year and infant mortality rates for all causes of death and prematurity, by employment status of mothers: Kensington, 1910-1914

	Observed births[a]			IMR	
Employment status of mother during pregnancy	**Survived**	**Died**	**Total**	**All causes**	**Prematurity**
Laundress/washing at home	849	172	1,021	168.5	23.1
Charwoman	299	66	365	180.8	
Other[b]	249	44	293	150.2	25.3
Total occupied	1,397	282	1,679	168.0	24.1
Not employed/at home	5,004	660	5,664	116.5	23.7
Total	6,401	942	7,343	128.3	23.8

Notes: a) The 7,343 births shown here are those occurring in households with a weekly income below 40s. Only such births were monitored by the Council's Health Visitors and represent less than half of the 16,219 births in the borough over the five year period. b) Because of the income threshold noted above, these occupations mainly consisted of cooks and other women in service plus hawkers and costermongers. For the prematurity IMR, the 'other' category also includes charwomen

Source: Sandilands (1910-14)

(Squire, 1924). Within six months they had registered and inspected 169 laundries in Kensington Town and 18 in Brompton (Dudfield, 1893: 157). The inspectors were able to order whitewashing of the laundries, improvements to drainage, water supply and sanitary facilities and ventilation in ironing and drying rooms. Given the nature of laundrywork, and its dominance in the poorer areas of Kensington such as Notting Dale, the public health department was especially keen to assess its impact on levels of infant mortality. Table 9.3 pools the results of the occupational information up to 1914, after which date the tables were discontinued. Although the information here relates to the employment status of women during pregnancy, a number of these mothers returned to work soon after the birth. The babies of charwomen appear to have had an even lower chance of survival that those of laundresses, and the babies of laundresses were no less likely to survive than the babies of all occupied women, yet it was laundrywork that claimed the attention of Dr James Sandilands, who was appointed Kensington's MOH following Dudfield's death in 1908. By 1911 he felt he had enough material to conclude that the engagement of mothers in laundrywork posed a serious threat to the survival chances of their babies. By demonstrating that the premature infant death rate was not influenced by occupation, he dismissed the idea that the arduous nature and long hours of laundrywork had an adverse effect on the child in pregnancy (Table 9.3). Although most of the women continued working until their seventh or eighth month of pregnancy, the problem lay not with ante-natal conditions and work *per se*, but with the post-natal child care and feeding regimes that going out to work entailed (Sandilands, 1911: 11).

To further strengthen his occupational hypothesis, Sandilands was obliged to deploy two further arguments or strategies: the first was to dismiss, or at least downplay, the role of factors other than women's employment; and second, to provide proof that working mothers were more likely to feed artificially than from the breast. In relation to other factors, his first report in 1909 did in fact list overcrowding as a major contributory element to high levels of infant mortality. While Newman himself had shown that much higher rates of infant mortality in Finsbury seemed to occur in families with resources meagre enough to only rent one room (Newman, 1907: 113; Marks, 1996: 96), Sandilands eventually concluded that there was no simple positive relationship between the tenement size and the level of mortality, the rates being highest for infants born into families occupying one and four room tenements and basement rooms (see Table 9.4).

Data on rates of survival and mortality were given separately for occupations and feeding, so an exploration of the interaction between occupation *after* birth and feeding was limited (cf. Holdsworth, 1997: 117). Nevertheless, artificial feeding and high levels of infant mortality were consistently linked to the employment of women. Two health visitors saw all babies born to families earning less than 40s per week; follow-up visits occurring only where the children failed to thrive. Between 1909 and 1911, more than 90 per cent of babies aged between 10 and 20 days were breast-fed. However, as the babies got older, this proportion dropped off significantly. In 1910, only 75 per cent of babies aged between three and four months were still at the breast. The following year, 87 per cent of babies aged between 20 days and three months were being breast-fed, and 68 per cent of those between five and seven months old (Sandilands, 1910: 10, 1911: 9). These data indicated to Sandilands that

Table 9.4 Legitimate births surviving to one year and infant mortality rates, by tenement size, Kensington: 1912-1914

Tenement rooms	Observed births				Total	IMR
	Survived		Died			
	N	%	N	%		
One	724	*18*	123	*21*	847	145.2
Two	2,078	*53*	293	*49*	2,371	123.6
Three	495	*13*	57	*10*	552	103.3
Four	245	*6*	48	*8*	293	163.8
Basement only	414	*10*	75	*13*	489	153.4
Total	3,965	*100*	596	*100*	4,561	130.7

Source: Sandilands (1912-14)

mothers weaned their babies upon returning to work and this assumed even greater significance when diarrhoeal mortality was analysed. Only ten per cent of the 71 infants dying from diarrhoea in 1910-11 were breast-fed immediately prior to the onset of their illness, compared to 53 per cent for all other causes of infant death; 21 per cent of diarrhoeal deaths were fed by a combination of breast milk with either cow's milk or condensed milk and the remaining 69 per cent were fed wholly by artificial means. The comparable percentages for all other causes were 13 and 34 respectively (Table 9.5).

Table 9.5 Mode of feeding at onset of illness for infant deaths[a] between two days and six months: Kensington, 1910-1911[b]

	Infant deaths			
	Diarrhoea		Other causes	
	N	%	N	%
Breast only	7	10	105	53
Mixed[c]	15	21	39	13
Artificial only[d]	49	69	112	34
Total	71	100	256	100

Notes: a) Deaths relate to those called upon by health visitors. b) The 1911 data referred to deaths between 2 days and 9 months c) Includes breast milk with either cow's milk or condensed milk. d) Includes cow's milk, condensed milk and other foods

Source: Sandilands (1910-11)

While women continued to be criticised for their methods of feeding infants (Sandilands, 1911: 10), conservative elements in Kensington local politics were reluctant to spend public money on infant welfare schemes that were thought to help remedy the problem (Marks, 1996). Yet neither of Kensington's two MOHs advocated the removal of mothers from the workplace in the light of their findings, unlike some other commentators (cf. Sykes, 1910). 'It appears from the statistics', wrote German Sims Woodhead, Chair of Pathology at Cambridge and public health campaigner, 'as though the loss to the nation in this way may be infinitely greater than if all the women who now act as wage-earners were fully pensioned and allowed to attend to their proper duties' (Woodhead, 1907: xiv-xv). In observing working-class mothers at close hand, Sandilands became convinced that there was 'no lack of maternal instinct among the poorer classes and no narrow limit to the amount of good which is certain to be effected by women who are so anxious to learn' (Sandilands, 1911: 14; cf. Dyhouse, 1978: 261). This quote may typify the patronising attitude of middle-class men to working-class mothers, but demonstrates restraint from the victim blaming that was so common at this time in the writings of observers in similar positions of administrative influence. As we have seen, Dudfield even went so far as to suggest that crèches should be funded by the local state (Marks, 1996: 134). In other British cities, the interdependence of the outer suburb and the central slum was crucial for the exercise of middle class power (Niemi, 2000: 125). Spatial relationships were no less significant in Kensington. The importance of laundrywomen (and, indeed, charwomen) residing in Notting Dale to the smooth functioning of the wealthier side of Kensington's economy and society probably served to preclude both Dudfield and Sandilands from suggesting mothers with young children should not be a part of that economy.

Crèche Provision in Kensington and London

The steady decline of infant mortality across the whole of London in the early years of the twentieth century was by and large credited to a succession of cool summers that reduced the toll of diarrhoeal mortality. Kensington saw the gradual narrowing of the gap between the borough's rates and those of the capital that accompanied this decline (see Table 9.2). That Kensington's improved performance predated the extension of the council's health visiting work in 1908, and that it post-dated the bulk of the borough's sanitary reforms by a number of years, indicated to the MOH that the decline 'must be attributed to indirect rather than direct administrative measures' (Sandilands, 1909: 9). Here, Sandilands was referring to the wide range of charitable organisations that existed in the area. The remainder of this chapter will consider one component of charitable provision that has tended to be overlooked in the literature on infant health and welfare provision, namely crèches (Ross, 1993: 205).

The rich history of charitable provision in Kensington can be seen from the selective sample in Table 9.6 (*Kensington Directory*, 1910). Charles Booth described the Kensington churches as giving out 'vast sums': St. Mary Abbots alone distributed an annual charitable income in excess of £20,000. The charitable efforts of the whole of Kensington seemed to be concentrated in Notting Dale where, Booth noted, '…

Table 9.6 Charitable provision in Notting Dale before 1914

Name	Type/service provided	Date
*Mary Bayley's**	*Mothers' club*	*1840-64*
London City Mission*	Domestic visiting	1847-
Carmelite Convent*	Sick and homeless	1859-
Rescue Society*	Temperance and moral rescue work	1859-66
Notting Hill Dispensary	Dispensary	1860-
*Latymer Rd Mission**	*Visiting, hall, crèche/infant day nursery (from 1880)*	*1863-*
Men's Institute*	Temperance work	1866-
St. Joseph's Home*	Old age home	1866-
Charity Organisation Society	Investigative	1869-
Salvation Army*	Hall	1889-
Harrow School Mission*	Boys' club	1882-
Rugby School Mission*	Boys' club	1887-
Rackham St. Infant Day Nursery	*Nursery*	*c1890-*
*Nurse Thompson's**	*Medical mission*	*1890-*
Congregationalist Mission, Bangor St.*	Mission hall	1893-
*St. James Mission**	*Lodging houses, youth club, creche*	*1893-*
Dr. Barnardo's Home and Mission*	Orphanage	c1894-
*Holy Cross Society of Trained Nurses**	*Nurses' school and home*	*c1895-*
Kensington Maternity Association	*Midwifery, maternal education*	*c1898-*
District Nurses Home	*District visiting*	*1903-*
Miss Merriman's	*Infant day nursery*	*c1904-*
Ladbroke Nursing Assoc. and Home	*Nurses' home*	*1905-*
*St. Clement's Day Nursery**	*Day nursery*	*1905-*
*Princess Mary Adelaide Crèche (Harrow Mission)**	*Crèche*	*1906-*
Notting Hill Day Nursery	*Nursery*	*1909-*
Women's Labour League Clinic	*Child welfare clinic*	*1910-*
Kensal Medical & Gospel Mission	*Dispensary, crèche, maternity nurse*	*c1910-*

*Note: * Denotes religious affiliation. Italics indicate a charity providing services for mothers and children*

Source: Kensington Directory, 1890-1914

the social agencies in operation are numerous …' (Booth, 1903: 155). There was a rapid increase in the number and type of charities at around the turn of the century, many of which were formed to fill the gap left by the Poor Law and the public health department in relation to mother and infant welfare. Childcare was a major problem for all women engaged in laundrywork, whether at home or in a workshop, for the work made great inroads into family life (McFeely, 1988: 38-40). Many babies were born to mothers who had worked until their confinement, and who returned to work before their infants were four weeks old. Breast-feeding was possible for those who worked from home (the minority by 1900), and babies could be brought to the workshop for nursing during the lunch break, but most mothers had to entrust their children to the care of others. As an alternative to baby farming or informal day care, the charitable crèche system proved important in this respect.[10] Crèches were notoriously uneconomic ventures (LCC, 1904: 7) and self-supporting crèches that were established in manufacturing districts were prone to financial failure (Hewitt, 1958: 158). With some notable exceptions, few factory owners demonstrated any interest in organising a day nursery for their employees' offspring (Hewitt, 1958: 164) and the argument that the state should not sanction such maternal irresponsibility held sway (Cooke Taylor, 1874). Almost all crèches were therefore in the charitable sector. There was an established tradition of crèche provision in north Kensington, where the great majority of laundresses lived. Most were begun by religious groups, and included Nurse Thompson's Free Medical Mission (opened in 1890), and the Latymer Road Mission (founded 1862), which ran an infant day nursery as early as 1881, charging mothers between 4d. and 6d. a day per child. The nursery manager, Miss Annie Rowan, described the impetus behind the establishment of the nursery:

> This part was known as Soap End's Island as it was the district of laundries. They were everywhere – in big buildings and private houses. Laundrying was carried on in almost every house. This meant that in the majority of cases the mothers were the breadwinners. As a result the babies had to take pot luck. So the crèche was started. I believe it was the first of its kind in London (*Morning Leader*, 15 April 1911).

The nursery accommodated 60 children under the age of five and was open from 7.30 a.m. until 10 p.m. (Gladstone, 1969: 157, 205; The Latymer Road Mission Scrapbook, 1 April 1888). The Thompson establishment (named 'Miss Thompson's Crèche' in a report of 1904) catered for an average of 25 infants, though it had capacity for 30. It was open 8 a.m. to 8 p.m. and charged 4d. per day, 3d. for the children of widows. As can be seen in Table 9.6, however, the majority of the area's infant day nurseries and crèches were created after 1903, with no less than six being opened between this date and 1909. According to a report made by Huddersfield's MOH, Dr Moore, in 1904, most crèches in England operated a similar regime to that of Miss Thompson's (Heath, 1907: 164-5). By this time, charitable crèche provision was seen as a way of providing a lifeline for the respectable working class mother. Agitating for the establishment of a crèche in Notting Dale as 'a form of aid that

10 The registration of day minders was a focus of the Infant Life Protection legislation (see Rose, 1986: chapter 12).

does the least harm', Mr Wheeler, the former chair of the Notting Dale investigative committee of 1896, argued that:

> In slum life … little can be done for the man or woman soddened with drink, the loafer or the hardened prostitute, something … can be done for the large class of women who are mothers and who are compelled to work often in the lowest and most precarious kinds of labour for the maintenance of themselves and [their] offspring (*The Times*, 26 December 1898).

Perhaps demand for charitable crèches increased as the Board of Education phased out school classes for under-threes by 1904 and was actively reducing the number of places for three to five-year-olds, but most crèches admitted children at one month of age, so they were important for infant care (Lewis, 1986: 110 [citing Whitbread, 1972]).[11] Just as crèche provision was about to take off in Kensington, a survey by the Public Control (later the Public Health) Department of the LCC showed that in the early twentieth century there were 55 crèches in 14 of London's 21 boroughs. Their services, described as being good 'as far as they go' (LCC, 1904: 4) were directed primarily at the respectable working-class mother: almost half refused to admit illegitimate children and the most successful were located 'near to where women work, near their homes, or between the two' (LCC: 1904, 6). A minority of crèches bathed their charges and provided clothing. It was feared in some quarters that the crèche system facilitated cross-infection, was attached to the stigma of pauperism (i.e. acceptance of charity) and threatened the unity of the family by freeing women from the responsibility of motherhood (LCC, 1904: 6; Inter-Departmental Committee on Physical Deterioration Vol. II, 1904: 312, 364). It was said that one paying crèche in the Potteries had been resisted because it took away the small income that some older people received for looking after young children (Inter-Departmental Committee on Physical Deterioration Vol. II, 1904: 338). Opponents of the crèche system also argued that in some cases the wages of the mother barely covered the costs of putting the child into day care (Sykes, 1910: 584), or that the presence of a nearby crèche would encourage mothers who might not otherwise work to do so and thus prevent breastfeeding (Pennybacker, 1995: 166). Alternatively, the supposed benefits of a well-administered crèche were outlined in the report as follows:

> 1. The crèche was usually a healthier place than the child's domestic environment, with better feeding regimes and more 'enlightened' care under the supervision of a matron
>
> 2. The standard of care was better than if the mother left the baby in the hands of a relative, neighbour or stranger
>
> 3. The child could be taught cleanliness and good habits, become stronger and have better life-chances
>
> 4. The medical supervision at the crèche allowed for prompt treatment of illness

11 Under the Factory and Workshops Act of 1891, it was illegal for employers 'knowingly to employ' any woman within four weeks of her confinement. The term 'knowingly' was clearly open to dispute (Hewitt, 1958).

5. Mothers were required to bring the child to the crèche clean, thus inculcating hygienic habits that would be adopted in the home

6. The crèche might be used for the education of school girls in the care of infants

(LCC, 1904: 9; Newman, 1906; Inter-Departmental Committee on Physical Deterioration Vol. II, 1904).

Perhaps with the exception of (5) and (6) – though (6) was seen as much a fault of the education system as a failure on the part of young mothers (Inter-Departmental Committee on Physical Deterioration Vol. II, 1904: 293-6) – it is striking how this list of 'advantages' resists blaming working mothers for poor infant health, or calling for mothers to come under direct supervision as they did under infant welfare schemes in their various guises (Sykes, 1910; Wainwright, 2003). In the early years of the day nurseries in Great Britain, supporters of the crèche system represented (albeit respectable) working mothers not so much as a social problem, but as a social reality of the modern industrial capitalist economy that should be accommodated (Chadwick, 1879; Cumming, 1874). It was long-emphasized that parents should be carefully screened before admitting their child as to the necessity of the mother working and that dropping off and picking up the infant offered an opportunity for educational intervention in the form of pamphlets and advice (Chadwick, 1879: 31).

The LCC report suggested that crèches were of the greatest benefit where there was most demand, namely amidst large amounts of female labour in factories and laundries (LCC, 1904: 9; see also Inter-Departmental Committee on Physical Deterioration Vol I., 1904: 91; Heath, 1907: 165). Perhaps the value of the crèche system is indicated by the fact that the LCC sought to bring that system under direct control and standardise provision across London as a Progressive instrument 'for taking charge' of the children when mothers were at work (LCC, 1904: 3; Pennybacker, 1995: 166). The report pointed to the desirability of providing a crèche place for each infant and child of every working mother in London, either by the Council itself in neighbourhoods where demand was greatest; or by Borough Councils and Boards of Guardians in association with public washhouses, workhouses and infirmaries; or through contributions to charitable crèches.

Conclusion

In the 1890s, the attention of Kensington's Vestry had been consistently drawn to the district's excessive IMRs by its own MOH. Finally stung into action by a damaging newspaper article, the creation of NDSA by the Vestry served to formally identify, in a very precise way, the streets that were believed to be at the root cause of the problem. While the extensive sanitary improvements carried out in NDSA in the late 1890s registered little immediate impact on infant mortality, in 1906 the borough decided that as it had fulfilled its duties according to the current public health legislation, it was no longer necessary to keep separate records for the five streets in Notting Dale. Nevertheless, the basic act of carving out and naming the area had irretrievably put infant health to the forefront of the public health department's research and health visiting agenda. As a result, poor infant health in Kensington came to be associated with a specific place, Notting Dale, and a specific segment of that place's population, mothers working in the laundry trade. To say a fundamental shift in thinking transplanted social rather than environmental causation would be misrepresentative (Armstrong, 2002; Niemi, 2000). The environmental, the moral and the social continued to be entwined in a complex relationship, in which the roots of the last two resided in the former. The chronic overcrowding of NDSA was said to engender immoral habits and it was from this slum environment that the respectable working mother and infant had to be saved. Given the dependence of South Kensington society on the laundresses who lived to the north in Notting Dale, it was unlikely that the middle class men who observed NDSA close up would paint the working mother as a pernicious social evil worthy of comparison to the drinking and sexual habits of a closely-packed, indigent underclass.

Kensington's MOH was given to state in 1910 that voluntary health care provision in the Borough 'may without exaggeration be described as extensive' (Sandilands, 1910: 15). He might also have said that the voluntary agencies had removed a great deal of pressure on the borough to provide its own infant welfare centres (Marks, 1996: 178). Sandilands was convinced that services such as Kensington Health Society's school for mothers 'could not fail to improve the physique of the rising generation' (Sandilands, 1910: 15; Prochaska, 1989: 395). In 1913 more than 800 mothers attended the Society's infant consultation classes and the total number of dinners supplied to expectant or nursing mothers exceeded 31,000 (Sandilands, 1913: 12), though he was unable to avoid the fact that declines in infant mortality had also taken place in districts with far fewer charitable agencies.[12] Notwithstanding increasing interest in the history of charitable provision in recent years, the relationship between infant mortality and maternal and infant welfare has tended to concentrate on schemes funded by local taxation, such as welfare centres, health visiting and milk depots. Perhaps because most day nurseries lay outside the realm of state funding and control, the system of crèche provision has not attracted a comparable amount of attention from social and cultural historians. It may be unwise to polarise the two systems, infant welfare schemes on one side and the crèche system on the other, since they both legitimated maternal education in some form and were predisposed to serving the 'respectable' working class. But it would appear that support for either alternative did indicate a significantly different world-view on the part of the protagonists. On the side of infant welfare schemes, health visiting was a means by which a mother could be assisted in bringing up the baby in her own home (Davies, 1988; Marland, 1993: 38). The organised regimes of infant welfare centres – such as dinners in the middle of the day and consultations

12 Despite the overall wealth of the borough, Kensington's IMRs consistently exceeded those of London as a whole after the First World War and well into the 1930s. This was probably the result of persistent poverty and overcrowding in parts of North Kensington, and the relative lack of maternal welfare services in the borough (and similarly wealthy districts), 'where provision was more limited and viewed as an individual rather than a municipal responsibility' (Marks, 1996: 290).

and classes in the afternoon (Sykes, 1910; Wainwright, 2003) – implicitly prevented fulltime participation in the labour force. Even mothers purchasing milk from some milk depots were assessed to ascertain whether they had a legitimate reason for not breast-feeding: one assumes being at work was not considered a legitimate reason (Wainwright, 2003: 169). Allowing for intense levels of medical surveillance of mothers, these predominantly rate-funded schemes perpetuated an ideology that sought to reduce a woman's participation in the workforce as much as possible after the birth of her child, an ideology that had been legislated for in the 1891 Factory and Workshop Act.[13] On the other side, the crèche system was far less invasive and this was one of the stated attractions to working class mothers. Indeed, it was reported in London that where daily bathing of the children was practiced at the crèche, mothers would object and the crèche was rarely full (LCC, 1904: 8). While crèches tended to accept the children of those mothers who were forced to work rather than those who chose to, support for the crèche system implied that obligations to the local labour market were at least as important as the obligations of motherhood.

13 54 & 55 Vict. C. 75. 1891, An Act to Amend the Law Relating to Factories and Workshops.

Chapter 10

Health Visitors and 'Enlightened Motherhood'[1]

Alice Reid

Newman saw the influence of the mother and home to be of crucial importance to the problem of infant mortality, and the issue of infant feeding and management to be the key to the mother's influence. His arguments have been explained in detail in Chapter 3 of the current volume, but it is worth re-iterating those points which are salient for this chapter. 'Infant mortality is a social problem concerning maternity', wrote Newman on page 221 of *Infant Mortality: a Social Problem*, concluding that 'we must turn in the last instance to the actual feeding and management of an infant by its mother'. He cites studies which clearly demonstrate the dangers of artificial feeding, exacerbated by practices such as long-tube feeding bottles and sugary comforters, and makes a connection with household circumstances and uncleanliness. 'The fact that some breast fed infants do die from diarrhoea means it is not just milk' which is the problem, but a lack of hygiene (Newman, 1906: 248). He concludes this section by stating: 'The problem of infant mortality is after all one of those elementary problems which depend more upon instinct and the physical faculties and functions which nature has provided in the mother than upon external environment' (Newman, 1906: 256).

The last three chapters of *Infant Mortality* deal with methods of prevention of infant mortality, which he placed into three categories: the mother, the child, and the environment. Environmental measures included improved sanitation in factories and homes, urban cleanliness, and the control of the milk supply. Those relating to the child covered birth registration, infant protection, and infant feeding, the latter including the establishment of infant milk depôts for those incapable of nursing their children. The preventive measures relating to the mother he undoubtedly thought were the most important:

> The problem of infant mortality is not one of sanitation alone, or housing, or indeed poverty as such, *but is mainly a question of motherhood.* No doubt external conditions as those named are influencing maternity, but they are, in the main, affecting the mother, and not the child. They exert their influence upon the infant indirectly through the mother (Newman, 1906: 257 [italics in original]).

---

1 The work for this chapter was funded by the Wellcome Trust under its History of Medicine Programme, which, in funding my PhD, facilitated the computerisation of 60 per cent of the data. The remainder of the data entry was funded using a British Academy Small Grant (SG:31626).

The measures he advocated to improve motherhood were themselves three-fold. First he recommended the reorganisation of the existing agencies for ante-natal and post-natal care, including baby clinics and food for nursing mothers. Secondly, he called for the education of the mother as to infant management through instruction of mothers using methods such as leaflets and circulars, the appointment of lady health visitors, and the education of girls in domestic hygiene. Thirdly, he was concerned with the occupation of the mother and suggested measures to prevent the employment of women within four weeks of childbirth, more uniform enforcement of factory sanitation, and the establishment and control of crèches.

The need for the education of the mother was a subject on which he expressed himself frequently and with no mincing of words:

> Few facts receive more unanimous support from those in intimate touch with this question than the ignorance and carelessness of mothers in respect of infant management. Such ignorance shows itself not only in bad methods of artificial feeding, but in the exposure of the children to all sorts of injurious influences, and to uncleanly management and negligence. Death in infancy is probably more due to such ignorance than to any other cause (Newman, 1906: 262).

Health visitors had been first introduced in 1862 by the Ladies' Sanitary Reform Association of Manchester and Salford, in order to overcome such maternal ignorance. The original visitors were volunteers, but the Association soon started to employ salaried visitors, and the service was taken over by the Manchester Medical Officer of Health (MOH) in 1890. By the time Newman came to write the 1906 volume he was able to pay tribute to the burgeoning numbers of lady health visitors already established in thirty-five to forty large towns across the country. The continued appointment of lady health visitors was the lynch-pin of his proposed programme of reform, and has proved to be one which endured to a considerable extent over the ensuing century.[2]

Health visitors would foster 'enlightened motherhood' through visiting new mothers, offering advice as to infant care and management, and 'carrying sanitation into the home' (Newman, 1906: 263-4). As the risk of infant deaths was highest in the weeks after birth and it was felt that good infant care practices had a better chance of success if established early, a visit as soon as possible after the birth was crucial, and a system was needed to alert each health visitor to recent births in her locality. Such a scheme of notification was pioneered by the Huddersfield health visiting scheme in 1905 (Marland, 1993), and its general adoption was one of the other measures advocated by Newman. Although he classed notification of births under preventive methods relating to the child, it was a corner stone of the successful operation of the health visiting movement. In 1907 the Notification of Births Act was passed, allowing local authorities to enforce the notification of every birth within 36 hours by the midwife, doctor or other person who was present. This Act was permissive legislation, but was widely taken up, and was made compulsory under the 1915 Notification of Births (Extension) Act. Both voluntary and municipal effort increased dramatically

2 Dwork (1987) provides a detailed investigation of the political and ideological antecedents of the maternal and child welfare movement. See also McCleary (1933), Dyhouse (1978) and Lewis (1980).

both before and after the 1915 Act, fuelled by fears of manpower shortage caused by the war. Between 1914 and 1918 the number of health visitors employed by local authorities quadrupled (Dwork, 1987: 208-14; McCleary, 1935: chapter II; Winter, 1977: 499). In 1918 the movement culminated in the Maternal and Child Welfare Act which empowered local authorities to make arrangements for attending to the health and welfare of expectant and nursing mothers and children under the age of five not attending school. This included establishing and maintaining a staff of health visitors to conduct visits and assist at infant welfare centres, distributing food and milk for mothers and children, and provision of a midwifery and medical service for pregnancy and confinement. While the schools for mothers and infant welfare centres were also advocated by Newman as opportunities for educating mothers, and they received considerable municipal investment, there was a feeling among MOHs that they were less effective at reaching those who would benefit most from instruction. In the opinion of Dr A.E. Thomas, Newman's successor as MOH for Finsbury, 'The attendance at the Welfare Centres is not satisfactory. The mothers who attend are the better class poor, who are already solicitous for the welfare of their babies. It is extremely difficult to get the slatternly, careless, dirty, indifferent, or negligent poor to come to these places' (Hope, 1917: 275). Dr D.G. Moore, the MOH for Huddersfield, was equally harsh:

> The woman who has sufficient energy and intelligence to attend a school for mothers probably has sufficient energy and intelligence to keep her infant alive anyhow. It is amongst the slothful, the ignorant, the indifferent, and the vicious that the heavy infantile mortality occurs. These cannot be relied upon to attend such institutions, and must be sought in their own homes (Hope, 1917: 295).

In sympathy with this view, the visiting of mothers in their own homes received the lion's share of Ministry of Health funding.[3]

Although the 1918 Act allowed for a uniform service, previously established schemes with a mix of voluntary and municipal effort in many parts of the country and the permissive nature of the legislation meant that even after 1918, local services were highly individual and differed markedly in character and emphasis (Hope, 1917; Peretz, 1992). This variety of provision is one reason for the difficulty in assessing how effective health visiting was in improving the health and survival of infants and children. Contemporary observers were optimistic that health visitors were able to reduce infant mortality. Dr Andrew Laird, the MOH for Cambridge, wrote that the appointment of health visitors (two in 1906 and a third in 1908) was,

> really the great outstanding event in the sanitary history of the Borough at this time … the steady decline in infant mortality from the time of the appointment of health visitors points very strongly to the work being a great, and … a prime factor in the saving of infant lives (Hope, 1917: 243).[4]

---

3 Health visiting received 31 per cent of the grants for maternal and child welfare made by the Minister of Health in 1919-20, and although such grants were reduced over the 1920s, they still received more than other maternal and child welfare services (Lewis, 1980: 105).

4 See also Marks (1996: 186) for similar contemporary sentiments.

Some more recent commentators have agreed that the coincidence of the fall in infant mortality with the growth of infant and child welfare services indicates success of the movement (Smith, 1979: 114). Others, however, have discredited the synchronisation of timing or argued that the simple temporal coincidence is not enough to prove a causal link (Marland, 1993: 45-8; Mein Smith, 1993).

This chapter aims to investigate the effect of health visitors on infant and child health using the notifications of birth and visiting records created by health visitors in the county of Derbyshire in the early 1920s. Although the choice of Derbyshire is primarily governed by the rare availability of rich data, the survival of the data for this county is serendipitous as it encompasses settlements of many different sizes and densities, with varying socio-economic structures and environmental settings, from textile towns and mining villages to spa-towns and both hill and arable farming.

Health Visiting in Derbyshire and the Derbyshire Health Visitor Data Set

In reflection of the permissive nature of the national legislation, health visiting in Derbyshire developed in a piecemeal fashion, some of the district councils having taken it up in response to the 1907 Act, and others only following suit after the county-wide adoption prompted by the 1915 Act.[5] It was not until 1919 that the county scheme was coalesced and unified, following a 1918 invitation to the district councils running health visiting schemes prior to 1915 to join the Derbyshire County Council administration.[6] By 1918 fifty health visitors were employed by Derbyshire County Council, with one full-time and one part-time maternal and child welfare medical officer, a superintendent of health visitors, an organiser of child welfare work, an ophthalmic nurse and a demonstrator of the infant welfare exhibition (Derbyshire MOH Report, 1918: 6-7). Under the Notification of Births Act midwives and doctors attending births were required to notify the local MOH within 36 hours, but in recognition of the fact that some births might slip through the net, the County Council agreed to pay local registrars 2d per sheet and 2d per entry for any registered birth which had not previously been notified (Derbyshire County Council Public Health Committee Minutes, 14 December 1915). Furthermore, they agreed to pay the same amount to the registrars for notifying deaths of children under one year of age to the local MOH (Derbyshire County Council Public Health Committee Minutes, 12 December 1916).

Once the notifications were received they were transcribed into printed ledgers, placed in approximate order of birth date, but with a different section for each urban or rural district. Such ledgers survive for the period 1917-22, for the whole county of Derbyshire, excepting County Boroughs and Municipal Boroughs. The areas (urban district and rural districts) covered by the data are shown on the map in Figure 10.1. Because different districts were brought under the county scheme at different times, the time periods covered varies from district to district, but full coverage was achieved by 1919.

---

5 Derbyshire County Council Public Health Committee minutes, 16 March 1915 and 8 June 1915. The three Municipal Boroughs of Chesterfield, Ilkeston and Glossop were not included in this initiative, nor was the County Borough of Derby.

6 Derbyshire County Council Public Health Committee minute, 10 September 1918.

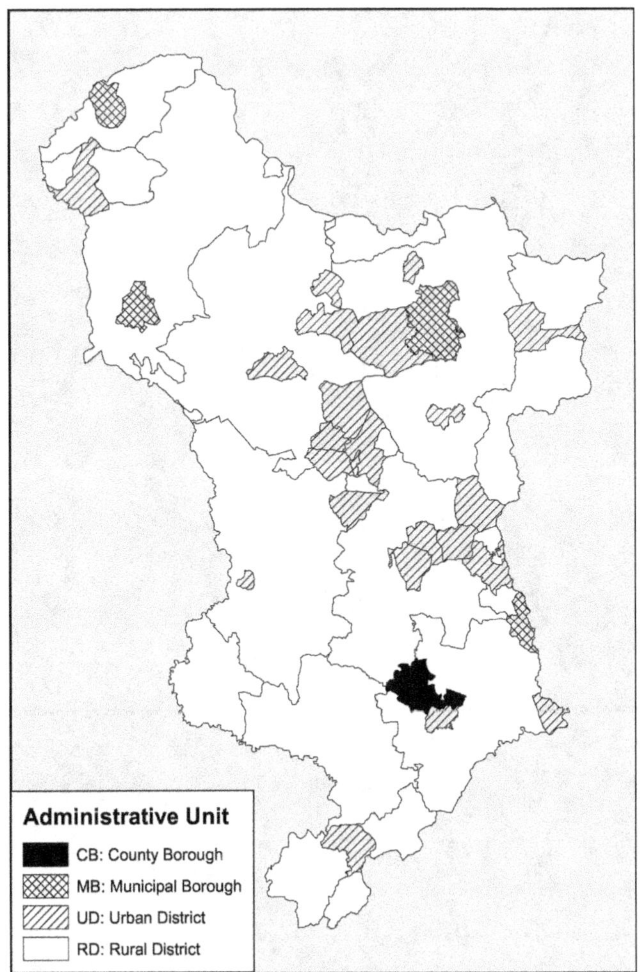

Figure 10.1 The administrative units of early twentieth century Derbyshire

Information derived from each visit was added to the record for each child at subsequent dates. The record for each child covers two consecutive pages, examples of which are shown in Figure 10.2, and contains the following information: surname, address, date and place of birth, name of doctor or midwife attending, sex, illegitimacy, multiple birth, dates of visits up to age five with feeding method at visits under one year, weaning and feeding at various ages over a year, illnesses and the ages at which they occurred, the occupations of mother and father and how long the mother worked during pregnancy, the number of bed rooms and living rooms in the house, the number of children born to the mother and how many of these had died or were stillborn. Stillbirths were identified and the date and cause of death was noted for children who died. The data set consists of a maximum span of six years, from 1917 to 1922. Previous use of this material

Figures 10.2 A and B: Example pages from the Derbyshire health visitor ledgers

Names hidden to anonymise data

B.F.—Breast Fed.
A.F.—Artificially Fed.
A. & B.F.—Artificial and Breast Feeding.

Belper R. SHEET No. 14.

	NAME.	ADDRESS.	Date of Birth.	Name.	Mother Working (Till / Occupation)	Father's Occupation.	Family Births	Family Deaths	House	Remarks.
1		Horsley Woodhouse.	9-8-18.	A. Webster.		Soldier. (Miner)	2		2 3	
2		Smalley Common.	11-8-18.	A. Webster.		Miner	1		2 3	
3		The Sycamores, Mottle.	11-8-18.	M.E. Rothwell	Fed	Farmer	3		4 5	
4		Crich.	12-8-18.	A. Roper.		Miner	1		2 2	Died Feb 1920 Enteritis
5		Mill House. Whatstandwell.	12-8-18.	C. Martin.		Millers Farmer.	7	1	3 3	
6		Crich Carr.	13-8-18.	A. Roper.	Machinist	Farm labourer.	1		2 2	Premature Dead 1-10-18
7		Crich.	14-8-18.	A. Roper.		Soldier	1		1 2	Moved to Bartsley 8-11-18.
8		The Hillside, Holloway.	16-8-18	C. Brooks.	Smith Hosiery worker.	Soldier	2		1 2	Removed—address not known. 16-5-19.
9		Lower Hilbourn	16-8-18.	M. Henshaw.		Miner	9	1+	2 2	
10		Langley Green	16-8-18.	Dr. Haslewood.		Soldier	4		1 2	
11		Longway Bank.	17-8-18.	A. Roper.		Farmer	5	2	4 4	
12		6 Duffield Rd. Allestree.	29-7-18.			Artificer (Navy)	1		2 3	[illegible] not to be found
13		The Common, Uttoxeter Rd.	17-8-18.	Mrs. Bennett.						
14		Church Walk, Allestree.	18-8-18.	C. Cooper.		Engine man. M.R.	6	1	2 2	
15		Golden Valley. Horsley Woods.	18-8-18.	A. Webster.		Miner	6		2 2	Mother died of Influenza
16		New Zealand Lane, Duffield.	18-8-18.	C. Sylvester.		Farm labourer	3		1 2	
17		Holbrook Moor.	19-8-18.	M. Henshaw.		Collier	4		2 2	
18		Black House, Quarndon.	19-8-18.	C. Cooper.		Farm man.	2		2 2	Died [illegible]
19		Mapperley Brook.	20-8-18.	Dr. Dobson.		Miner	3		2 2	Dead 21-8-18. Difficult labour.
20		The Pottery, Denby.	23-8-18.	Dr. Feroze.		Soldier (Prisoner)	1		0 3	Left for Lincoln
21		[illegible] Farm	20-8-18.	R. Butler.		Farmer.	4		2 2	

BELPER R. SHEET No. 14a.

	DATES OF VISITS. 2nd Year.	3rd Year.	4th Year.	5th Year.	Weaning and Diet.	Illnesses.	Date of Vaccination.	Teeth.	Attendance at Clinic.	Home Conditions.	General Development—Walking, Talking, etc.
1							Objects.	8/12 2 teeth. 10/12 4 teeth			
2	25-9-19.	2-3-21. 20-6-21.	[illegible]	[illegible]	Weaned about 3/12. [illegible]	7/12 Influenza. Had Operation 2 10/12 Disch. ears		1 1/12 7 teeth.	Allestree Welcome Stanley Common		1 1/12 walking.
3			22-11-22	27-3-23.	Weaned about 12/12.			7/12 2 9/12 5			Walked at 12/12
4	21-1-20					Now at School		[illegible] 20. 8/12 9 teeth. 1 3/12 12.			
5	5-11-19. [illegible]	5-10-20.	20-3-22	29-9-23.	A.F. from birth:—G.M.	2 Measles. Had Influenza		2 1/12 16.			1 3/12 Walking.
6						School Age.					
7	17.12.19.	10.12.20. 25.5.21.	7.2.22	29.9.22.	Weaned at 13/12	4 1/2 Had Pneumonia some weeks ago	Objects.	9/12 4 teeth. 9/12 7 teeth.	2 3/12 Full teeth		
8											
9	14.5.19. [illegible] 14-1-20.	11-1-21.	23-1-22.		Weaned at 1 3/12.		Objects	8/12 6 teeth. 1 1/12 16. 11/12 11 teeth.			
10	18-12-19.		21-9-21.		Weaned at 1 1/12.	1 1/12 Measles.		4/12 6 teeth. 1 1/12 8. 3 1/12 Teeth decaying.			
11	9-10-19. 16-12-19. 19-2-20.			4-12-22.		14/12 Had Adenoids. Now ready for School.		14/12 5.			14/12 Creeps.
12	12-7-19.							1 4/12 2 teeth.			
13											
14											
15					A.F. from 4/12 [illegible] W.			4/12 5 teeth.			4/12 walking
16	30-10-19.	8.11.20. 14.6.21.	22-6-22.	5-9-23.	Weaned at 9/12	[illegible] School		1 3/12 9. 2 4/12 16.			
17								4/12 4 teeth. 4/12 5 ".			
18	12-1-19.	16-2-21.			Weaned about 1 1/12.			4/12 4 teeth.			
19					1 Stillborn + 1 Died 20-8-18.						
20					A.F. from birth—B.G.						
21	15-7-19.										

(Reid, 1999, 2001a, 2001b, 2002, 2004) employed only a sub-set of the data, but the full data set of 51,376 births is analysed here (see also Reid, 2005).

Targeting Visits

Systematic analysis of the effect of health visiting on individuals is complicated by the fact that health visitors, in their attempts to improve survival, concentrated on those they felt were at highest risk of death; the poorest in particular. In Kensington in 1909, for example, only those whose incomes were less than 40s. a week were visited; nevertheless this covered a large proportion of the population (Marks, 1996: 174). Across health visiting services in different parts of the country, there was considerable variety in the percentage of newborns receiving a visit, reflecting differences in the policies of the local services (Peretz, 1992: 107; Newsholme, 1913: 106-117). In some areas infants needing visiting were selected using criteria provided by the notifications, resulting in a minority of babies receiving even one visit. For example, only 28 per cent of babies in the Scottish districts studied by Buchanan (1983) were visited. Other local authorities aimed to visit all births at least once, and used the first visit to identify births that needed repeat visits, thus 90 per cent of babies in the Durham districts studied by Buchanan (1983: 285-286) were visited. Derbyshire followed the latter model, aiming to visit all infants born in the area. Accordingly 89 per cent of newborns received at least one visit. Those that weren't visited belonged to one of two groups: the first were generally better-off families, whose doctor had stipulated did not need visiting; the second were those who could not be traced by the health visitors, having either moved away or provided an incorrect or inadequate address (Reid, 2001a).

The Local Government Board (LGB) clearly felt that efficacy was proportional to the number of visits: repeated contact with the mother would give health visitors a chance to ensure that women were remembering and acting on their advice and to identify new areas where advice might be needed: a letter from the LGB was received by the Derbyshire County Council urging them to increase the number of health visitors so as to permit an increase in the number of visits (Derbyshire County Council Public Health Committee Minutes, 13 March 1917). Previous analysis of the sub-set of the Derbyshire health visitor data found that twins, children who were artificially fed from a very young age, those living in urban districts or mining areas or areas classed by the MOH as 'unhealthy' were visited more frequently, as were higher birth orders, illegitimate children, those with mothers who had worked during pregnancy, whose mothers had had a previous child death, and those with an 'unsatisfactory home'. Children whose fathers worked in non-manual occupations and those born in institutions received fewer visits. Children who were suffering from specific neonatal conditions such as *ophthalmia neonatorum* were visited very frequently in the first months of life (Reid, 2001a). These children were also the ones most likely to die, and health visitors focused on them as offering the greatest potential saving of life. Because the health visitors were targeting their visits on the higher risk children, however, it proved impossible to link the frequency of visits to the survival of the infants, any beneficial effect on survival being masked by the higher risk of death among such babies. Even controlling for the characteristics on which health visitors appear to have targeted, a higher frequency of visiting was still linked to a higher chance of death. Assuming that health visitors did not actually increase mortality, this suggests that the instruments available for us to measure the targeting by health visitors are much cruder than the indications of increased risk that they used. Without the ability to control for the reasons why health visitors visited some infants more than others, it has proved impossible to assess the effect of the number or frequency of visits on survival.

Another fundamental aspect of health-visiting success, however, was the facility to visit mothers while their babies were still very young: this was the very reason for notification of births legislation, the corner-stone of health visiting. Therefore, another way of assessing the effect of health visiting on mortality is to look at the time since birth at which health visitors paid their first visit. While it is likely that health visitors were prioritising their first visit to those in most perceived need at earlier ages, it is also probable that the information that they were using to do so is available in the records, and it should therefore be possible to control for such prioritisation. It is likely that more subtle targeting was only able to occur after the first visit, when the health visitor could assess the condition of the child, the resources available and state of the household, along with the attitude and knowledge of the mother.

Notification of Births and Age at First Visit

Arthur Newsholme, in the Forty-Second Annual Report of the Local Government Board for 1912-13 published a report on 'conditions as to visiting under the Notification of Births Act' based on returns from large towns, small towns and metropolitan areas (Newsholme, 1913: 98-105). Notification of births within 36 hours was designed to allow a health visitor to visit as soon as possible after the birth, although the desirability of very prompt visiting depended on whether the attending midwife or doctor was continuing to visit and dispensing useful advice. 'Where the midwives can be relied upon to encourage, by the best modern methods, the breast-feeding of infants, and to report all cases of opthalmia or other disorders, it will be regarded by most as advisable to defer the health visitor's first visit until after the tenth day, at which the date the midwife usually ceases to attend the mother' (Newsholme, 1913: 100). In practice the timing of the first visit varied considerably, local MOHs not being of a common mind as to whether midwives could be trusted to report problems and encourage breast-feeding, and thus whether the timing of the first health visitor's visit would affect breast-feeding. Of the comments published, five thought early visiting by health visitors (overlapping with a midwife's continued visits) was advantageous for feeding, and the same number thought the timing of the health visitor's first visit was immaterial. Newsholme's more detailed perusal of the full returns, however, led him to write that 'in many districts the discontinuance of breast-feeding is permitted under circumstances in which more intelligent and painstaking care would have enabled it to be continued' (Newsholme, 1913: 100-101).

Newsholme's report also found that births attended by doctors were, as a matter of policy, often visited later. Of the 41 urban areas which detailed the timing of the first visit made to midwife-delivered cases, 85 per cent intended to visit within the first ten days, whereas in only ten per cent of the 21 places giving the timing of the visits to medically-attended cases were visits planned for the first ten days. With the continuing improvement in the training and supervision of midwives following the 1902 Midwives Act, Newsholme offered the opinion that in a few years the standard of midwifery might have risen sufficiently to allow a delay in the first visit of the health visitor. The five to ten years which elapsed between Newsholme's report and the creation of the records analysed here may account for the rather later age at first visit in the rural districts and small towns of Derbyshire.

Overall, the mean age at first visit was 39 days, but the distribution was skewed towards the younger ages, the median being twenty days and thirty per cent of babies having received a visit by the time they were fourteen days old. These calculations, and the following analysis, are based on the 40,247 live-born infants who were born in the area covered by the data set and for whom a date for the first visit was recorded. Babies whose births were not notified, generally because they had not been born in the area, have been excluded because they were not 'at risk' of being visited from the day of their birth, and in most cases the date of their notification or transfer of notification to the district is unknown. Another group of babies has been excluded because (in some districts) the actual dates of the early visits in 1919 were not recorded, simply a tick in the relevant column for the age in months of the child.[7] There is still considerable variation between different groups of infants: Table 10.1 gives mean ages at first visit for different groups and Figure 10.3 shows the mean age at first visit for the different districts in the data set.

Table 10.1 Mean age of first visit for different groups of infants born in Derbyshire, 1917-1922

		Number	Age in days at first visit	95% confidence intervals
All infants		40,247	39	39-40
Rural or urban district	Urban	16,803	**34**	**33-35**
	Rural	23,444	**43**	**42-44**
Mining or non-mining district	Mining	28,246	**33**	**32-34**
	Non-mining	12,001	**55**	**53-56**
Multiplicity of birth	Twin	939	**28**	**23-33**
	Singleton	39,308	40	39-40
Sex	Male	20,571	40	39-41
	Female	19,419	39	38-40
	Sex not given	257	**63**	**54-72**
Legitimacy	Illegitimate	1,006	34	29-39
	Legitimate	39,241	40	39-40

7 The mothers of many still-born infants were also often visited, and the average age at first visit for those that were was very similar to live-born infants, but these have not been included in the analysis as there was no immediate scope for preventing an infant death.

Place of birth	At home	40,209	39	39-40
	In institution	38	**82**	**58-106**
Birth attendant	Doctor only	9,318	**48**	**46-49**
	Midwife only	27,084	**35**	**34-36**
	Both doctor and midwife	3,845	**52**	**49-54**
Birth order	1	11,684	41	39-42
	2-4	17,177	38	37-40
	5+	10,323	**35**	**33-36**
	Not known	1,063	**84**	**80-89**
Mother's previous experience of child death	None	32,183	41	40-42
	At least one	8,064	**33**	**32-35**
Social class	I	132	**67**	**54-80**
	II	3,782	**57**	**55-60**
	III	23,088	**34**	**33-35**
	IV	5,810	**43**	**41-45**
	V	4,165	37	35-39
	Not given or not known	3,270	**54**	**51-57**
Mother occupied during pregnancy	Occupied	1,021	41	36-45
	Not occupied	39,226	39	39-40
Father's occupation	Agriculture	2,725	**66**	**63-68**
	Bricks/Potting	574	**30**	**24-36**
	Coal mining	17,567	**32**	**30-33**
	Lace	322	**23**	**15-31**
	Cotton	375	**62**	**54-70**
	Other occupations	16,407	41	40-42
Number of rooms in the house	1-3	5,840	41	39-43
	4-6	29,145	**36**	**35-37**
	7+	1,559	**46**	**43-50**
	'Large house'	349	**78**	**70-85**
	Not given	3,354	**62**	**60-64**
Unsatisfactory home		313	32	23-40
Unhealthy area		90	**66**	**50-81**
Opthalmia neonatorum		63	20	2-39

Note: Figures in bold are significantly different to the value for all infants.

It is clear from the variation in the mean age at first visit shown in Table 10.1 that health visitors were selecting some infants to visit sooner than others. Twins, for example, were visited nearly two weeks earlier than singleton births, and mothers who had had a previous child death were visited a week earlier than those who had not suffered in this way. Better-off people, as identified by both the social class of the father (derived from his occupation) and those living in large houses, were visited later. Some of these differences are likely to have been the result of conscious prioritization: the Derbyshire MOH obviously felt that more infant welfare work was needed in mining districts where rates of infant mortality were higher: '[The miners'] wives, on whom the health of the children largely depends, have in the past received no instructions in home management' (Derbyshire MOH, 1918: 38). This focus can also explain the earlier visits to the infants of miners, and to those with moderately sized (as opposed to small) houses, the MOH having observed in 1915 that in coal mining districts wages were higher, housing was better, and women were less likely to have gone out to work (Derbyshire MOH, 1915: 9). Overt prioritisation was given to babies suffering from *opthalmia neonatorum*, a condition of the eyes resulting

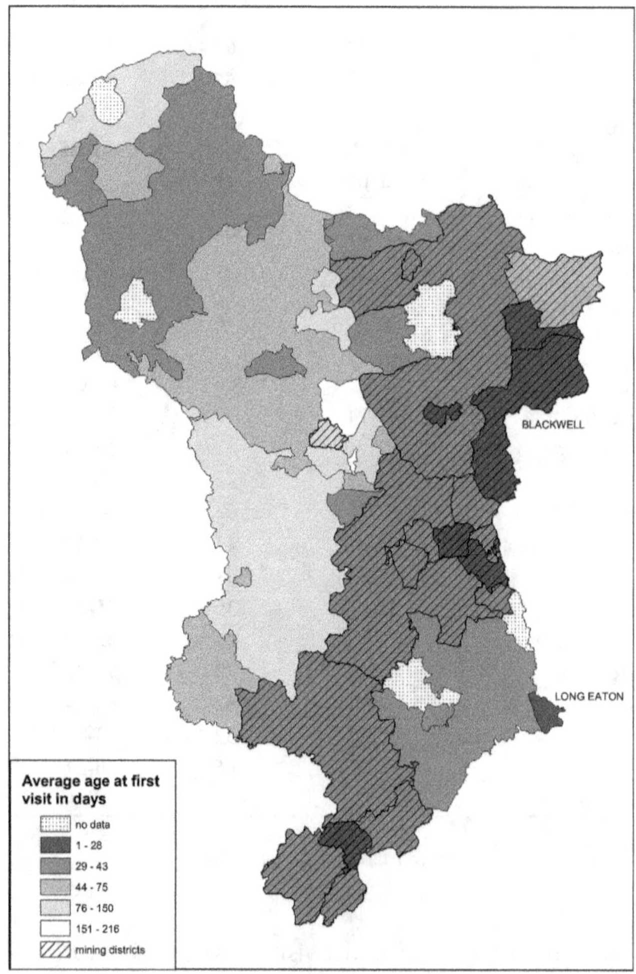

Figure 10.3 The mean age of infant at first visit by a health visitor, by Derbyshire administrative unit, 1917-1922

from inadequate cleaning of mucus from the eyes after birth and associated with maternal venereal disease (Buchanan, 1983: 259; Cugalj, 1984). It was responsible for 25 to 50 per cent of infantile blindness in the early twentieth century, and there was a campaign to reduce it in Derbyshire, according to which midwives were required to notify all cases, and paid one shilling per case for notifications received immediately (Buchanan, 1983: 259; Derbyshire County Council Public Health Committee minutes, 14 September 1915). The sight of such infants was dependent on the establishment of good hygienic practices, and they were visited substantially earlier than average, although the difference was not highly statistically significant.

Other differences between districts may be a function of the ratio between health visitors and births, but unfortunately there is no information on the number of health visitors assigned to different areas. However some district councils, notably Long Eaton Urban District (UD) and Blackwell Rural District (RD), were anxious for more health visitors than they were allocated, and agreed to pay for the services of an additional health visitor (Derbyshire MOH, 1916: 6). Long Eaton was a centre of the lace industry, and the additional resources put into health visiting there may help to explain the low age at first visit to the children of lace workers. Lower densities of population and therefore longer travelling distances in rural districts may have been the reason that infants in rural districts had to wait longer until their first visit than those in urban areas. Figure 10.3 shows, however, that there was considerable variation from district to district which did not fall along rural-urban or mining-non mining lines.

Other differences are likely to be a product of the generation of the data set and the notification process. The first visit to births at which doctors were present was considerably delayed, being about a fortnight later than those to midwife notified births. Although as noted above, visits to doctors' cases were often consciously postponed, partly due to the tendency of the doctors to have attended better-off families who may not have been prioritised for visits, and possibly also because doctors were more trusted than midwives to report problem cases and to promote breast-feeding. However, there is some evidence that doctors were less conscientious about notifying births, which might also have led to a delay in the first visit. The Maternity and Child Welfare Committee minutes record no instances of midwives failing to notify births, but one doctor was censured for failing to do so (Derbyshire County Council Maternity and Child Welfare Committee *Minutes*, 6 March 1923) and many of the births out of chronological order in the ledgers were attended by doctors. Similarly, although births in institutions in Derby, Chesterfield and Sheffield have already been excluded, there were a number of small maternity homes in the area covered by the data set, and a long time elapsed before the first visit to babies born at these. Infants for whom information on sex, parental occupation, birth order, and housing were not available, also tended to have waited longer until a visit.

We might have expected health visitors to try to visit illegitimate infants and first births more quickly, but in fact higher birth orders were visited earlier, however this may have been a function of the higher fertility of coal miners who were also prioritised. The lack of a difference for illegitimate births might also be disguised by correlations in the data set, but could suggest that legitimacy status was not provided at notification preventing health visitors from targeting their first visit at illegitimate infants. Similarly, although previous analysis showed that infants living in homes identified by health visitors as 'unsatisfactory' received more visits, they did not receive significantly earlier visits, and health visitors appear to have been more reluctant to pay early visits to infants born in unhealthy areas, although this may be a function of the places in which these were located.[8]

Performing a multivariate hazards analysis of the time to the first visit takes account of such correlations. It reveals that in comparison to a hypothetical child (one who is a singleton birth, delivered by a midwife only, of manual social class, living in a small to medium house in a non-mining, rural area), twins and those in urban or mining districts were likely to have been visited sooner, and those delivered by doctors, in non-manual classes, and in large houses had to wait longer for their first visit. Once other factors are controlled, it emerges that illegitimate infants were also more likely to have received an early visit. Significant hazard ratios for the risk of being visited are shown graphically in Figure 10.4. This analysis calculates an age-dependent risk function for the risk of being visited, and hazard ratios correspond to a scaling up or down of that risk function. In other words, once a pattern of age at first visit is established (reflecting a higher likelihood at young ages), groups of infants with hazard ratios lower than one are less likely to have been visited, and those over one are more likely to have been visited. Analysis of the risk of death shows that the characteristics distinguishing the infants who were more likely to be visited were also among the factors associated with early death, making it difficult to identify any beneficial effect of early visiting (Reid 2001a). However, because the information which the health visitors were using to decide who to visit is also

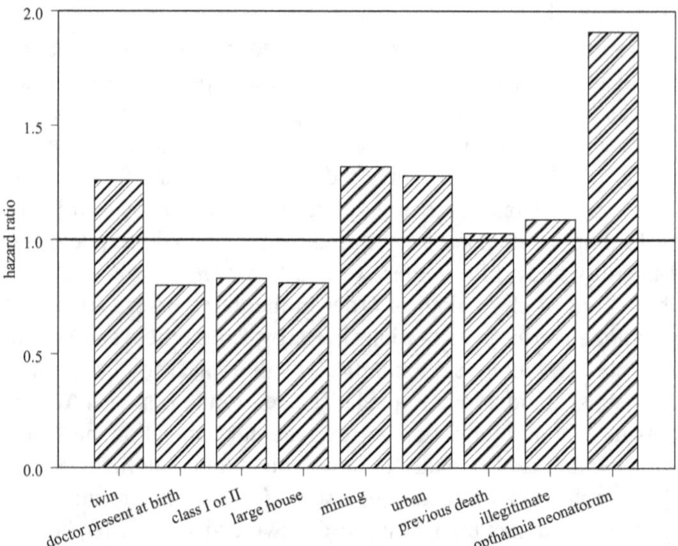

Figure 10.4 Hazard ratios for the risk of being visited: Derbyshire, 1917-1922

8 Homes labelled by health visitors as 'unsatisfactory' were also described as 'dirty', and mothers as 'indifferent' or 'careless', suggesting that 'unsatisfactory' referred to the standard of the housekeeping rather than the accommodation.

available to us, it should be possible to control for these factors and to try to identify whether the timing of the first visit affected future survival.

Hazard analyses were performed to examine the effect of having received a visit in the first fourteen days of life on the chance of death, while controlling for the other characteristics associated with early visiting.[9] Initial results are slightly disappointing, suggesting that even when controlling for the factors which induced health visitors to visit early, children receiving their first visit within the first fourteen days were still more likely to have died. This suggests that the health visitors knew more about the individuals than is available in the transcribed ledgers, and were able to target their visits on high risk or sickly babies more effectively. We know that health visitors were being officially informed of babies with *opthalmia neonatorum*, and it is possible that midwives also communicated other facts about vulnerable infants to the health visitors. Health visitors may also have been using prior knowledge of the mothers in their areas derived from visiting their previous babies, and were focussing early visits on those mothers they felt would have more sickly babies. It is possible that this explains some of the variation in the age at first visit, but if it were also explaining the correlation between early visits and likelihood of dying, we would expect no such connection when first births only are considered. However, when the analysis is performed for just first births, the results are the same, indicating that even if health visitors were using their prior knowledge of mothers to target visits, this was not how they identified the sickly babies they managed to visit at early ages.

Nevertheless, it is still possible to identify some of the effect of health visitors by exploring other avenues, such as examination of the influences on the change from breast to artificial feeding, as one of the most important roles of the health visitor was to improve survival by encouraging breast-feeding. A hazards analysis on the risk of being artificially fed was therefore performed using the first time the infant was observed to have been artificially fed, or the reported age at weaning, whichever was earlier, as the dependent variable, and other characteristics as the independent variables.[10] Figure 10.5 shows that twins and illegitimate infants were over twice as likely to have become artificially fed, and those delivered by a doctor or whose father belonged to a non-manual occupation were also more likely to have been weaned early. Early weaning was less likely for higher birth orders, however, and of particular importance for this analysis, for those who received an early visit by a health visitor.

The reason for the imperative, observed in the LGB directives and among local health officials, to encourage breast-feeding was that artificial feeding had only relatively recently been recognised to have an association with an elevated risk of death. Infants fed with inappropriate milk substitutes were at particular risk

9 The effect of a visit before different dates was experimented with, but differed little. Thirty per cent of infants received a visit before fourteen days.

10 This is not an entirely accurate way of capturing the transition to artificial feeding, as the date at which the infant was first fed artificially is likely to have been before the date of the first visit at which they were reported as having been so. In this analysis, we are looking at the effect of an early visit (before 14 days) on the risk of changing to artificial feeding. If those artificially fed from birth were also targeted, this might obscure any relationship, so to capture an association between visit and feeding, the risk of being artificially fed is calculated from the date at which it is noted if a visit has occurred before 14 days).

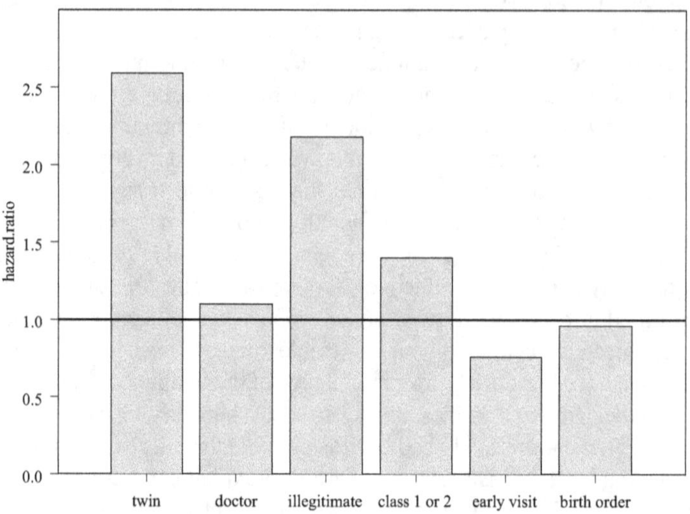

Figure 10.5 Hazard ratios for the risk of being artificially fed: Derbyshire, 1917-1922

from diarrhoeal disease, but were also prone to wasting and debility (Coutts, 1911; Davies, 1913; Newsholme, 1913: 84-5; Coutts, 1914, 1918; Woodbury, 1922). Health visitors' records, such as those from the city of Derby in the early years of the twentieth century, provided information on the feeding method of deceased infants which was instrumental in establishing the connection between feeding and mortality (Howarth, 1905), and in Derbyshire one of the health visitors' most important briefs was to encourage breast feeding for an ideal period of around nine months. The first exhortation on one of the leaflets ('How to rear an infant during the first year of life') distributed to new mothers by health visitors in Derbyshire, stressed the importance of breast feeding:

> Every infant ought to be fed at the Mother's Breast for the following reasons:
>
> Because there is not and never can be any nourishment so good as the Mother's own milk.
>
> Because artificial feeding in the first year is difficult, expensive, troublesome, and dangerous.
>
> Because an enormous number of children artificially fed die before they are 12 months old of sickness and diarrhoea, and of those who live, many are feeble and unhealthy.
>
> Because it is easier and healthier for the mother to feed her child at the breast.
>
> Because the breast-fed child is happy and contented, and the bottle-fed baby is cross and fretful and gives the mother no peace.

(from leaflet N.B.6. reproduced in Derbyshire MOH report 1914: 12).

Health visitors appear to have been successful in delaying weaning among infants they visited in the first days and weeks of life. However, Medical Officers and Health Visitors also recognised that not all mothers were able to continue breast feeding for the recommended 6 to 9 months. The leaflet on how to rear an infant in the first year of life recommends that 'if the mother is losing flesh the child should be weaned', and a further leaflet (N.B.7) provided detailed instructions on feeding infants by hand, including the storage of milk, the best feeding bottle, and preparation and quantity of milk for different ages of infant. From 1918 health visitors were also one of the ways in which mothers unable to breast feed could gain access to free or subsidised milk under the Milk (Mothers and Children) Order of 1917, having authority to give priority certificates (see Derbyshire County Council Public Health Committee minutes, 12 March 1918). Under this scheme in 1920 free milk was given to those with family incomes under 30 shillings (after 4 shillings was deducted for each child under the age of 14 years) per week, and half price milk was available to those with incomes between 35 and 40 shillings (following deductions) (Derbyshire MOH report, 1918: 38-9).[11] In the year ending March 1920, the County Council spent £2,571 16s 11d on cows' milk and £4,057 14s 1d on dried milk, although much of this was recovered from those who were expected to pay half price (Derbyshire MOH report, 1919: 73). The Milk Order had only come into place in 1919, but the Derbyshire County Council Maternity and Child Welfare Committee minutes (25 March 1919) show that £800 18 0d had already been spent on provision of milk and virol (an infant food) in contrast to only £193 4s 2d spent on infant welfare centres. Did this supply of a more suitable milk product and careful instruction on its use and preparation improve prospects for artificially fed children?

Figure 10.6 shows the results of a hazard analysis of the risk of dying in the post-neonatal period, treating feeding as a time-varying variable.[12] Infants who were known to be already artificially fed at the time of a visit were around six times as likely to have died than those who were not. Those who were visited in the first two weeks were also twice as likely to have died, but an interaction between an early visit and a change to artificial feeding shows that artificially-fed infants were substantially less likely to have died if they had received an early visit from a health visitor. In more intuitive terms, the predicted post-neonatal mortality for an infant with average values of each variable would be a mortality rate of 34 per thousand between the end of the first month and the end of the first year. For infants who

11 The qualifying incomes and allowances were changed in 1920 and again in 1922 (Derbyshire County Council Maternity and Child Welfare Sub-Committee minutes, 1 April 1920; Derbyshire County Council Maternity and Child Welfare Committee minutes, 21 March 1922).

12 As explained above, using the first known date at which a child is artificially fed is not the most accurate way of identifying those who were artificially fed as some infants may have been already fed in this way before this date. It is, however, a much more sensitive way of measuring the likelihood of being artificially fed than previous attempts using this data set which have looked at whether infants were artificially fed by a certain age, such as one month (Reid, 1999, 2002).

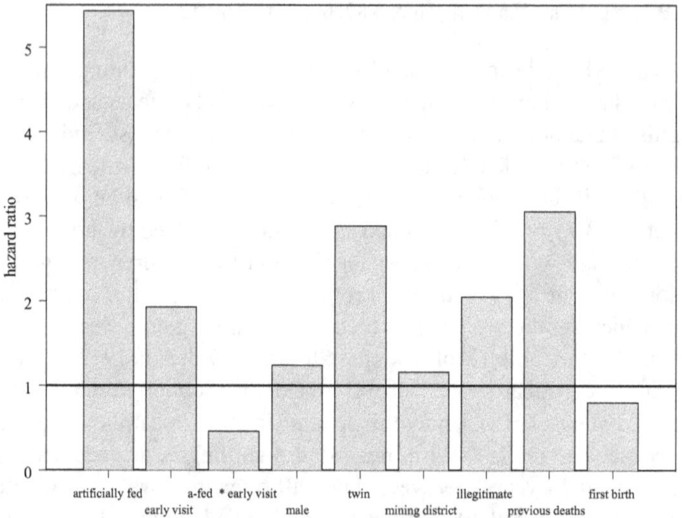

Figure 10.6 Hazard ratios for the risk of post neonatal mortality: Derbyshire, 1917-1922

were otherwise average, but were artificially fed, mortality was 73 per thousand among those who were not visited before fourteen days, but 65 for those who were visited within a fortnight of birth. This suggests that the milk substitutes or the advice on feeding given by health visitors was instrumental in reducing the risk of death among artificially-fed infants in Derbyshire, although the risks associated with such behaviour were still high. Available milk substitutes, even those recommended by health visitors, were far from ideal for infant digestive systems, and women who hand-fed their infants were often hard-pressed to carry out the health visitors' recommendations on hygienic preparation in circumstances without running water and safe waste disposal. The MOH for Derbyshire recognised this in his 1918 report: 'given all possible instructions in home management it is impossible to bring up healthy children in a house with an unpaved common yard and a privy midden' (Derbyshire MOH report, 1918: 38).

The circumstances which allowed safe hand-feeding of infants may help to explain the exemplary performance of the town of Long Eaton in infant mortality terms, where the infant mortality rate was fifteen percent lower than that in all the urban areas covered by the data.[13] In 1915 96 per cent of the houses in the district were supplied with mains water, and by 1920 the district had converted all but twenty of its privy middens into water closets. The MOH report for Derbyshire wrote in 1920 that 'The sanitary administration of this District is excellent – the energetic manner in which the conversion of privies and pail closets has been tackled is an

13 Using published births and infant deaths, the infant mortality rate for Long Eaton for the years 1917-1923 (excluding 1921 for which the numbers of infant deaths were not published) was 64 per thousand, whereas that of all the urban districts combined was 76 per thousand.

example to the rest of the county'. As mentioned earlier, of course, Long Eaton was also well-staffed with health visitors, and managed the second lowest average age at first visit, but in the absence of information about the sanitary circumstances of individual houses, it is impossible to tell whether it was the sanitation or the health visiting which contributed to the survival of artificially-fed babies in Long Eaton.

Conclusions

Although health visitors in Derbyshire had a policy of near universal visiting, there is evidence that they were trying to target their visits on the most vulnerable babies. The dates of the first visit varied dramatically: around thirty per cent of infants were visited before they were a fortnight old, but some were visited only a day or two after birth and some were much older. The information received with notifications of births enabled health visitors to focus on particular groups of high risk infants, and can explain the earlier visits to twins, illegitimate infants, those with *opthalmia neonatorum*, and those living in mining areas. Even after controlling for these factors, however, babies visited young were still more likely to die, indicating that health visitors were in possession of information which enabled them to identify sickly or more vulnerable babies even before the first visit. At the first visit, they would have been able to make further assessments as to which families would benefit most from visiting, and give general advice on feeding and infant care.

One of the most important forms of advice given by health visitors regarded infant feeding: their brief was to encourage women to breast-feed, and it is clear that they were at least partially successful in this: mothers who were visited before their baby was fourteen days old were more likely to persist with breast-feeding. Health visitors also advised on safe methods of artificial feeding, and helped artificially feeding mothers to acquire subsidised safe fresh or dried milk. In this area, too, an early visit to establish good practice was crucial: among babies who were artificially fed, those who had been visited before they were fourteen days old were much less likely to die. Therefore, although targeting on sickly babies renders it impossible to establish a statistical link between earlier visiting and higher survival, it is clear that the policy of early visiting paid dividends in delaying weaning and improving artificial feeding practices and therefore reducing the risks of mortality in the first year of life. This provides an answer to the MOHs who debated the merits of whether early visits by health visitors would affect feeding; the first few weeks having now been shown to be crucial in establishing successful breast-feeding, and endorsing Newsholme's view that 'intelligent and painstaking care' would often be rewarded by higher rates of breast feeding (Newsholme, 1913: 100-1).

Health visiting was not an isolated initiative, and its success was also dependent on other measures. The registration of births was crucial in enabling early visiting and probably also in alerting health visitors to more vulnerable infants. The encouragement of breast feeding and dissemination of knowledge about safe artificial feeding was a major part of the health visitor's brief, and the latter was aided by other developments such as the control of the milk supply, and better sanitation in the home which limited the spread of disease. The success in terms of infant survival of places such as the town of Long Eaton in Derbyshire, may be because of their integrated approach to the problem of infant mortality: putting resources into a broad range of Newman's suggested preventive measures relating to the mother, the child, and to the environment.

PART III

Chapter 11

Infant Mortality and Social Progress in Britain, 1905-2005

Danny Dorling[1]

Introduction

In Britain by the end of the nineteenth century it became evident that birth rates were falling and infant mortality was rising. This lead to a rapid decline in 'natality' and consequently to a fall in the supply of 'infants for Empire'. By 1905, and in a remarkably apposite observation, the medical officer for health in Battersea wrote that within Britain:

> England is now regarded as the nucleus of a great Empire, with colonies which, though vast in extent are poor in population, and the fact must be faced that in view of our declining natality, the stream of emigrants that formally left our shores cannot be expected to continue (McCleary, 1905, quoted in Dwork, 1987: 6)

A century later, in 2005, there was great concern among some without imagination over a stream of immigrants coming to our shores. McCleary did not see that far ahead but in the century since he wrote it has become clear that fertility, mortality, natality and migration are all intimately linked over the course of lifetimes. However, society in Britain is still arranged very much in a hierarchy from those who 'think they know best' to those who are 'not to be trusted'. For instance, medical officers still treat much of the population with suspicion as became evident when the Department of Health, in 2005, said that, rather than increase benefit levels, mothers should be paid to eat, and give their infants to eat, 'healthy start food'; however, mothers could not be trusted to do this without vouchers requiring a statutory instrument laid before parliament. In the latest such instrument it was stated that within Britain:

> A person … is entitled to benefit in accordance with these Regulations with a view to helping and encouraging her to have access to, and to incorporate in her diet, food of a prescribed description. The benefit to which a person described … is entitled is Healthy Start food to the value represented by a voucher… (Department of Health, 2005: 13)

For the ten possible categories of persons entitled to 'healthy start food' in 2005 see the draft regulations laid before Parliament under section 13(10) of the Social Security Act 1988 for approval by resolution of each House of Parliament (2005:

1 I am grateful to Tiffany Manting Tao for help with drawing an earlier version Figure 11.5, to Graham Allsopp and John Pritchard for help redrawing all figures and to Eilidh Garrett and Nicola Shelton for comments on two earlier drafts.

Part II). However certain types of person were excluded, for example category 3(b) is:

> a woman under the age of 18 who has been pregnant for more than ten weeks, provided that she is not a person to whom section 115 (exclusion from benefits) of the Immigration and Asylum Act 1999(b) applies.

Clearly pregnant teenagers who were claiming asylum were not to eat too healthily by 2005. Progress is a strange concept.

This chapter addresses the questions of how, in the space of just four generations, concerns of 'natality for Empire' moved on to 'healthy start foods'; why so many parents are still prevented from access to the resources needed to even properly feed their infants in Britain, and thus how much further still has to be gone. The main underlying cause of infant deaths in Britain was poverty at the beginning of the twentieth century and it remains poverty now. Infant health, and in particular, the unprecedented decline of infant mortality, is used here to begin to answer the questions of how far we have come and how far we have to go; starting with what was just beginning to be widely realised at the turn of the last century:

> It is difficult to escape the conclusion that this loss of infant life is in some way related to the social life of the people (Newman, 1906: vi).

The death of an infant, a child in the first year of its life, is extremely painful to imagine. It is even more painful when it is considered that in every year in the last century, and continuing into this, most of those youngest of deaths were caused by poverty. 'Healthy start food' is being introduced to replace the Welfare Food Scheme established in 1940 (Department of Health, 2005: 5) which itself was a response to the work of Newman, McCleary (whose concerns over Empire were quoted above) and others who pioneered better health for infants a century ago. McCleary (1905) argued that the provision of clean milk for infants through depots was needed to reduce infant mortality. He recommended the establishment of the first milk depot in Battersea in 1902 (Dwork, 1987: 105). It then took forty years for a means tested right to clean milk to be established and sixty more years for that right to be extended to other forms of food so that poorer mothers who choose to breast feed are not discriminated against, as they do not need free milk; one rationale for the 'healthy start foods' initiative. Perhaps in another century there will be no need for such targeted and inflexible 'benefits'. One day a substantial number of infants and children will not have to be fed by the state through various forms of benefit and free meals. However, in order to understand how this might be achieved, we need to first understand how the realisation, a century ago, that the social lives of people mattered led to the poorest – excluding asylum seekers – being allowed to choose a little cheap fruit in place of milk powder in 2005.

Background

For all of the period (1900-2005) considered here, poorer people have been much more likely than most to see their babies die, and the rich the least likely. A century

They walked softly over and stood by the cradle side looking at the child; as they looked the baby kept moving uneasily in its sleep. Its face was very flushed and its eyes were moving under the half-closed lids. Every now and again its lips were drawn back slightly, showing part of the gums; presently it began to whimper, drawing up its knees as if in pain.

"He seems to have something wrong with him," said Easton.

"I think it's his teeth," replied the mother. "He's been very restless all day and he was awake nearly all last night".

"P'r'aps he's hungry."

"No, it can't be that. He had the best part of an egg this morning and I've nursed him several times today. And then at dinner-time he had a whole saucer full of fried potatoes with little bits of bacon it in."

Again the infant whimpered and twisted in its sleep, its lips drawn back showing the gums: its knees pressed closely to its body, the little fists clenched, and face flushed. Then after a few seconds it became placid: the mouth resumed its usual shape; the limbs relaxed and the child slumbered peacefully.

"Don't you think he's getting thin?" asked Easton. "It may be fancy, but he don't seem to me to be as big now as he was three months ago."

"No, he's not quite so fat," admitted Ruth. "It's his teeth what's wearing him out; he don't hardly get no rest at all with them."

They continued looking at him a little longer. Ruth thought he was a very beautiful child: he would be eight months old on Sunday. They were sorry they could do nothing to ease his pain, but consoled themselves with the reflection that he would be all right once those teeth were through.

"Well, let's have some tea," said Easton at last.

... [later] ...

The woman did not reply at once. She was bending down over the cradle arranging the coverings which the restless movements of the child had disordered. She was crying silently, unnoticed by her husband.

For months past - in fact ever since the child was born - she had been existing without sufficient food. If Easton was unemployed they had to stint themselves so as to avoid getting further into debt than was absolutely necessary. When he was working they had to go short in order to pay what they owed; but of what there was Easton himself, without knowing it, always had the greater share.

extract from Tressell, 1914: 53-58

Figure 11.1 1905: The meaning of poverty

ago the majority of people would have had first hand experience of infant mortality and of living in poverty. That is no longer the case, so a fictional account of a couple living at the turn of the last century likely to lose their baby is reproduced in Figure 11.1. It is worth reading this account before turning to the argument below as it is all too easy to forget what and who is involved when short lived lives are turned into digits. The account shows too why people such as McCleary were arguing for clean milk, not just for infants but to reduce the chances of mothers going hungry, and by 1905 for education on breast feeding (termed 'nursing' in the account). The contents of Figure 11.1 also, superficially, illustrate the problems of trying to imagine the change that has occurred over the course of the last century and how rapid that change has been.

One way to imagine change is to compare the fortunes of recent generations of one's own family. You are only reading this because your parents and their parents survived their first year of life. What were their chances of doing so?

My two oldest grandparents were both born in 1905, the year in which George Newman wrote the book this volume draws on and in which McCleary's *Infantile Mortality and Infants' Milk Depôts* was published. My parents were both born during the Second World War. My partner and I were born around 1971 and our two children in the first three years of the present century. There will hopefully never again be four generations in Britain who experience such different chances of surviving their first year of life. Rates are traditionally expressed as proportions of live births. The respective infant mortality rates (IMRs), per 1,000 live births, of the cohorts born in the same years as the pairs of my most immediate relations were 151, 60, 18 and 5 per 1,000. These numbers translate to the following brief account of progress as applied to the cohorts corresponding to the four generations of my family and all other families born in Britain around the years 1901, 1941, 1971 and 2001.

In England and Wales in and around 1901, one in just over six babies died in their first year of life. The average chance of any pair surviving to age one was 72 per cent and most prospective pairs of grandparents did not survive, as a pair, from birth until they were old enough for it to be socially acceptable for them to have children of their own. Rates varied widely between poor and rich geographical areas and social classes; terms which were often synonymous with very small areas such as affluent streets or poor ones. In 1901 at the age of thirty, Benjamin Seebohm Rowntree reported that amongst the worst off employees of his father's chocolate factory, 247 of their babies were dying for every 1,000 born. Thus a quarter of the children of the poorest working classes in York died before their first birthday; whereas, for Rowntree's servant keeping classes only 94 babies were dying for every 1,000 born (Newman, 1906: 189). The highest infant morality rate Newman reports is 289 per 1,000 for illegitimate infants in London around the same time (Newman, 1906: 17). Thus depending on to whom and where a child was born, its chances of surviving to its first birthday ranged from around three quarters to better than nine tenths and infant mortality was the major influence on life expectancy at birth. Within towns, where health was often poorer than in parts of the countryside, life expectancy in total varied from 30 years in central Liverpool to 50 years in the Clifton district of Bristol by 1900 (figures from Szreter and Mooney, 1998: 90).

By 1941 one in just over sixteen babies died in their first year of life. The chances of an 'average' pair surviving to age one were 88 per cent. Great improvements in health during the childhoods of these children meant that the large majority were surviving to be old enough for it to be socially acceptable for them to have children of their own. Local statistics were hard to come by during the years of the Second World War. Much changed in the years during and immediately following the War for the 1941 cohort, weaned more equitably on rations and eligible for free milk if needed. By the earlier 1950s decennial reporting of mortality statistics by area had begun again. The Registrar General's *Decennial Report* covering the period 1950-53 found that, by area, IMRs varied from being 68 per cent above the national average in Port Glasgow to being 68 per cent of the national average in Oxfordshire. Furthermore, and coincidentally, for 68 per cent of the population, local IMRs were no more than a quarter above or a fifth below the national average. Note that 5/4 and 4/5 are comparable reciprocals for comparing distributions around unity. Hence the ratio of the worse off infants by area being 2.5 times more likely to die in their infancy than the best off did not vary markedly over this time period (247/94 = 2.62 in 1901 and 168/68 = 2.47 in 1951). In other words for every 5 infants who died in the worst off areas in both 1901 and 1951, only 2 died in the best off areas.

By 1971 fewer than one in fifty six babies died in their first year of life, the chances of a pair surviving to age one were 96 per cent. Thus there was only one chance in twenty five that both my partner and I would not live to see our first birthday (and hence later be able to meet if we ignore our much smaller chances of dying in childhood and younger adulthood). Whether children would survive long enough to have children of their own had largely stopped being an issue. What mattered now, for most that had a choice, was whether and, if so, when they choose to have children. At the extremes geographically 64 per cent more baby boys died in Coatbridge, Lanarkshire, than the national average and 57 per cent less in the rural districts of Buckinghamshire (Dorling, 1997). Although this variation may appear a little wider than before, some four fifths of the population lived in areas where rates did not exceed a quarter nor were a fifth below, the national average. Put another way, no matter where my partner or myself had been born, or to whom, we had a better chance of reaching age one than had almost anyone born in 1941 no matter how privileged their social or geographical circumstances. In contrast, those born into the worst situation in 1941 did 'only' as well as the most privileged babies born in 1901. By 1971 for every 2 babies dying in the best off areas, 6 were dying in the poorest places (as compared to 5 above).

By 2001 less than one in 186 babies died in their first year of life, the chances of a pair of children surviving to age one were more than 99 per cent. IMRs had become so low that they were no longer routinely published at local authority level for individual years (ONS, 2005). For the year 2002 IMRs reported for very large areas were lowest at 3.8 per 1,000 live births in the Norfolk, Suffolk and Cambridgeshire Strategic Health Authority (SHA) and at 3.9 per 1,000 in Thames Valley SHA. Rates were highest at 7.0 per 1,000 in West Yorkshire SHA and 7.7 per 1,000 in Birmingham and the Black Country SHA. The extreme ratio is thus lower if these large areas are used but is wider when measured for cities (see below). However, three quarters of births, and a similar proportion of the population as in 1971 lived, in 2002, in areas with rates no more than a quarter above nor a fifth below the national average for England and Wales. A child born into the worst off Strategic Health Authority in 2002 had a much better chance of reaching age one than did a child in the best off district of the early 1970s. In many of the best off districts as defined in the 1970s there are now years in which no infant dies in their first year of life. If progress were to continue at its present rate then this is what many of the currently poorest districts will experience, in just a generation's time. However note that as very low rates are achieved inequalities have risen over 6 fold (10.3/1.5) between the largest of English cities by 2001 (see Figure 11.3 below). This translates to more than 6 infants dying in the worst-off cities in 2001 for every one dying in the best off cities, as compared to local authority areas in 1901 (2½:1), 1951 (2½:1), and 1971 (3:1). Ratios of probabilities are difficult entities to compare when the probabilities are reducing so quickly. However it is clear that as infant mortality has become more rare its association with poverty rather than chance (otherwise known as the more general environment) has widened. Probabilities are also difficult concepts to apply to individuals.

As any statistician will gleefully tell you, in hindsight your chances of being born are 100 per cent and thus also are those of all your past relations. However, far more of us are the product of more affluent parents and grandparents than would have been the case had these chances of death been more equal. The eugenicists of 1905 would be pleased by this outcome (see Dwork, 1987: Chapter 1, entitled 'Infant mortality and the future of the race'). Conversely, that most of us are largely the product of poor grandparents and great grandparents is testament simply to how extensive poverty in the past was. There were very few, very affluent people living at the start of the twentieth century in Britain. They almost completely monopolised the telling of the past, certainly its quantitative recording, and only a tiny minority of them were interested in the poor or the idea of infant mortality as a social problem. The account given in the extract from Tressell above is a very rare example of a non-elite publication (Davey Smith *et al.*, 2001: 135). To give an example of how enlightened health professionals approached these issues at the time, five years following Seebom Rowntree's report on York, George Newman reported on the work of Dr Niven, the Medical Officer of Health of Manchester:

> By means of a number of beer-traps Dr Niven contrived to count the flies in some dozen houses in Manchester during the summer months of 1904, and from these data he concluded that the advent of the house-fly in numbers precedes by a short time the increase in the number of deaths from diarrhoea. In the fortnight ending August 13th, for instance, the number of flies caught in these traps was 37,521, the maximum in any fortnight, and in the fortnight following the maximum number of deaths from diarrhoea occurred – namely, 192 (Newman, 1906: 168-169; see also Dwork, 1987: 48-49).

Diarrhoea is now one of the major killers of infants worldwide but also no infants die as a result of untreated diarrhoea in Britain now. One possible account of the geography and recent history of infant mortality in Britain would begin with the Battersea milk depot of 1902, the Manchester beer-traps of 1904, and other key events, such as the 1906 election, and work progressively forward through ever increased experimentation, autopsy, argument, realisation and intervention through to the situation in 2006 to produce a story of political, social, medical and technological achievement. Although such a story would be interesting, by concentrating on the nuts and bolts of what occurred to lower infant mortality so dramatically, it is possible that we would miss something quite remarkable in all the detail. That something we might miss concerns progress itself and what often has to be sacrificed in order to achieve it: the growth in the short term wealth of the richest. The remainder of this chapter thus concentrates on what happened that was coincident with the most and least rapid periods of the fall of infant mortality, to suggest why infant mortality fell when and where it did beyond the necessary but not sufficient improvements in public sanitation, private hygiene, state care, general finance and public medicine which occurred over this period.

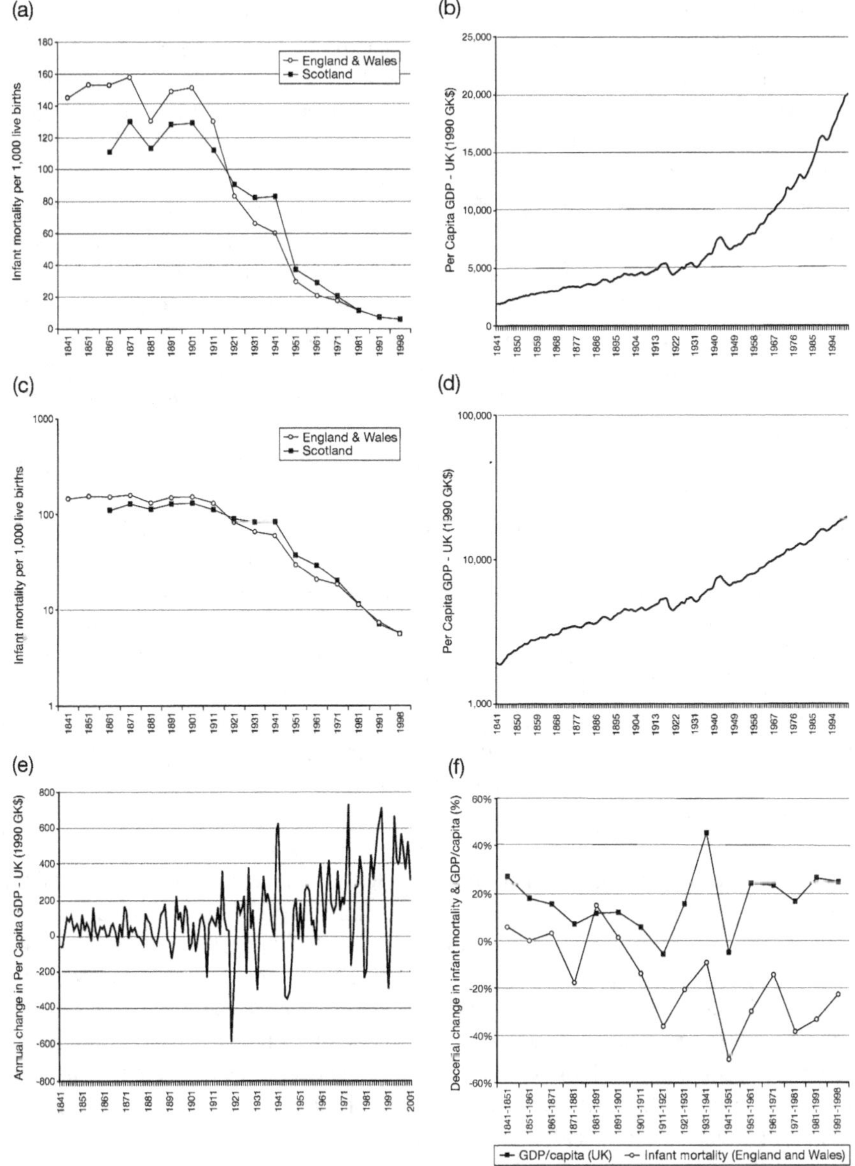

Figure 11.2 Infant mortality and affluence in the UK: six views of 160 years (1841-2001)

Social Progress

Figure 11.2 contains 6 graphs. The first, Figure 11.2a, of the decline of infant mortality 1841-1998 has been reproduced in many forms numerous times (this version is taken from Davey Smith *et al.*, 2001: xxiii). The second, Figure 11.2b, should be equally familiar but is now of exponential growth rather than decline, in this case of Gross Domestic Product per person (GDP per capita derived from Maddison, 2005). The GDP per capita figures are produced such that they can be compared over time assuming equal purchasing power for a universal dollar at each point in time. When these two graphs are compared it would appear that as monies have risen mortality has fallen. The British population became wealthier as their incomes rose and they were able to afford and produce better medical, environmental and social care for their children, progressively more and more of whom survived to their first birthday. This improved care took numerous forms: cleaner milk, fewer flies, less exhausted mothers (and fathers), a decline also in parental mortality and family size, a rise in health visitors, maternity units, paediatric specialists, income support payments and so on. These advances were some of the first expenditures made with the excess of monies, and investment in child care continues, encompassing the wider care of families with children as a result. As standards of living rose exponentially, IMRs fell exponentially.

There is, however, a problem with this story and that problem is hinted at in Figures 11.2c and 11.2d which are simply the earlier two figures with the vertical axis drawn on a log scale. Log scales ensure that comparable rates of change are comparable lengths on the graphs. In Figure 11.2c it is easier to see that IMRs did not begin to fall continuously until after 1901 and fell fastest after 1941 with the most improvement being experienced in the two decades in which world wars were fought (Dwork, 1987). In Figure 11.2d it is evident that, in real terms, GDP per capita rose steadily from 1841 to 2001 with no great break in slope around 1941. Most importantly it actually fell significantly during the final and immediately subsequent years of war – at the very times when the greatest progress was being made in reducing aggregate infant mortality. If improvements in the health of infants are so closely linked to rising living standards then why did infant health improve fastest when living standards, as measured by GDP per capita, fell?

Figure 11.2e shows the annual change in GDP per capita with local minima (measured using consistent international purchasing power dollars and in descending order of magnitude) in the years to 1919, 1945, 1931, 1991, 1980, 1908 and 1926, and 1973. The graph is simply the first derivative of Figure 11.2b and shows how economic fortunes oscillate in an ever more chaotic fractal pattern. What matters here is that these periods when GDP per capita actually fell the most coincided in aggregate far too often with the fastest proportionate decreases in infant mortality as shown in Figure 11.2f. The two lines in Figure 11.2f show the decennial proportionate increase in GDP per capita and in infant mortality (which in most cases is a negative figure as the rate was decreasing). The most rapid decrease in infant mortality coincides with the decade 1941-1951 centred around the 1945 realisation (at least by those who held the purse strings) that Britain had emerged on the winning side of the war but had been 'bankrupted' by it. The next most progressive decade in terms of

IMR decline, 1971-1981, included the start of the 1980s recession. Thus, in general, infant mortality fell fastest in Britain when GNP per capita rose most slowly. What then occurred? As a clue take one paragraph written at the start of the decade of most improvement – in infant mortality and much else – and ask: Were the reforms that were then introduced, introduced through altruism or supposed necessity?

> … facts … should dominate planning for [the] future … the low reproduction rate of the British community today: unless this rate is raised very materially in the near future, a rapid and continuous decline of the population cannot be prevented ... [which] makes it imperative to give first place in social expenditure to the care of childhood and to the safeguarding of maternity (Beveridge, 1942: point 15 of introduction).

The declines in infant mortality, through better infant health, through the 'safeguarding of maternity' and better 'care of childhood' were achieved as a result of a decision made to implement a plan. The plan itself was justified on the almost reverse Malthusian grounds that without such safeguarding and better care Britain would soon run out of people, or at least fit and able bodied people. Forty years earlier the concern was that the nation was running out of people to fill an Empire, now it was running out of people to fill a geographically tiny island! That, at least, was how the plan was sold to those with the traditional power over such plans, and to the population as a whole who had not had a vote for some ten years.

The landslide election of the 1945 Labour government allowed the plan's implementation, probably more completely than had been realised when the plan was proposed (Bevan, 1947). Four decades before, a landslide Liberal party victory followed an earlier national debate on poverty, infant mortality and the physical degeneration of the population that arose from the earlier war in South Africa (Dwork, 1987). When those in power in Britain thought they might begin to lose their Empire – they used the resources of that Empire, the national wealth arising from it, to improve domestic infant health. Wealth may be necessary but it is certainly not all that is sufficient to reduce infant mortality. This is most obviously seen in poor countries where just a few individuals now hold almost all the wealth. It even appears to be the case, in England and Wales at least, that infant mortality falls fastest when national government are least effective or interested in maximising a more even overall wealth distribution.

Where Most Infants Die

IMRs today remain relatively high in much the same places that they were high a century ago. Figure 11.3 shows the map of the cities with the largest populations in England at the beginning of the 21st century. Bar the slight effects of the locations of medical facilities that care for very sick infants – and therefore see elevated mortality rates – the map is mainly a north-south divide. Figure 11.4 shows that divide again, but charts when rates of infant mortality by area in 1921 and 1931 are compared with age/sex standardized rates of mortality for all people under age 65 in the early 1980s and early 1990s respectively. A naïve, epigeneticist interpretation of these two graphs might be that high rates of infant mortality in the past are indicative of

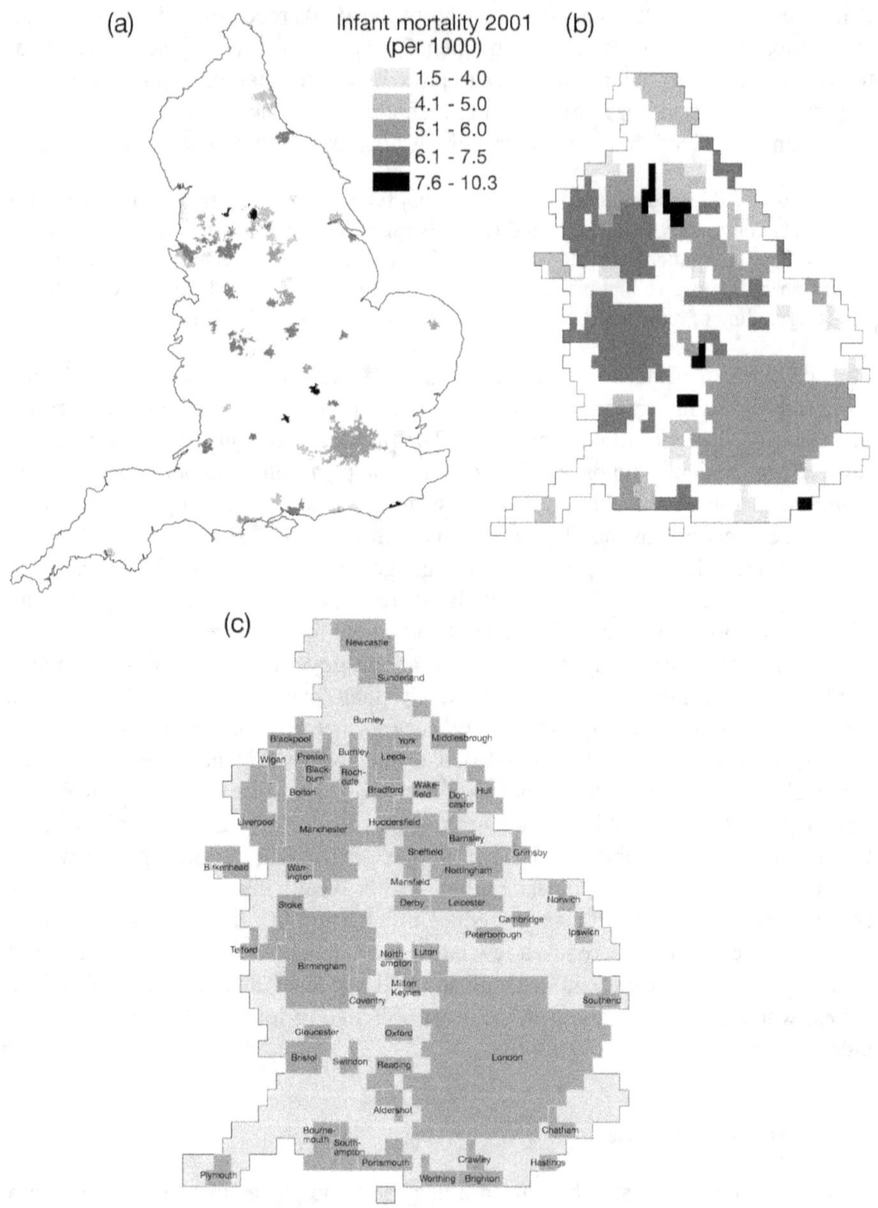

Figure 11.3 **(a) and (b) Infant mortality for the largest cities of England, 2001. (c) A key to the location of the major cities in England – by population space**

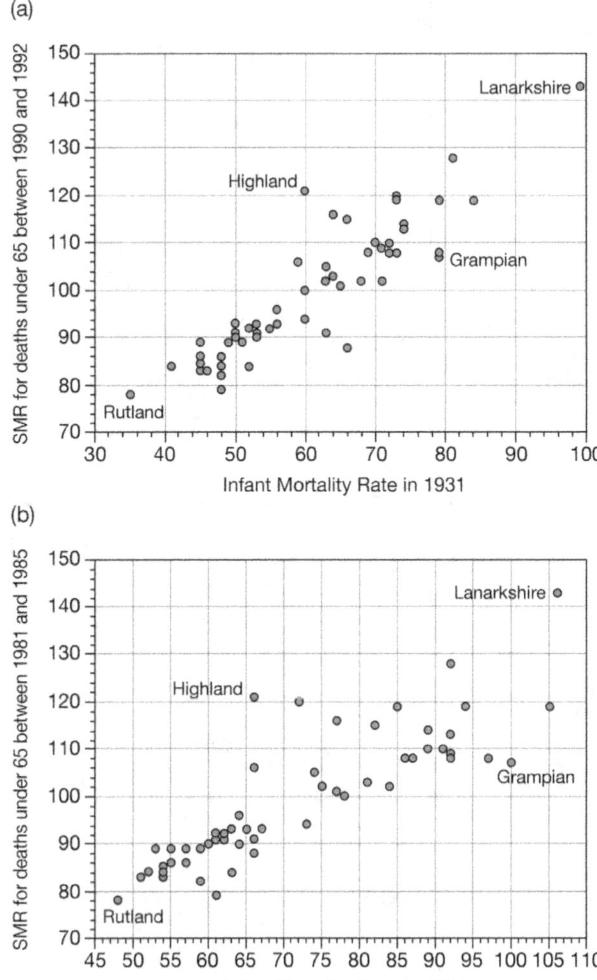

Figure 11.4 (a) The relationship between IMR around 1931 and SMR under age 65 by 1990-1992. (b) the same relationship for 1921 and 1981-1985

a poor inter-uterine environment in gestation which is reflected by relatively high rates of mortality amongst the population of those areas some sixty years later – a consequent relatively higher mortality of their grandchildren's infancy. Of course, the majority of people living in Rutland county – the best off area in health terms in the figure – were not born there, let alone will most of their children or grandchildren stay there. What the graphs instead show is that areas which are affluent tend to have remained relatively affluent over time whilst areas which were poor at the start of the last century remain so even today. In many cases the people currently living in the worse social conditions in some of our poorest northern mill towns bear almost no hereditary connection with those who lived there in the past. It is not

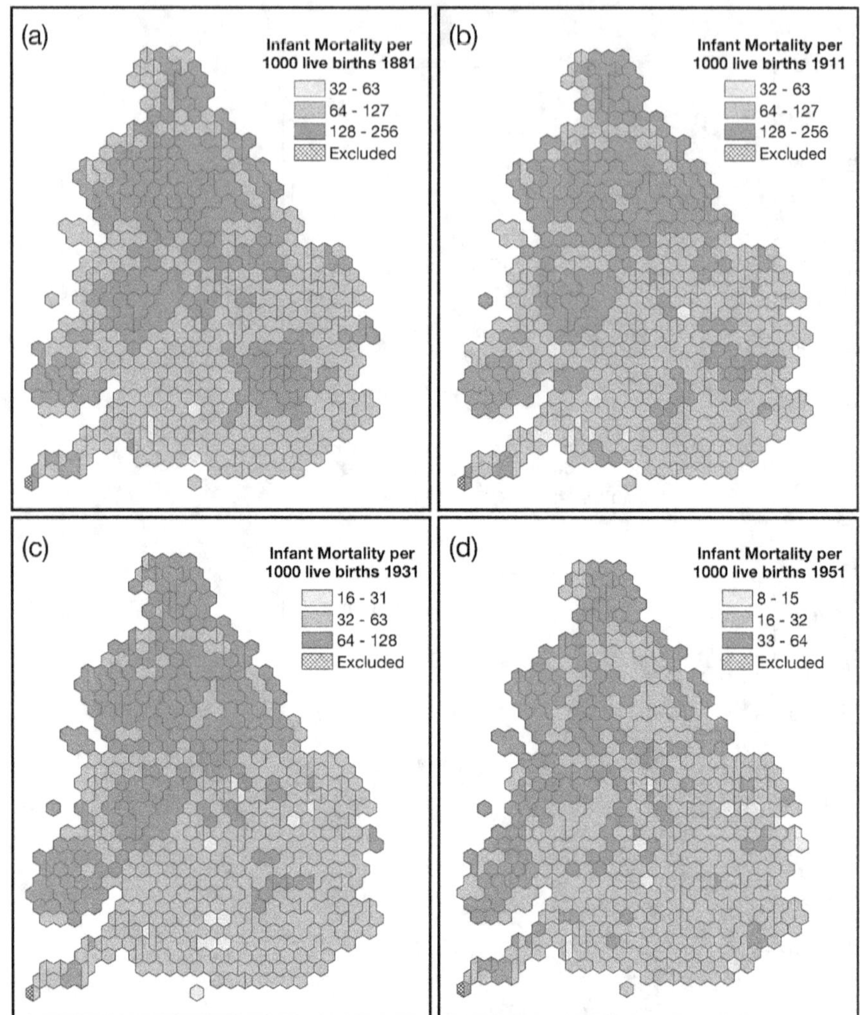

Figure 11.5 (a-d) Infant mortality in England and Wales, 1881-1951

just that they are not the same people, they are not even closely related to the same people but instead to people, say, in the Indian sub-continent. Some Northern mill towns suffer relatively high infant mortality not because the current population is related to people who used to live there, nor because the large numbers of the current population are related to people living in what was India, but because of the social conditions of life in those towns. Life in towns in Britain is mainly determined by where those towns are located, largely in relation to London, and the fact that British society is apparently organised mainly to maximize profit in London.

The areas shown in Figure 11.3 are the built up areas of the cities of England as defined in 2001. The geographic areas used in Figure 11.4 were 'historic' counties of

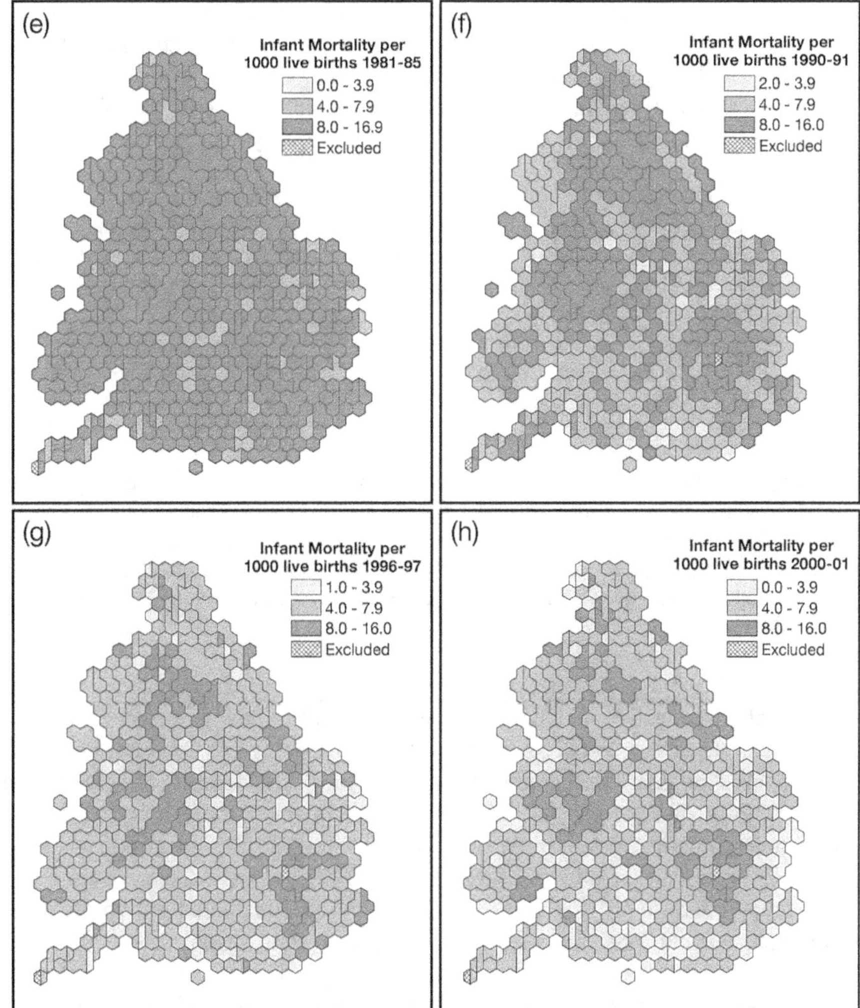

Figure 11.5 (e-h) Infant mortality in England and Wales, 1981-2000

Britain (as they are now termed) for which IMRs were published before the Second World War. Contemporary data can be re-aggregated to historic areas. In contrast the areas shown in Figure 11.5 are the contemporary local authority districts of England and Wales to which historic data has been approximated (see source of figure: www.visionofbritain.org.uk). Here rates are shown in 1881, 1911, 1931, 1951, 1981-85, 1990-91, 1996-97 and 2000-01. Care should be taken over interpreting these last three periods as by the end of the twentieth century, in some small districts, there were areas where no infant deaths occurred in some years. Figure 11.3 is a much better guide to the contemporary geography of IMR than Figure 11.5h. However, even with this caveat in mind it could be argued that Figure 11.5 suggests that the north-south geography of mortality (which was always partly an urban-rural geography) is becoming more evidently urban-rural over time. High relative rates of infant mortality have also returned to London after a century of better than usual improvement. However, in the main the relative picture is one of stability – as Newman himself noted at the start of the twentieth century:

> The chief [infant] mortality occurs in the North and North-Midland districts, and the lowest mortality in the counties below a line drawn from the Wash to the mouth of the Severn. Each decennium shows some minor variation, but practically it may be said that during the whole last century this was in the main the general distribution (Newman, 1906: 22).

Conclusions

> Last year (1905) there was a loss to the nation of 120,000 dead infants in England and Wales alone, a figure which is almost exactly one quarter of all the deaths in England and Wales in that year. That is to say, that *one quarter* of the total deaths every year is of children under the age of twelve months. And this enormous sacrifice of human life is being repeated year by year and is not growing less (Newman, 1906: 2, emphasis as in original).

In 2005 roughly 3,000 infants died in their first year of life in England and Wales, which is almost exactly one half of one per cent of all deaths in England and Wales in that year. Thus one in two hundred of the total deaths is of children under the age of twelve months. This sacrifice has been falling, with no exception, every year since 1951 and is expected to continue to fall. However, the toll is increasingly concentrated amongst the children of the poorest areas of Britain. For the sacrifice to continue to fall requires further social progress. We now have the means to reduce IMRs to near zero in affluent areas of Britain, but for the infants of poor areas the government consults on the costs of 'healthy start food' (Department of Health, 2005).

It was not until 1946, forty years after George Newman published his findings, that sufficient measures were enacted to improve the health of new born infants and the number dying simultaneously – and thereafter without exception – fell year on year. Government officials often talk about time lags in health policy and how many years it might be before an improvement might be seen as a result of a particular measure being implemented but, in the case of mid twentieth century infant mortality, there clearly was no such lag. It is possible that improvements in health in their childhood of the women who gave birth in 1946 might have led to 3.1 fewer infants dying per 1,000 in that year as compared to 1945, but the timing just appears a little too fortuitous with the ending of the War and the small but immediate improvement in living standards which that engendered. Furthermore the largest drop ever recorded in infant mortality, of 7.5 fewer infants dying per 1,000 born in 1948 as compared to 1947 coincides far too conveniently with the introduction of the National Health Service and the rights and access to better care which that gave prospective and new mothers.

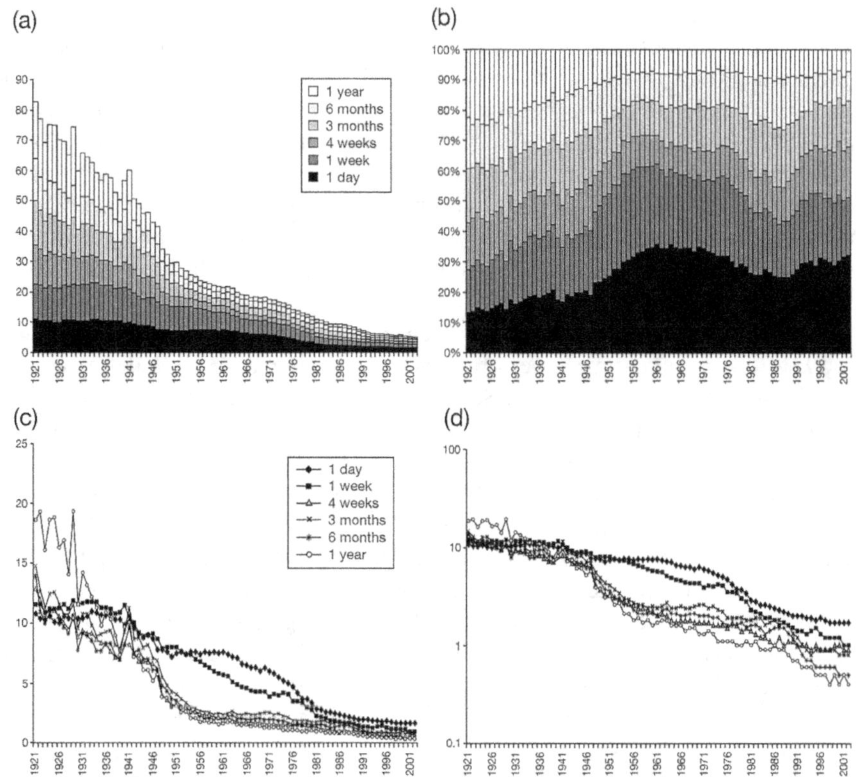

Figure 11.6 Infant mortality 1921-2002 in England and Wales by age of death
(a) IMRs per 1,000 births per year
(b) deaths by age as a proportion of all infant deaths per year
(c) deaths by age per 1,000 births per year
(d) deaths by age per 1,000 births per year

In the story of the fall in infant mortality the most important time-lag was not between policy change and result, but the forty years between the turn of the century escalation in calls for social justice, of which Newman's book was just a small part, and the winning of enough hearts and minds to prevent the continuation of the enormous sacrifice. That winning of hearts and minds was not just over the elite who had most power over resources but of people in general growing up in a much more infant health conscious society informed by the work conducted at the start of the last century and its propagation through books and magazines over the course of four decades. Of course, medical advances were made in the interim but, as seen above, by 1941 the poorest of children only had the chances of the best off in 1901. What occurred after 1941 was possible with the medical knowledge of the previous century, but it required people to vote in 1945 and for them to know what they were voting for and why. Until it is made clear that infants die because of poverty rather than fecklessness, why vote against poverty, as happened in 1945, 1964 and 1997? No amount of medical knowledge, as is obvious from worldwide infant mortality figures today, is sufficient without the will and funds to implement it. The 'vast array of small human beings that lived but a handful of days' that Newman found 'so difficult to make real to the mind' (Newman, 1906: 3) has grown by millions worldwide in the century since he wrote. In 2005, the 'make poverty history' year, pop stars clicked their fingers on television screens every three seconds to signify the death of an infant through poverty. Infant mortality was first accurately measured globally in the 1970s. Worldwide since that date it has fallen fastest where it was lowest to begin with.

Back in Britain, the final piece of evidence presented here, Figure 11.6a, shows the 1921-2001 national time series of infant deaths in England and Wales by age of the infant at their death in days, weeks, and months. Although there has been some change in the distribution of the ages at which infants are most likely to die, it has not been as dramatic as might have been the case were specific medical interventions the key. Figure 11.6b shows that in 1921 13 per cent of infant deaths occurred in the first day of life, by 2002 that figure had reached 32 per cent. Of the far fewer babies who could not be saved by 2002, more as a proportion could not be saved in that first day and far fewer now die after living for six months. Congenital diseases, associated most with the earliest deaths, decline more slowly in the graphs than do the infectious diseases associated with deaths later in the first year of life. Figure 11.6c shows that the decline in the latter occurred fastest a generation before the decline in the former, but Figure 11.6d suggests that rates of decline of death at all ages are following much the same long term trajectory. Each generation demands overall improvement. It occurs for each but different proximal mechanisms are at play. For the mothers of the 1940s it was diseases that their infants were freed from most quickly, perhaps partly due to increased availability of antibiotics. For the mothers of the 1970s deaths in the first days and weeks of the lives of their infants fell fastest; perhaps greater access to incubators helped. For the mothers of the start of the present century deaths after a number of months became rarer again more quickly, perhaps associated with increased and better information on preventing Sudden Infant Death Syndrome and other once largely unknown causes. However, for most generations the best off have enjoyed access to the kind of care that the worse off only achieve a generation later. Medical and almost all other innovations appear to take roughly a generation, some twenty five years, to diffuse down social hierarchies in Britain. To achieve this for a fifth generation will require social progress equivalent to the ending of child poverty in Britain within the next two decades. Progress of a similar magnitude has been achieved before.

Chapter 12

The Health of Infants at the Beginning of the Twenty-first Century[1]

Yvonne Kelly

Introduction

Health status during infancy has profound implications for health and social trajectories throughout the lifecourse. This chapter gives a brief overview of markers of infant health during the twentieth century and presents a descriptive account of current social and ethnic differences in markers of infant health: birthweight and gestational age, infant feeding and the occurrence of common illnesses in the first nine months of life using data collected in the first survey of the Millennium Cohort Study (MCS), a prospective study of infants born at the beginning of the twenty-first century.

Infants born with weights at the lower end of the birthweight distribution are at increased risk of adverse outcomes, and there is a distinctive reverse-J-shaped relationship between birthweight and mortality during the first year of life. There is a large body of work on the possible links between birthweight and a range of childhood outcomes such as cognitive performance, behaviour and respiratory health. Moreover during the last two decades a great deal of work has been done to explore the effects of growth and development during foetal and infant life on consequent biological programming of organs and corresponding lifetime capacity for function throughout the lifecourse. Whilst questions remain about the magnitude of such effects and about the mechanisms through which they might operate, work has been done that suggests that the effect of birthweight on health in childhood and later life is contingent on the social environment.

A large body of work on the benefits of breastfeeding for health and development exists alongside research on the possible detrimental effects on morbidity and mortality associated with artificial feeding in infancy. The potential benefits of breastfeeding are wide ranging with decreased risk of infections and illness for infants, and in developing country settings mortality which can be attributed to these causes. In mothers breastfeeding is associated with a reduced likelihood of the development of some cancers. The World Health Organisation (WHO) currently recommend that infants should be breastfed exclusively for the first six months of life and that after the introduction of complementary feeding that breastfeeding should continue until children are aged two years (WHO, 2002). In 2003 the UK Department of Health adopted the recommendation on exclusive breastfeeding for the first six months of life (Department of Health, 2003). However, concerns remain

1 This work was funded by a grant from the ESRC ref no. RES-000-23-1191.

about the implementation of this among the population as survey work carried out just prior to this recommendation estimated that only about one per cent of infants in the UK were exclusively breastfed for the first six months of life.

Over the last 60 years technological and medical advances have led to major changes in curative health practice. The advent of the NHS in 1948 brought with it major changes in preventive health care and health promotion strategies that recognise that many of the causes of ill health lie outside the direct influence of the medical sector (Ranade, 1997). These have been accompanied by changes in the scientific understanding of risk, public knowledge of health and awareness of individual risk and wider societal factors. Health improvements over this time have resulted, in part, from a marked decline in the occurrence of major illness and mortality during childhood. For example mortality due to infectious diseases among 0-4 year olds has fallen from about 20 per cent of all deaths in this age group in the early 1930s to less than 1 per cent in the early 1990s (Ranade, 1997). Hence the occurrence of illnesses that are common but in the main not serious can be used as markers of health.

It is clear that health and illness show strong social patterning and this is a major source of scientific interest, and understanding the underlying causes of these inequalities in a range of contexts is of great importance. Attempts to tackle the question of how disadvantage affects health have been made from a range of viewpoints with research exploring how material conditions, psychosocial factors and health related behaviours explain social gradients in health outcomes. Work has also been done on how social experience in childhood and adult life influence health, these include elements such as life satisfaction, autonomy and control, reward and effort, prejudice and racism. Investigators have also focused on the importance of stressors such as social cohesion, poverty, wealth and resources, and opportunity as influences on health. Furthermore, work has been done on the identification of underlying biological mechanisms through which these stressors may operate and manifest clinically.

The Use and Meaning of Birthweight and Length of Gestation

The regular weighing of babies with the primary aim of monitoring growth began in the latter half of the nineteenth century and the term 'prematurity' was used to describe babies who were born with low birth weights (Macfarlane and Mugford, 2000). The criterion of birthweight less than 2,500 grams as part of the definition of prematurity was first used in 1869 by Nikolay F. Miller who was at that time Physician-in-Chief of the Moscow Foundling Hospital (Cone, 1980). In his landmark work Sir George Newman (1906: 78) tried to further refine the definition of prematurity and in doing so included babies who were born either too small, too early or both. Newman's definition of prematurity included: infants that were born after the twenty-sixth completed week of gestation but born before 36 weeks; and infants who were born weighing less than 2,500 grams, (those born before the twenty-sixth completed week defined as either abortions or miscarriages). Work by Arvo Ylppo in Finland in 1919 further popularised the use of the 2,500 grams threshold and endorsement by the WHO in 1948 led to universal use of this cut-off point.

In order to separate the effects of birthweight and length of gestation out from the notion of 'prematurity' it was suggested in the early 1970s that gestational age be categorised as follows: pre-term (babies born at less than 259 days), term (259-294 days) and post-term (greater than 294 days). This definition did not specify whether gestational age should be estimated from the date of the last menstrual period or from a paediatric assessment. The estimation of gestational age in the ninth and tenth revisions of the *International Classifications of Diseases* is based on the date of the mother's last menstrual period (WHO, 1992).

Newman's work also documented the backgrounds and characteristics of mothers and their infants and identified that low birth weights were related to a number of adverse circumstances including infections in the mother, poor housing conditions, the nutritional status of the mother and 'the activities and pleasures of society' (Newman, 1906: 177-220). Work carried out in more recent decades has identified and characterised a broader array of factors that influence birthweight, these encompass physical and behavioural characteristics of the mother, social and economic position and features of the wider social context such as residential area.

The exact meaning of weight at birth in terms of adverse outcomes varies within and across populations and there remains intense speculation about the use and interpretation of birthweight data (Wilcox, 2001; David, 2001; Hertz-Picciotto, 2001). For example there is no particular increase in the risk of mortality below the 2,500 gram cut-off point, and the lowest risk of mortality in any population, also known as the 'optimal' weight, is typically 300-400 grams more than the average birthweight for a given population. Use of the 2,500 gram cut-off point has led to confusion in the interpretation of birthweight data among infants with non-White ethnicities. One consequence of this in early studies of ethnic differences in birthweight was the paradoxical finding, that small US born Black infants had lower mortality rates than their White counterparts despite overall higher mortality rates for Black infants (North and MacDonald,1977). In a series of papers Rooth (1980) and Wilcox and Russell (1983a; 1983b; 1986) proposed the use of population specific standards for birthweight distribution. When this is done for different ethnic groups the apparent advantage in survival for low birth weight US born Black infants disappears and it is seen that mortality for Black infants is higher than for White infants across the birthweight distribution. Thus, it has been proposed that the most useful birthweight measures to employ in studies of ethnicity and birthweight are mean values, standardised scores and if the use of a cut-off is desirable that this should equate to a point a certain distance from the mean, for example minus two standard deviations (Wilcox, 2001).

Despite the rationale for using context specific definitions of low weight at birth, the proportion of infants born weighing less than 2,500 grams is still used as a marker of health across and within countries. Worldwide there are wide variations in the proportions of babies born weighing less than 2,500 grams with about one quarter of babies born in India, Pakistan and Bangladesh born below this threshold. In the Caribbean 14 per cent of infants are born weighing less than 2,500 grams. Overall rates across Europe, Australia and the US are comparable, ranging between four and eight per cent, though there are wide variations within

Europe with rates in Sweden and Finland at four per cent which is half that of the UK (UNICEF, 2004).

Data collected in the 1946, 1958 and 1970 British Birth Cohort studies suggest a decline in average birthweight, with a drop of approximately 110 grams over this period (Wadsworth *et al.*, 2003). In contrast, analysis of national data presented in 500 gram categories, from the mid 1950s to the late 1980s showed a temporal increase in the proportions of babies in higher weight categories and a decrease in the proportion of low weight births, suggesting an overall shift to the right in the birthweight distribution and a corresponding increase in average birthweight (Albermann, 1991; Power, 1994). In the 1990s a small upward trend in the prevalence of low birth weight was seen, and it has been suggested that this is a result of increased survival of very small babies, an increased number of multiple pregnancies and changes in socio-demographic factors such as increasing age of childbearing (Macfarlane and Mugford, 2000).

In contrast to measures of weight at birth there are no accurate worldwide estimates of gestational age though it is estimated that the prevalence of pre-term birth ranges from five to 15 per cent, and that most of the variation is due to demographic and geographical factors (Slattery and Morrison, 2002). With the exception of Scotland, data on gestational age are not collected routinely in the UK. Over the last decade in Scotland there has been a small, but steady increase in the prevalence of pre-term birth, from 5.5 in 1990-1992 to 6.0 per cent in 2000-2002 (NHS Scotland, 2005).

Infant Feeding

A wealth of scientific evidence has highlighted the benefits of breastfeeding for infants and mothers. Infants who are breastfed are at reduced risk of gastrointestinal infections, respiratory illnesses and allergies (Howie *et al.*, 1990; British Paediatric Association, 1994; Wilson *et al.*, 1998; Kramer *et al.*, 2001), and there are recognised beneficial effects for cognitive performance, development and behaviour later in childhood (Lucas *et al.*, 1998; McCann and Ames, 2005; Sacker *et al.*, 2006). For mothers breastfeeding is associated with reduced risk of breast and ovarian cancers (Gwinn *et al.*, 1990; Collaborative Group on Hormonal Factors in Breast Cancer, 2002).

The feeding of infants with non-human milks dates back some 5,000 years and the formulation of artificial milk on a commercial basis began in the latter half of the nineteenth century (Fildes, 1986). Wet nursing was very popular among socially advantaged groups in the eighteenth and nineteenth centuries but this practice was largely replaced by the use of formula milks in the early part of the twentieth century (Golden, 1996). The social patterning of formula milk use from birth reversed between the 1940s and 1970s and became most common among families living in disadvantaged circumstances. Data from the British Birth Cohorts have been used to illustrate the dramatic decline in the proportions of infants fed on breast milk over this time period. In 1946 rates of breastfeeding initiation were 75 per cent, in 1958 this had fallen to 70 per cent and by 1970 breastfeeding was initiated for only 38 per cent of infants. The same pattern of decline was seen for the proportion of infants who continued to be fed breast milk at one month of age (Wadsworth *et al.*, 2003).

In response to the low breastfeeding rates seen in the early 1970s and the recognition of the widespread benefits for population health the British Government commissioned the Infant Feeding Survey (IFS) which began in 1975 with the primary aim of monitoring infant feeding patterns. In 1975 the breastfeeding initiation rate was 51 per cent. Since that time there have been substantial increases in rates of breastfeeding initiation, in 1990 63 per cent of infants were breastfed at least once and by the year 2000 rates had risen to 70 per cent (Hamlyn *et al.*, 2002). These increases are due at least in part to public health campaigns, baby friendly hospitals and the availability of trained breastfeeding counsellors.

Current Perspective on Inequalities in Infant Health

Britain has a strong tradition of Birth Cohort studies and evidence from these has been extensively used as a base for health and social policy. The massive economic and demographic transitions that have taken place in the UK since the last of the previous birth cohorts in 1970 prompted the setting up of the MCS. The MCS provides a multi-purpose dataset that describes the diversity of backgrounds from which children born at the beginning of the twenty-first century are starting out in life. It has detailed data on an extensive array of social and economic factors, and so provides an opportunity to answer questions about the precursors of early life health.

The sample of 18,553 households was drawn from all births occurring in England and Wales between September 2000 and August 2001 – and in Scotland and Northern Ireland between November 2000 and January 2002 – who were alive and still living in the UK at age nine months. The sample was drawn from child benefit registers and has a random clustered design. Clustering was at the electoral ward level, with over-representation of disadvantaged wards and areas with high proportions of Black and South Asian residents. First sweep interviews with the mother and father (where resident) took place when the infant was aged nine months. At interview questions focused on the time leading up to birth, the birth itself and the nine months that followed. Topics covered at interview included the following:

- Before birth: mother's height, weight, age, number of previous pregnancies, antenatal care, illnesses during pregnancy, prior infertility, cigarette smoke exposure, family composition and socioeconomic position.
- Around the time of birth: details about the labour and delivery, birth complications, exposure to cigarette smoke, family composition, and parent's socioeconomic status.
- Around nine months of age: common illnesses such as wheezing and diarrhoea, infant feeding practices, growth, parents' psychological and physical health, cigarette smoking and alcohol consumption, family composition, ethnicity,

socioeconomic status, social support and networks, parental relationships, domestic arrangements and neighbourhood characteristics.

Social class by occupation was measured using the National Statistics Socio-Economic Classification (NS-SEC). This social class schema was first used in the 2001 census and is based on the following principles which relate to markers of working conditions: the timing of payment (monthly, weekly, daily or hourly); presence of incremental pay; job security; degree of autonomy; promotion opportunities; degree of planning own work tasks (Coxon and Fisher, 1999). In this chapter data are shown using the five category version of the NS-SEC as follows, Managerial and Professional, Intermediate, Small employers and Self-employed, Lower Supervisory and Technical, Semi-Routine and Routine. Data on ethnicity for infants and their parents were collected using the 2001 Census categories and in this chapter data are shown for the following groups: White, Indian, Pakistani, Bangladeshi, Black Caribbean, Black African, and Other.

In the MCS birthweights were reported by mothers at the time of interview and gestational age was estimated by the difference in the expected date of delivery reported by the mother and the actual date of birth. It is not known whether the reported expected delivery dates were based on information about the mothers last menstrual period, from ultrasound examination, both of these, or some other estimate.

The weight of an infant at birth is influenced by a wide range of factors, including those that are fixed such as gender and mother's stature, age and reproductive history. However, characteristics such as stature and age at motherhood are themselves heavily influenced by the social environment. Moreover, a wide range of socio-economic factors have been shown to be associated with birth weight including family composition, economic status (Shiono *et al.*, 1986; Gould and Leroy, 1988; Parker *et al.*, 1994), local area characteristics and measures of local area deprivation (Wilcox *et al.*, 1995; Jarvelin *et al.*, 1997; O'Campo *et al.*, 1997; Roberts, 1997; Fang *et al.*, 1999), cigarette smoking (Rush and Cassano, 1983), and access to and use of antenatal care. Social class gradients are evident for birth weight; routine hospital episodes' statistics data for England and Wales show that low birth weight is approximately 60 per cent more common among infants born to fathers in routine and semi routine occupations compared to those in managerial and professional occupations (Mortality Statistics, 2003). Birthweight data from the MCS are shown for boys and girls separately by mothers' occupational social class and by the infants' ethnic group (see Table 12.1). Girls were, on average 118g lighter than boys and compared with boys have a birthweight distribution that is shifted to the left. There were clear social class differences in birthweight with infants born to mothers in managerial and professional occupations being on average 104g heavier than those born to mothers with semi-routine and routine occupations. There were greater absolute and relative birthweight differences for girls compared with boys by mother's occupational class. Girls born to mothers with semi-routine and routine occupations were 5.0 per cent lighter than girls born to mothers in managerial

Table 12.1 Birthweight of MCS cohort members by mother's occupation and ethnic group

		Mean birthweight (kg)			% of births < 2,500g		
	N	Boys	Girls	All	Boys	Girls	All
Mother's occupation							
Management & Professional	4,822	3.499	3.349	3.426	4.7	5.8	5.2
Intermediate	3,094	3.455	3.331	3.394	5.8	6.7	6.3
Small employer & Self-employed	648	3.485	3.311	3.400	6.8	6.1	6.3
Lower supervisory & Technical	998	3.385	3.315	3.352	8.8	7.8	8.2
Semi-routine & Routine	6,762	3.365	3.280	3.324	7.4	8.4	7.9
Never worked	2,166	3.285	3.160	3.222	10.4	11.6	11.0
Total	18,490	3.422	3.304	3.365	6.6	7.4	7.0
Ethnic group							
White	15,257	3.448	3.331	3.391	3.2	3.2	3.2
Indian	464	3.054	2.965	3.011	5.5	3.8	4.7
Pakistani	896	3.157	3.075	3.116	2.3	3.1	2.8
Bangladeshi	366	3.138	2.996	3.065	3.6	4.5	4.1
Black Caribbean	243	3.291	3.147	3.228	5.1	2.6	4.0
Black African	373	3.404	3.172	3.291	3.1	7.3	5.2
Total	17,599	3.422	3.304	3.364	3.1	3.2	3.2

and professional occupations, whereas boys were 3.8 per cent lighter. The lightest infants were born to mothers who had never had a paid occupation. Low birth weight was least common for infants born to mothers with managerial and professional occupations and was some 50 per cent more common among infants born to mothers in semi-routine and routine occupations. The highest rate seen was for infants born to mothers who had never had a paid occupation and this was more than twice that for infants born to mothers in managerial and professional occupations.

It is well documented that infants born to mothers from non-White ethnic minority groups are generally lighter than those born to White women. There is a shift to the left in the birthweight distribution, and, as discussed earlier, this results in a higher proportion of babies from ethnic minority groups being classified as low birth weight. This has been shown in the US for babies of Black, Hispanic and Asian women (Friedman *et al.*, 1993; Martin *et al.*, 2002), while in the UK babies born to mothers from India, Bangladesh, Pakistan, the Caribbean and Africa are more likely to be classified as low birth weight compared with their White counterparts (Dawson *et al.*, 1982; Moore *et al.*, 1995; Collins *et al.*, 1997). The highest rates of low birthweight are seen for infants born to mothers who were born in India, Bangladesh and Pakistan (12.1, 11.2 and 11.1 per cent respectively) (Registrar General, 2003). Clearly, the use of mother's country of birth as a proxy measure for ethnicity in UK Government reports may mask birthweight differences for infants born to the growing numbers of second and third generation mothers. However, when ethnicity is assessed by individual perception of ethnic origin as it is in the MCS we still see marked differences in birthweight (see Table 12.1). White infants had an average birthweight of 3.391 kilograms and were the heaviest and Indian infants at 3.011 kilograms the lightest. The same patterns of birthweight differences were seen for girls and boys separately. To take account of the differences in birthweight distribution across ethnic groups the proportion of infants weighing less than two standard deviations from the ethnic group specific mean value are shown in Table 12.1. Infants with birthweights below these cut-off points were overall most common in the Black African and Indian groups (5.2 per cent and 4.7 per cent respectively) and least common in the Pakistani (2.8 per cent) group. These patterns varied by infant sex and the highest proportions were seen for Black African and Bangladeshi girls (7.3 per cent and 4.5 per cent respectively), and Indian and Black Caribbean boys (5.5 per cent and 5.1 per cent respectively).

For gestational age there are clear overlaps with the precursors of birthweight, and major influences include features of the pregnancy and the mother's reproductive history, for instance previous premature delivery. Other features of reproductive health that influence the length of gestation include health behaviours and infections and these in turn are socially patterned (Kramer, 1987; Albermann, 1994; Kogan *et al.*, 1995; Kramer *et al.*, 2000; Bukowski *et al.*, 2001).

In the MCS no differences were seen in mean gestational age by mother's occupation and there were small differences in pre-term rates with 11 per cent more infants born before term to mothers in semi-routine and routine occupations compared with those in managerial and professional occupations (see Table 12.2).

Length of gestation has been reported to be shorter for African, Caribbean and Indian women in the UK (Wilcox *et al.*, 1993), and non-White women in the US and Norway (Migone *et al.*, 1991; Vangen, 2002). In the MCS, the longest average gestational ages were seen among White and Black African mothers (277 days) and the shortest for the Black Caribbean group (274 days) (see Table 12.2). Pre-term birth was most common for Black Caribbean and Indian infants (10.3 and 10.2 per cent respectively). Overall pre-term birth was more common for boys than girls (8.3 and 7.1 per cent respectively) but this gender difference varied across ethnic groups and pre-term birth was more common among Pakistani and Bangladeshi girls (7.2

Table 12.2 Gestational age in days and proportion of preterm births in the MCS by mother's occupation and ethnic group

	N	Mean gestation (days)			% <37 weeks gestation		
		Boys	Girls	All	Boys	Girls	All
Mother's occupation							
Management & Professional	4,828	277.2	277.4	277.3	8.4	6.1	7.3
Intermediate	3,096	277.2	277.5	277.3	7.9	6.9	7.4
Small employer & Self-employed	651	278.0	276.7	277.4	7.0	7.8	7.4
Lower supervisory & Technical	1,002	276.1	277.9	276.9	9.2	6.4	7.9
Semi-routine & Routine	6,779	276.4	277.1	276.7	8.1	8.0	8.1
Never worked	2,196	275.6	277.3	276.5	8.8	7.0	7.9
Total	18,552	276.8	277.3	277.0	8.2	7.1	7.7
Ethnic group							
White	15,257	276.9	277.5	277.2	8.3	7.0	7.6
Indian	464	273.6	276.0	274.8	11.6	8.9	10.2
Pakistani	896	276.2	276.3	276.3	4.5	7.2	5.7
Bangladeshi	366	276.1	274.8	275.4	4.8	8.0	6.4
Black Caribbean	243	274.1	274.3	274.2	11.2	9.2	10.3
Black African	373	276.8	277.6	277.2	8.2	6.3	7.3
Total	17,599	276.8	277.3	277.0	8.3	7.1	7.7

and 8.2 per cent respectively) compared with Pakistani and Bangladeshi boys (4.5 and 4.8 per cent respectively).

The UNICEF Global Database Breastfeeding Indicators (2001) pools data on breastfeeding prevalence and duration and covers approximately two thirds of the world's infant population. From these data it is estimated that worldwide one third of infants are exclusively breastfed for the first four months of life. The highest rates of exclusivity at four months are seen for South and South East Asia at 49 per cent and the lowest overall rates are seen across Europe with 16 per cent of infants breastfed exclusively for the first four months of life. There are wide variations across South Asia with rates in Bangladesh, Pakistan and India at 53, 16 and 55 per cent respectively. In sub Saharan African and Latin America rates of exclusive breastfeeding for the first four months of life are 33 and 38 per cent respectively. In Europe the highest observed rates are seen in Sweden with 98 per cent initiation and 61 per cent of infants being exclusively breastfed at four months.

The promotion of breastfeeding in the UK has been highlighted in a range of policy developments as a means of improving health, and in particular as a way of reducing health inequalities among mothers and children (Department of Health, 1999). Recently the Department of Health have set a target for Primary Care Trusts for breastfeeding to 'deliver an increase of two percentage points per year in breastfeeding initiation rates, focusing especially on women from disadvantaged groups' (Department of Health, 2002). In 2003 the Department of Health adopted WHO recommendations that exclusive breastfeeding should whenever possible be undertaken for the first six months of infant life. The evidence for this recommendation emerged from a systematic review of infant feeding studies. This report showed that the risk of respiratory illness and gastrointestinal morbidity and mortality, particularly in developing countries, was substantially reduced for infants exclusively breastfed for six months compared with those breastfed for shorter durations (Kramer and Kakuma, 2002).

Despite the recognised benefits and the implementation of a range of initiatives to promote the initiation and duration of breastfeeding, rates in the UK remain low compared to many other parts of the world. The UK Infant Feeding 2000 survey (2000 IFS) reported initial breastfeeding rates at 70 per cent in England and Wales, 63 per cent in Scotland and 54 per cent in Northern Ireland (Hamlyn *et al.*, 2002). By six months 21 per cent of babies in the total sample were still being fed breast milk. However it was not possible to estimate precise figures on exclusive breastfeeding from these data. Since the previous Infant Feeding survey in 1995 little change had occurred in breastfeeding rates in England and Wales, although significant increases took place in both Scotland and Northern Ireland.

Data collected in the early 1990s by the Avon Longitudinal Study of Pregnancy and Childhood (ALSPAC) showed that 26 per cent of mothers in the least educated group were still breast-feeding at four months, compared with 74 per cent of mothers in the highest educational group (Emmett *et al.*, 2000). A recent population based study of an economically deprived area in North East England showed that mother's education was the strongest predictor of breastfeeding continuation at six weeks and that infants living in materially disadvantaged circumstances were more likely to be weaned at younger ages compared with those from advantaged backgrounds (Cassiday *et al.*, 2004; Wright *et al.*, 2004). Thus, marked inequalities are apparent in breastfeeding. In the UK mothers most likely to initiate breastfeeding are those who reach higher educational levels, are in advantaged social classes, are aged over 30 years, and are feeding their first as opposed to subsequent babies (Hamlyn *et al.*, 2002; Kelly and Watt, 2005).

In the MCS breast feeding initiation was assessed by the question 'did you ever try to breastfeed?' and continuation thereafter in months was computed on the basis of questions about the age of the baby when last given breast milk and when formula or other types of milk or solids were first given to the baby.

The proportions of babies for whom breastfeeding was initiated and the duration of breast feeding by mothers' occupational social class are shown in Figure 12.1. Seventy one per cent of the total sample were initially breastfed. A clear social class gradient in breastfeeding initiation was evident, with 86 per cent of mothers from the Managerial and Professional occupational group initiating breastfeeding compared with 56 per cent of mothers in Semi-routine and Routine occupations. Not surprisingly, a social gradient in breastfeeding continuation was also apparent. At four months 50 per cent of the infants born to mothers in Managerial and Professional occupations were being breastfed compared with 20 per cent of infants born to mothers in Semi-routine and Routine occupations; and by six months the rates were 37 and 14 per cent respectively. Multivariate analysis of these data has shown that mothers in occupations with the least favourable working conditions were more than four times less likely to initiate breast feeding and twice as likely not to exclusively breast feed their babies at one and four months of age compared to women working in occupations with the most favourable conditions (Kelly and Watt, 2005).

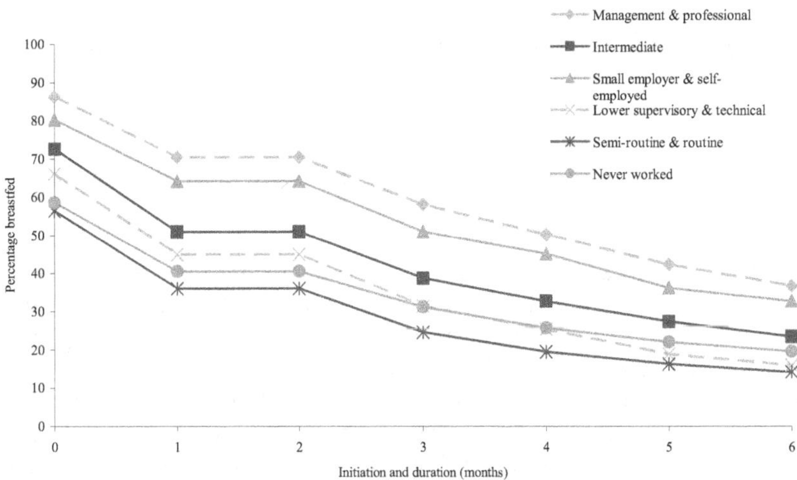

Figure 12.1 Breastfeeding initiation and continuation rates by mother's occupation

At one month 34 per cent of infants were fed exclusively on breast milk and by four months only 3 per cent of the sample were exclusively breastfed (in line with contemporary recommendations to introduce solids at three to four months). At six months only 0.3 per cent of MCS infants were fed exclusively on breast milk (Kelly and Watt, 2005). At one month a clear occupational class gradient in exclusive breastfeeding rates is seen with 47 per cent of mothers in Managerial and Professional occupations exclusively breastfeeding compared with 22 per cent of those in Semi-routine and Routine occupations and 22 per cent of those who had never had a paid occupation. At one month, of those mothers who were breastfeeding about two thirds, across all occupational groups, were doing so exclusively. Whereas of those who had never had a paid occupation and were breastfeeding at one month, just over half of them were exclusively breastfeeding their infants. At four months the proportions of mothers exclusively breastfeeding as a proportion of all those breastfeeding varied little by occupation, although interestingly the highest rates were seen for mothers who had never had a paid occupation (Figure 12.2).

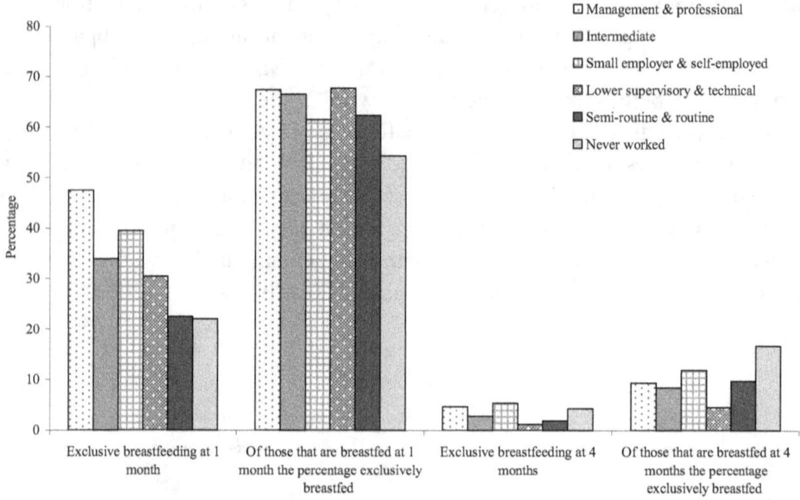

Figure 12.2 Exclusive breastfeeding rates at 1 and 4 months by mother's occupation

The 2000 IFS showed that the highest breastfeeding initiation rates were among Black (95 per cent) and Asian (87 per cent) compared to White mothers (67 per cent) (Hamlyn *et al.*, 2002). In contrast, data from the US demonstrate the lowest rate of breastfeeding amongst non-Hispanic Black mothers (Ryan *et al.*, 2002) and the most recent US data from the 2002 National Immunization Survey (NIS) indicates that the highest rates of breastfeeding initiation are found amongst Hispanic mothers (80 per cent), compared with White (72 per cent) and non-Hispanic Black (51 per cent) mothers (Li *et al.*, 2005).

In the MCS there were wide differences in the initiation of breastfeeding across ethnic groups (Figure 12.3). Breastfeeding initiation and continuation over time shows a pattern of consistently higher rates for Black Africans, and consistently lower rates for Whites and Pakistanis, while exclusive breastfeeding shows Black Caribbean and Others with consistently high rates, Indians in the middle, and all of the other groups with low rates (Figure 12.4).

It can be seen that exclusive breastfeeding as a proportion of any breastfeeding varies markedly by ethnic group. At one month, of those mothers who breastfed their infants two thirds of the Black Caribbean and White compared with two fifths of the Black African group did so exclusively. And at four months, of those mothers breastfeeding their infants 20 per cent of the Pakistani group did so exclusively, in contrast to 6.3 per cent of Black African and 8.5 per cent of White mothers.

Multivariate analysis of these data has shown that when a range of social, demographic, economic and cultural factors were taken into account, Black African, Black Caribbean, Bangladeshi, Pakistani and Indian mothers were ten, nine, five, two and two times respectively more likely to initiate breastfeeding compared with White mothers (Kelly *et al.*, 2006).

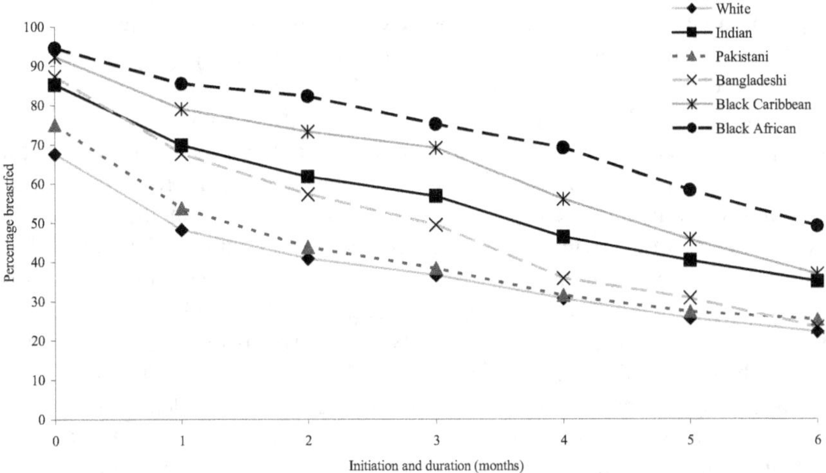

Figure 12.3 Breastfeeding initiation and continuation rates by mother's ethnic group

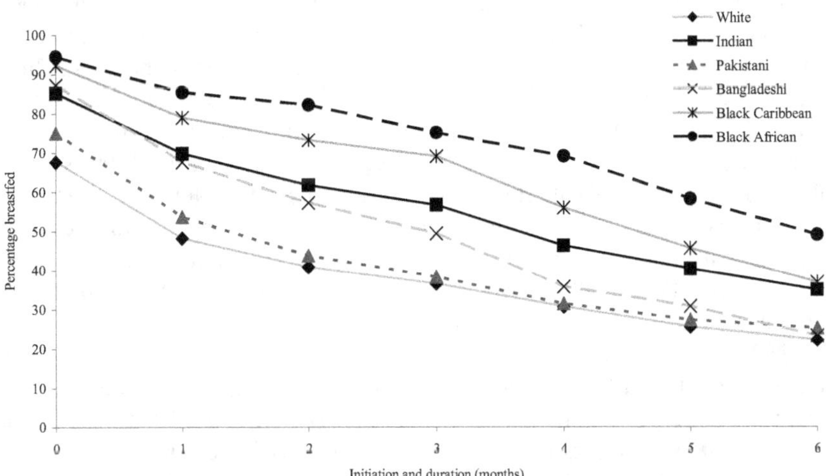

Figure 12.4 Exclusive breastfeeding rates at 1 and 4 months by mother's ethnic group

The initiation and maintenance of breastfeeding is determined by a range of clinical, personal, social, cultural and environmental factors (Maclean, 1989; Hoddinott and Pill, 1999). To meet the revised Department of Health breastfeeding recommendations will require a radical shift in practice and policy in the NHS and across society as a whole (WHO, 2002). In light of the evident inequalities in infant feeding practices and their consequent effect on infant health there need to be public policies and supportive health services aimed at promoting breastfeeding particularly amongst socially disadvantaged groups. However, as ethnic minority groups in the UK appear to have a breastfeeding advantage compared with the ethnic majority, future policies aimed at increasing breastfeeding rates need to pay attention to the different social, economic and cultural profiles of all ethnic groups.

In Scandinavia, where 30 years ago about two thirds of infants were initially breastfed multi-faceted interventions were implemented at a national level. These included: problem-based information about breastfeeding, more peer support groups, improved management skills among health workers, an increase in the collective sharing of breastfeeding experience due to the rising numbers of women who successfully breastfed, and improved maternity leave and conditions on return to work. However, it is not clear which, if any, of these individual features were most effective in achieving the almost universal (98 per cent) breastfeeding initiation rates seen at the beginning of the twenty-first century (Protheroe *et al.*, 2003).

There are no nationally available data in the UK on the prevalence of common morbidities such as respiratory illnesses, skin and ear problems and diarrhoea in the infant population. Studies that have been done on General Practitioner (GP) consultation rates report that during the first year of life the most common reasons for visiting the GP are respiratory complaints including chest infections and wheezing illness, skin problems including eczema and rashes, ear infections and gastrointestinal illnesses including gastroenteritis (Macfarlane and Mugford, 2000). These data are limited as they reflect only those cases that are presented to the GP and it is not possible to ascertain morbidity rates for the general population. Other routine data sources based on GP consultation rates include infants but cover a wider age range (0 to 4 years) and again because of their nature only reflect the tip of the clinical iceberg (McCormick *et al.*, 1995). A paper by Saxena *et al.* (1999) showed small but significant social patterning of GP consultation rates. Moreover, a report on these GP consultation data show clear social patterning for visits due to infections, gastrointestinal illness, respiratory and skin problems (Key Health Statistics from General Practice 1998, 2000). However these data are aggregated for the 0-15 year age group and may not be relevant for the infant population.

ALSPAC collected data on common illnesses and a report from this study estimated that in the early 1990s approximately 20 per cent of infants had wheezing symptoms, and 25 per cent had bouts of diarrhoea (Baker *et al.*, 1998). Unlike the MCS these data were not restricted to primary care consultations and this probably explains the large differences in observed rates. There have been reports of ethnic inequalities in health which indicate that differences are seen throughout childhood with higher rates of morbidity in Black and Asian children (Raleigh and Balarajan, 1995; Nazroo *et al.*, 2001). However these reports have concentrated on older children. Furthermore, because of the predominantly White ethnic make up of the ALSPAC study population, it is not known whether the same level of morbidity seen for White infants exists among those from ethnic minority groups.

In the MCS the most common reason in the first nine months of life for cohort members to visit the GP, a health centre, a health visitor, a casualty department or to have phoned NHS Direct was for a chest infection and this occurred for 29 per cent of the sample. Consultations for skin problems, ear infections and vomiting and diarrhoea occurred for 19.5, 10.1 and 13.8 per cent of infants respectively (see Table

Table 12.3 Visit to GP, health centre, health visitor or phone call to NHS Direct and hospital admission during the first nine months of life among MCS cohort members by mother's occupation (NS-SEC) and ethnic group

		GP visits					Hospital admission		
	N	Chest infection %	Wheezing or asthma %	Ear infection %	Skin problem %	Vomiting or diarrhoea %	Chest infection %	Wheezing or asthma %	Gastro-enteritis %
Social Class									
Management & Professional	4,828	0.3	5.4	11.2	23.4	15.8	2.9	0.5	1.0
Intermediate	3,095	9.3	5.6	10.2	19.1	14.9	3.4	0.6	1.2
Small employer & Self-employed	650	6.9	5.6	11.8	18.5	13.5	4.0	0.6	2.2
Lower supervisory & Technical	1,002	4.4	8.8	11.2	19.1	13.7	3.5	1.0	1.6
Semi-routine & Routine	6,777	9.3	7.3	9.4	17.5	12.4	4.8	1.2	1.6
Never worked	2,170	6.6	7.5	6.8	15.0	9.4	5.6	1.8	2.1
Total	18,522	9.0	6.4	10.1	19.5	13.8	3.9	0.9	1.4
Ethnic group									
White	15,288	29.9	6.4	10.6	19.8	14.3	4.0	0.8	1.4
Indian	465	21.4	2.5	6.8	19.3	9.0	1.9	0.9	0.6
Pakistani	899	28.3	5.9	5.9	14.8	8.4	4.5	2.3	1.7
Bangladeshi	367	18.0	4.7	5.3	13.4	7.6	4.7	1.8	0.0
Black Caribbean	243	20.7	11.5	6.9	19.0	8.6	0.0	0.0	0.6
Black African	381	17.2	7.6	2.7	16.0	9.2	1.9	0.0	0.4

12.3). Consultations for wheezing illnesses were least common and these occurred for 6.4 per cent of infants. Hospital admissions were rare and for chest infections, wheezing illness and gastroenteritis rates were 3.9, 0.9 and 1.4 per cent respectively (Table 12.3).

There was little difference in GP, health centre, health visitor, casualty or NHS Direct consultations for these common ailments by mothers' occupation. However for hospital admissions there was clear social patterning with admission rates for chest infections being 66 per cent, wheezing illnesses and asthma 140 per cent and gastroenteritis 60 per cent higher for infants born to mothers in Semi-routine and Routine compared with Managerial and Professional occupations (see Table 12.3).

Marked differences were apparent for proportions of GP, health centre, health visitor, casualty or NHS Direct consultations during the first nine months of life by ethnic group. White infants were most likely to have consultations for chest and ear infections (30 and 11 per cent respectively), skin problems (20 per cent) and vomiting and diarrhoea (14 per cent). Bangladeshi infants were the least likely to have such consultations for skin problems (15 per cent) and vomiting and diarrhoea (8 per cent) and second least likely for chest and ear infections (18 and 5 per cent respectively) and wheezing illness (5 per cent). However Bangladeshi infants were the most likely to be admitted to hospital due to a chest infection (5 per cent) and second most likely to be admitted for wheezing illness (1.8 per cent). Pakistani infants were the most likely to be admitted to hospital for wheezing illness (2.3 per cent) and gastroenteritis (1.7 per cent) and second most likely for chest infections (1.8 per cent). Black Caribbean infants were the most likely to have GP, health centre, health visitor, casualty or NHS Direct consultations for wheezing illness (12 per cent) and the least likely to be admitted to hospital for this cause.

Causes of Ethnic Inequality in Infant Health

The magnitude of the effect for explanatory variables such as cigarette exposure or household income on early childhood health outcomes might be consistent across ethnic groups, or they might vary, or interact differentially. For example, a report from California showed differential effects of neighbourhood deprivation across ethnic groups with reduced birthweight in Black and Asian women, but not White and Latino women (Pearl *et al.*, 2001). These findings from the US may have some relevance for understanding the patterning of inequalities in early childhood health among UK ethnic minorities. For example it might be that the 200-300 gram differences in average birthweights seen across ethnic groups in the UK are explained to varying extents by different socio-environmental factors.

As well as acute behavioural exposures – for example there is marked variation in the prevalence of cigarette smoking across ethnic groups (Primatesta and Prior, 2001) and current differences in material circumstances – markers of infant health such as birthweight may also reflect differences in accumulated disadvantage over successive generations, for example poor nutritional status (Cohen *et al.*, 2001). It has been suggested that increasing birthweight might help to improve the health of ethnic minority children. There is evidence from the US that second generation

Table 12.4 Socio-demographic characteristics of the mothers of MCS cohort members

	Ethnic Group					
	White	**Indian**	**Pakistani**	**Bangladeshi**	**Black Caribbean**	**Black African**
N	14,949	425	756	286	249	326
Age (years)	28.3	28.5	26.5	26	29.4	30.4
SD[a] of age	6.0	5.1	5.3	4.8	6.8	5.8
Height (metres)	1.6	1.6	1.6	1.6	1.6	1.7
SD[a] of height	0.1	0.1	0.1	0.1	0.1	0.1
3 or more children (%)	20.7	21.3	39.1	47.0	28.2	38.7
One parent household (%)	13.6	4.6	7.0	5.2	46.7	38.4
Resident grandparent (%)	4.7	30.5	25.5	28.7	11.2	4.6
Smoking during pregnancy (%)	36.8	7.0	4.3	2.9	38.1	6.5
Degree or higher qualification (%)	17.9	26.5	8.3	3.5	12.8	27.1
Semi-routine & routine occupations (%)	34.5	29.7	31.6	25.0	28.9	28.0
Household income below poverty line (%)	20.1	20.5	43.2	51.2	43.9	42.0
Living in rented accommodation (%)	29.5	15.7	21.8	49.7	58.2	75.0

Note: a) SD = Standard deviation

Black immigrants have heavier babies compared with first generation immigrants (Chike-Obi, 1996). However, as yet there is no clear evidence of increases in average birthweight in first and second-generation Asian immigrants in the UK (Dhawan, 1995; Draper *et al.*, 1995), although there is some evidence of upward social mobility for second compared with first generation migrant groups (Nazroo, 2004) and there is some evidence suggesting that both upward and downward social mobility of ethnic minority adults within a generation is related to health outcomes (Harding and Balarajan, 2001; Harding, 2003).

The prevalence of risk factors for low weights at birth, pre-term birth and other markers of infant health such as the occurrence of respiratory and gastrointestinal infections, differ across ethnic groups with wide variations in family composition, level of academic qualifications, income, age at motherhood and cigarette smoking during pregnancy (see Table 12.4). Thus, there remains the challenge of determining whether differences in the predictors of birthweight across ethnic groups can be quantified for the purposes of targeted interventions. Furthermore, because of the accumulation of disadvantage in previous generations it may be that it takes several generations before the benefits of consequent policy changes are seen.

It is known that a number of factors, including birthweight and gestational age, influence health during infancy and childhood. Of these factors, exposure to tobacco smoke and feeding with formula milk increase the risk of wheeze and diarrhoea (Eaton-Evans and Dugdale, 1987; Baker *et al.*, 1998; Gergen *et al.*, 1998; Lux *et al.*, 2000). From analysis of MCS we know there are marked social differences in breast feeding initiation and continuation (Kelly and Watt, 2005). We also know from these data that infant feeding practices differ markedly across ethnic groups and broadly speaking there are higher rates of breastfeeding among ethnic minorities compared with Whites. Family composition, housing, maternal education and overcrowding, are also important predictors of infant health outcomes (Pope *et al.*, 1993; Glass and Bern, 1994; Baker *et al.*, 1998; O'Connor *et al.*, 2000). Low income, poor housing and so on are all known to be associated with smoking and non-optimal diet, and these factors also affect birthweight. People from certain ethnic minority groups are more likely to experience low income, job insecurity or lack of employment and poor housing (Modood *et al.*, 1997). The work of Nazroo (1998; 2001; 2003) has shown that apparent ethnic differences in adult health can often be explained in terms of social and economic disadvantages when these are appropriately measured, although other factors may also be important.

There are differences in the distribution of birthweight across ethnic groups but the implications of this for future health are not entirely clear. A large body of work points to the long lasting effects of size at birth and growth in infancy on health in later life (Barker, 1994; 1995). Size at birth also influences employment opportunities in adulthood but the pathways for these effects remain unclear (Bartley *et al.*, 1994). And whilst questions remain about the magnitude of such effects and about the mechanisms through which they might operate (Huxley *et al.*, 2000) work has been done that suggests that the effect of birthweight on health in childhood and later life is contingent on the social environment (Barker *et al.*, 2001; Kelly *et al.*, 2001; Marmot, 2001). Furthermore, in the UK where large proportions of particular ethnic minorities are living in economic disadvantage it may be that differential effects of birthweight for future health across ethnic minority groups will be seen.

Conclusions

There are clear inequalities in markers of infant health such as birthweight, feeding and common illnesses as described above. Such inequalities have important consequences and are amenable to change as health during infancy influences health and well being throughout the life course and will therefore impact on the likelihood of social and health inequality in adult life. A range of economic, psychosocial and physiological factors are known to affect birthweight and other markers of infant health and some of these factors, such as cigarette smoking and maternal stature have a social distribution. Ethnic inequalities for infant and early childhood health have been observed internationally. However, the extent to which these can be explained by differences in the distribution of recognised risk factors in different ethnic groups is not clear. Furthermore it is not known how far any ethnic variations in risk factor distribution may be purely a result of socio-economic differences. If differences in health across ethnic groups are solely a function of social and economic circumstances then this will point to the need to address deprivation among these groups to reduce the likelihood of subsequent poor health. Even at the beginning of the twenty-first century, issues with which Newman struggled a century before still haunt those trying to ensure that infants, whatever the circumstances into which they are born, have the best possible chance of surviving their first year.

Chapter 13

Conclusion: The Social Dimension of Infant Well-being

Nicola Shelton

When Newman began writing *Infant Mortality: a Social Problem* he would have been unaware that the sustained decline of infant mortality in England and Wales was probably underway and that levels would never return to those he was familiar with in the 1890s. Newman believed that infant mortality was a social problem. Rather than identifying the direct epidemiological causes of infant mortality, he focused on the social conditions that led to the production of diseases. His thesis was that three sets of factors caused infant mortality, attributable to: the mother, the environment and the child. Chapter 13 returns to this thesis and considers it in the light of the research reported in other chapters and elsewhere, but especially the agenda for research set out in Chapter 3. Newman's approach involved social science in several ways. He did not believe in looking for a single cause for high infant mortality and as a social epidemiologist he was concerned to look at the national picture, though he was aware that infant mortality varied greatly over space. By moving away from a mechanistic approach to public health medicine Newman reintroduced people and society into the equation, but this both humanised public health and demonised the people.

Newman's Thesis Revisited

Infant Mortality was organised into eleven chapters (see also Chapter 3 in this book). In Chapter I[1] Newman was concerned with the present position and incidence of infant mortality. The worry was that the infant mortality rate (IMR) was rising at a time of falling general death rates and this despite immense improvements in the social and physical life of the people. In Chapter II, which was concerned with the geographical distribution of infant mortality in Great Britain, Newman pointed out that industrial areas in the North of England (including mining and textile districts) experienced higher than average infant mortality. He argued that the high IMR in the North was not due to climate and topography or habits and customs as these had 'like the climates much in common' and as people 'on the whole treat their infants the same' (Newman, 1906: 26). He then argued that 'mere density of population' was also not one of the main causes of infant mortality, as London did not have the highest rate, and that 'density is an effect of the towns only' (Newman, 1906: 31). On

1 Here again Newman's chapters are referred to using Roman numerals, while chapters of the current volume are denoted by Arabic numerals.

one level Newman was right, there was more to IMRs than mere population density alone (see Chapter 5). It was unclear at this point in the book what he believed to be the problem. In Chapter III Newman concentrated on the fatal diseases of infancy. He was concerned by increasing levels of diarrhoea, pneumonia and bronchitis and prematurity despite the introduction of vaccination, antitoxin and more general improvements in sanitation and public health. Ante-natal influences on infant mortality were considered in detail (Chapter IV). Much blame was placed on the poor health of parents caused by infections, such as tuberculosis and syphilis, and exposure to toxaemias through heavy metal poisoning (e.g. lead), and alcoholism. Long hours at work were also criticised and he recommended 'no laborious toil' for women during the last two months of pregnancy. In Chapter V Newman investigated the role of the occupation of women and infant mortality. He drew on commentary from Sir John Simon that blamed mothers: 'infants perish under the neglect and mismanagement which their mothers' occupation implies'. Newman reported that the Factory Act of 1891, which prohibited mothers from returning to work within a month of giving birth, was not being observed or enforced. Newman also drew on work that would now be called feminist, such as that of Clara E. Collet (1898) who blamed the environment, not the mother.

It was often unclear whether Newman's own beliefs supported or opposed those of the other reformers he mentioned. For example, he reported that the type of work undertaken, especially by pregnant women, was unsuitable. But then he also sympathised with working women appreciating that they did so because their husbands were too sick to work or their wages too low. He criticised the length of the working day for women, the number of days they worked, and the fact that there was no regulation of the housework or homework that they did. He reported that there were problems with sanitation in factories, direct and indirect industrial injury and poor housing. Yet he blamed the high IMR on bad feeding quoting the claims of other reformers that there was 'blind adherence to customs long since deservedly condemned'.

At the end of the nineteenth century there had been a series of hot summers and the numbers of infant deaths (especially diarrhoeal) had soared, so epidemic diarrhoea was an unsurprising focus of Chapter VI (see Chapter 8). Newman blamed the inflated IMR on the combinations of temperature, water supply and sewerage problems aggravated by flies and dirt floors in houses. He reported that the milk supply was contaminated at source, during transport and in storage. He noted that more infants born into higher social classes survived even when bottle-fed; he believed this was due to better hygiene. It was in Chapter VII, on the influence of domestic and social conditions, that Newman revealed his true opinions about the causes of infant mortality. Essentially, he believed that it was not urbanisation, but urbanism that was to blame for the high IMR and he argued that the State needed to protect people from themselves. Perhaps, unsurprisingly given his Quaker origins (see Chapter 2), Newman feared that drink was to blame, especially the proximity to alcohol in urban settings.

In his discussion of infant feeding and management (Chapter VIII) Newman returned to the topic of carelessness and alcoholism. Again, the effects of the environment were dismissed; he reported that overcrowding was just as bad in

Scotland and Ireland, but the IMR was lower and that it was infant feeding that made the difference. (It should be noted that much of the overcrowding in Scotland and Ireland would have been in rural areas.) Newman clearly stated that 'Infant mortality is a social problem concerning maternity' and that ignorance and carelessness directly caused a large proportion of infant deaths (Newman, 1906: 221). He emphasised this with the point that in some families in the worst districts all their children survived and in some families living in the best areas not all their children survived. He believed this was due to differences in knowledge, attention and care. This was clearly in contradiction to his claim at the beginning of the book that behaviour and custom were the same everywhere.

Newman certainly believed that international, ethnic/religious and cultural differences in behaviour existed. He discussed the benefits of prolonged nursing. He suggested that the higher IMR in England due to convulsions was brought on by solid food being introduced into an infant's diet at three months compared to nine months in Scotland. (The complexity and diversity of experiences in Scotland are illustrated in Chapter 7). Newman extolled the virtues of Norway and Sweden whose mothers used extended breastfeeding as a birth spacing strategy and, consequently, also had lower infant mortality, but admitted that they also had the advantages of lower summer temperatures, a smaller urban population and dry soil. The higher IMR in France was blamed on higher rates of wet nursing and infant feeding of pap (flour and water) in combination with poor accommodation and maternal occupations preventing breastfeeding. In Austria, the much higher IMR was blamed on geography and the use of brandy in crèches. In Russia, where rates were much higher again, Newman blamed a variety of problems: early marriage, poor hygiene, a lack of separate beds for infants, a lack of medical aid combined with ignorance and superstition and no breastfeeding because mothers were out at work.

Having more or less dismissed regional variations in this first chapter, Newman made comparisons between Derby, Finsbury and Brighton in Chapter VIII. In Derby the higher IMR was blamed on the use of modified milk, including the use of sweetened condensed milk, which was skimmed. Newman recommended its replacement with diluted cow's milk. In Finsbury, he found that the infants who were healthy at one month were mainly those who were breastfed, and fewer artificially-fed infants had survived to nine months. Newman suggested that diarrhoeal deaths in breastfed infants were due to dust and dirt on their comforters and their mothers' hands. He recommended the use of floor tiles and dustbins. In Brighton he reported that there were large numbers of diarrhoeal deaths due to artificial feeding with condensed milk, which gave less nutrition and was more liable to contamination. In Dorset the IMR and diarrhoeal rates were lower; Newman suggested this was due to the higher breastfeeding rates and that there was no factory employment. He also acknowledged that there had been sanitary improvement in the area and there was little overcrowding; wages were good and there was low unemployment; there was lower alcohol consumption, a good milk supply and medical aid was available.

In his final three chapters, Newman made recommendations to reduce infant mortality. These were based around the mother, the child and the environment. In Chapter IX the focus was the mother. He recommended that there be a re-organisation of the care of mothers. He cited the lack of maternity insurance compared with

Germany, which offered a post- and ante-natal occupational maternity fund. France had a society for nursing mothers for the period of childbirth and first year post-natally. The policy was to 'feed the mother' in dining rooms (Newman, 1906: 260) partly to improve her breastmilk supply, but also because this was cheaper than supplying infant milk. Returning to the argument that ignorance and carelessness were the causes of infant mortality, Newman also recommended the education of mothers. Education should involve the instruction of mothers, the appointment of medically-trained lady health visitors, who would act as friends and not inspectors, offering advice and sympathy with tact (see Chapter 10 for a discussion of their work), and the education of (school) girls in domestic hygiene. He thought that written instruction was not successful due to the low literacy levels. Despite having criticised maternal employment, in his recommendations he described it as 'a necessary evil' suggesting that mothers in employment should be offered support rather than being prevented from working. He recommended extending the Factory Act of 1861 to improve workplace sanitation and increase the provision of crèches. He also recommended increasing the period of maternity leave from four weeks to three months post-natally, and one month ante-natally to bring Britain into line with common practice in Switzerland and Germany. Newman also suggested the introduction of maternity pay.

In Chapter X Newman focused on the child. He recommended improved birth and other child-related registrations, including childcare, and improved feeding. He commented that artificial feeding occurred because mothers were unable or unwilling or because they found breastfeeding impractical. As an alternative, he recommended pure sterilised (cow's) milk and improvements in the cleanliness of milking. He gave examples of the positive benefits of improving artificial feeding for infant mortality. Though sterilised milk did not prevent diarrhoea, weight loss was less than with other methods of artificial feeding.

Despite having focused much on the behaviour and attitude of mothers, Newman did recommend improvements to the environment in his final chapter. He recommended urban cleanliness and, once again, the control of the milk supply, improved sanitation in factories and homes, and the supervision of mothers' employment in dangerous trades, with prohibition of employment before and after confinement. He drew upon the *Report on the Health of Sheffield* (1904) suggesting improvements for urban cleanliness. These included abolishing privy middens, providing water closets, emptying dustbins, repairing defective drains and sewers, paving of streets and yards, and increased street cleaning. To control the milk supply, he recommended that milk production should be clean and sanitary, dairy cows should be healthy and that they should undergo veterinary inspection. He recommended improved standards in the sale and distribution including the use of bottling and that, as in the United States; pasteurization should be replaced by sterilisation.

Factors that Led to the Improvement in Infant Survival

It is important to appreciate the extent of infant mortality decline since 1906 and the changing set of factors that were probably responsible. Newman's trinity of mother, child and environment certainly helps to begin and order the account, but it will not be sufficient to complete the story. We need to remind ourselves of the remarkable improvement in the fortunes of the newly born. In 1901, and throughout the Victorian era, the average infant born alive in England and Wales had an 85.0 per cent chance of surviving to the end of his or her first year. In 2001 the equivalent chance was 99.5 per cent. This can be expressed in another way; the risks facing Victorian infants were 30 times worse than those experienced by today's babies (see Chapter 11). Both of these perspectives on such a remarkable transformation are illustrated in Figure 13.1. It must also be recalled that England and Wales was not alone in Western Europe, most other countries followed roughly the same trajectory and now have IMRs of about 5 per 1,000 live births, or less (Schofield *et al.*, 1991; Bideau *et al.*, 1997).

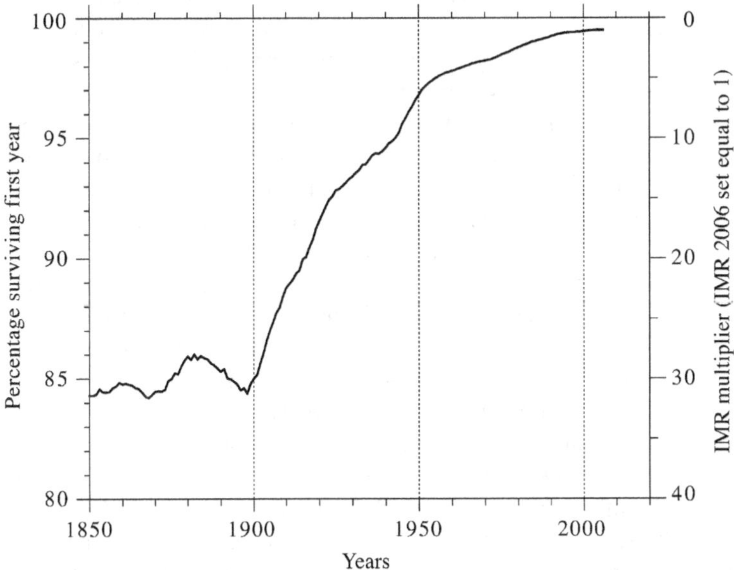

Figure 13.1 Infant survival percentages and the relative risk of dying in infancy: England and Wales, 1850-2006

At first glance the post-1901 infant survival percentages illustrated in Figure 13.1 appear to form a continuous, almost straight-line series. But on closer examination, some differences can be detected. Certainly the rate of improvement slowed down during the second half of the twentieth century. During the first half of the century it was rapid and seems not to have been affected either by the First World War or the Great Depression of the 1930s. It should also be noted that both maternal and late-foetal (stillbirth) mortality did not begin their secular declines until the late 1930s; that is nearly forty years after infant mortality began its decline (Loudon, 1992). Also, the decline of infant mortality lagged some forty years behind the decline of early childhood mortality (ages 1 to 4) and that of mortality among children in general. With Newman, we need to ask why infant survival remained low and relatively unchanged (although geographically and socially variable) in the Victorian era, and then how could it have changed so dramatically in the twentieth century.

Although Newman gave relatively little attention to the environmental factor, it seems most likely that the poor (especially urban) sanitary environment was responsible for the maintenance of high infant mortality (via the post-neonatal component) during the nineteenth century. As far as infants were concerned, the benefits of the sanitary revolution came rather late but once in place, they had a profound and sustained impact. Newman was probably right to direct the attention of his Edwardian contemporaries away from the obsession of earlier generations with sewerage and water supply (see Chapters 6 and 8 for example).

Partly due to Newman's social rather than biological-medical approach, there were several areas where he did not focus on potential changes to improve infant welfare. Many things that have improved could have caused the changes he was hoping for. The biggest of these were improved medical care and the decline in poverty (see Chapters 3 and 11). Especially since the introduction of the National Health Service, many medical interventions have been aimed at maternal and infant welfare. These include immunisation and antibiotics for the control of infectious disease, and prescription medication and dental care made free of charge to children, pregnant women and new mothers. Improved maternal care has also come in the form of obstetric care, ante- and post-natal screening and monitoring. The reduction in absolute poverty during the late twentieth century came combined with technological change, made washing facilities and heating accessible and affordable. Improved maternal and infant diets with access to safer, better quality food led to increased disease resistance and decreased disease exposure. Targeted financial assistance for families with children, such as family allowance introduced in 1945, would have helped to reduce poverty (Department for Work and Pensions, 2002).

However, apart from antisepsis it is difficult to see exactly what new forms of effective medical care were made available before 1940. Chapter 10 has, for example, shown how difficult it is to evaluate the impact of health visiting on infant survival chances, although it is hard to believe that such programmes were anything other than positive in their influence. It may not be a coincidence that the life chances of general practitioners only became superior to those of the general public around 1900. During the nineteenth century the medical knowledge of doctors was insufficient to guarantee them higher life expectancies than other men of equivalent ages (Woods, 1996). But in the twentieth century the benefits became cumulative. Infants were delivered with more skill, they were better fed, better housed, better cared for, their progress was more effectively monitored and, what is more, there were, on average, substantially fewer of them per mother. All of these factors combined to drive survival chances upwards. However, with the secular decline being such a multi-factor phenomenon, the task of researchers seeking to unravel the various causes is far from straightforward, as is witnessed in the varied contributions to this volume. At least Newman's focus on mother and child appears to have been partially justified.

Newman Today: The Inequality Problem and its Implications for Policy

We have seen that there have been considerable changes in the level of infant mortality during the twentieth century. This has occurred in all social groups. Levels in the United Kingdom are now amongst some of the lowest in the world. Does inequality still remain? In order to answer this question we must decide how to measure inequality across time, space and between social groups. The challenge in measuring inequality across time is two-fold. Due to the massive decrease in IMRs between the nineteenth and the twenty-first century it is difficult to argue that the differences between IMRs of 200 and 100 per 1,000 could be found in a society with half the inequality than that between rates of say 4 and 1 per 1,000. Also the absolute numbers of infants experiencing the worst levels of mortality has declined, partly due to the significant reduction in birth rates in the last century. Statistically as death rates tend toward zero, inequalities can tend towards infinity.

Table 13.1 helps to illustrate just how complicated, yet extremely important, can be these distinctions between perspectives focusing on absolute mortality rates and those employing mortality differentials measured by ratios (Woods and Williams, 1995; Woods, 2004). It deals with the age components of infant mortality, especially neonatal and post-neonatal, but it also considers late-foetal mortality (stillbirths) and average birthweight. The data on occupations (mainly father's) have been coded and combined into the standard six social classes used in Britain. The final column gives the percentage of total live births in each of the social classes. The bottom row shows the general average for birthweight (in grams) and the various mortality rates (all per 1,000). These are for the coded population, a sample of no more than 10 per cent of the total. (Hence the difference between the IMR where the father's occupation has been coded and that where the mother's has.) The ratios in the body of Table 13.1 compare social classes II to V with social class I, which has been set at 1,000. The mortality rates are all extremely low by historical and current international standards, yet they are not negligible. The differentials appear considerable, however. Why should infant mortality be 66 to 71 per cent higher in social class V (unskilled manual workers) than social class I (professionals)? Although Table 13.1 helps to confirm that inequalities in life-chances to still persist, it also reveals that social class V is something of an anomaly. In terms of health and survival risks it has come to be seen as a 'trailing class'. It comprises a relatively small group of people who are especially disadvantaged as adults and whose children also suffer in consequence. There is a widening gulf between IIIM, the largest class, and V, but the gap between IIIM and I continues to narrow as overall mortality declines further. This 'trailing class' phenomenon has particularly important implications for the targeting of health policy and for social exclusion in general.

Regardless of the level of inequality, the differences between mortality rates in the best and worst areas as shown in Chapter 11, or the richest and poorest groups, (these may not always be interchangeable), highlight the fact that potentially avoidable infant deaths still occur due to social inequity 100 years after Newman identified infant mortality as a social problem. The remainder of this chapter discusses Newman's legacy in terms of the ways in which contemporary policy relates to his agenda for prevention, and how contemporary debate on the subject of the social

Table 13.1 Early-age mortality differentials by social class: England and Wales, early 1990s

Social class	Birthweight	Stillbirth	Neonatal	Post-neonatal	Infant (father)	Infant (mother)	%
I	1,000	1,000	1,000	1,000	1,000	1,000	8
II	994	1,000	1,129	1,000	1,067	1,043	26
IIIN	988	1,200	1,258	1,231	1,222	1,000	11
IIIM	978	1,200	1,323	1,385	1,311	1,106	34
IV	971	1,378	1,484	1,538	1,467	1,255	15
V	968	1,733	1,581	2,154	1,711	1,660	6
Total	3,370	5.5	4.1	1.8	5.9	6.5	100

Note: For explanation see text
Source: Derived from Botting (1997): Tables 7.1, 7.2, 7.3, 7.4, 7.5

problem of infant mortality (more widely infant health inequalities and morbidity) differ in their focus. It considers whether blaming mothers/parents is supposed to help, by virtue of making them responsible, and why it is still seen as the first step to educating them. Further, it asks why knowledge is perceived to be so partial and so divided by social class.

If Newman had written his book today, how would his approach to the social problem that is infant mortality have been different? Certainly, his audience would have changed. Newman was concerned about pregnant women and nursing mothers' exposure to environmental hazards. Conditions in the workplace have significantly improved and Government legislation to protect the health of childbearing women is extensive (Health and Safety Executive, 2003). Women are no longer at the mercy of their employer. Exposure during pregnancy to lead and mercury is harmful, as Newman rightly pointed out. Risk assessment in the workplace now covers this and other hazards, and changes are made to the work of women to prevent their exposure. Newman might be less pleased about the current policy towards alcohol consumption as it is more tolerant than the official attitude in his day. Mothers are warned about the dangers of alcohol and smoking in pregnancy. In Britain today, pregnant women are advised to stop smoking altogether (Department of Health, 1998), but Department of Health policy recommends that they may drink in moderation, defined as 1 to 2 units per week (that is 2 glasses of wine, 1 pint of beer or 2 measures of spirits) (Department of Health, 2006). These limits are set despite the fact that all alcohol goes into the baby's bloodstream and there is no evidence that this is not harmful (Konovalov *et al.*, 1997). In several other countries the policy is one of zero alcohol consumption for pregnant women. In the United States, for example, there has been a more stringent proposal that recommends complete abstinence for women who may become pregnant, as well as those who already are (Surgeon General, 2005; International Center for Alcohol Policies, 1999).

Newman wanted mothers to be educated through the appointment of lady health visitors and domestic hygiene teaching for girls. Both of these recommendations were adopted nationally. Marriage and parenting skills and personal hygiene now form part of the school National Curriculum for both boys and girls, although domestic science in schools was aimed primarily at girls until the 1970s. Newman encouraged local authorities to increase health-visiting schemes. Though these are still run at local level they are now funded centrally. Newman would probably be pleased that interventionist models of medical provision that beleaguered his work towards the end of his career (see Chapter 2) have been supplemented, if not replaced, by a return to the provision of public health information and education. The problem remains, how well is this information targeted at and received by different social groups since the education and instruction of (especially lower social class) mothers remains a thorny issue.

All pregnant women in Britain are offered free ante-natal education, but this concentrates mainly on labour and the initiation of breastfeeding. Mothers are visited for 10 to 14 days by a community midwife and then regularly by a health visitor for up to six months (National Collaborating Centre for Women's and Children's Health, 2003). Plans to increase post-natal midwifery support to between one and up to three months have been suggested in the Maternity Standard National Service Framework for Children, Young People and Maternity Services (Department of Health, 2004). Though the Acheson Report (Acheson, 1998) identified health visitors as the key workers providing services to mothers and babies, that evidence stressed the importance of this early intervention to outcomes in later life. The regulatory body for health visitors (CPHVA) has criticised that the recommendation made in the Acheson Report to expand health-visiting services, has not been taken on board and incorporated into policy and that health visiting is being marginalised by midwifery in the National Service Framework for Children, Young People and Maternity Services (CPHVA, 2005).

Recent plans in Scotland to stop health visiting after six weeks to higher social class mothers who are not experiencing problems, have also been heavily criticised (Hardie, 2005). The idea that high social class families do not need health visiting after six weeks suggests a view that they are more competent to deal with problems that might occur after this period than those who are poorer, not just that they will have fewer problems. The charge of ignorance among poor mothers still persists. A similar example is found in the guide for health professionals produced by the Maternity Alliance with funding from the Department of Health, on how to give effective healthy eating advice to disadvantaged pregnant women, as part of the Government's 'Healthy Start' scheme (Maternity Alliance, 2005). Again the suggestion is that ignorance remains in the lower social classes.

Newman recommended increased maternity leave (to give mothers a chance to breastfeed) and maternity pay (so they could afford to stay at home and continue to breastfeed). He had been trying to persuade employers to introduce maternity benefits and maternity leave schemes whereas these are now largely the responsibility of central government. Maternity leave was introduced in England and Wales by the time Newman was writing, and in 1919 the ILO Convention on Maternity Protection led to the establishment of maternity benefits for working mothers (International Labour Organization, 1919; Kammerman, 2000). Since 2003, six months of paid leave (at 90 per cent full pay for 6 weeks and £106 per week thereafter) are available to women in the United Kingdom who have been with an employer for at least a year before the birth of the child and six months of unpaid leave for those with an employer less than one year (Department of Trade and Industry, 2005). Contemporary international comparisons have shown that an extension in paid maternity leave is associated with lower infant mortality, whereas an extension in unpaid leave does not have a beneficial effect (Ruhm, 2000; Tanaka, 2005).

The benefits of breastfeeding have been discussed in Chapter 12. In 2003 a WHO policy was introduced that recommended increased exclusive breastfeeding (that is no artificial milk, water or other food) for a duration of from four to six months for all infants (World Health Organisation, 2003). This was adopted by the Department of Health for England and Wales in 2004, though not as yet in Scotland (Shelton, 2005). This, however, is not currently matched by six months paid post-natal maternity leave. At the request of the mother, maternity leave can be commenced from 11 weeks before the expected date of delivery until the day of delivery. But if the mother takes one day or more of maternity-related sick leave in the last four weeks prior to the expected date of delivery, maternity leave commences from that day (Department of Trade and Industry, 2005). In this situation, a mother would have her post-natal maternity leave reduced to just 22 weeks. Plans to increase paid maternity leave to nine months are due to be implemented in 2007 (Department of Trade and Industry, 2005a). Only three weeks of post-natal maternity leave is compulsory and this despite recommendations that women having a caesarean section (currently over 20 per cent of deliveries (Department of Health, 2005)) should refrain from strenuous activities until they are fully recovered (National Institute for Clinical Excellence, 2005). The short period of compulsory post-natal leave may encourage women (perhaps especially those who are self-employed) to return to work before they are fully fit and at the expense of the best interests of the baby. It has been found in the United States that early maternal return to work (within six weeks of birth) has negative effects on child health outcomes (Berger et al., 2002).

Nor does the WHO's recommendation on the extension of exclusive breastfeeding necessarily help to reduce social inequality. Specific breastfeeding initiation interventions aimed at lower income/social class mothers are usually required to do this (Dyson, McCormick and Renfrew, 2005). No mother can be breastfeeding at six months who did not even initiate it after birth. In 2000, 91 per cent of mothers whose partners were in social class I initiated breastfeeding compared to just 57 per cent of mothers whose partners were in social class V, and 50 per cent of mothers who did not have a partner or whose partner's occupation was unclassified (Infant Feeding Survey, 2000; see also Table 13.1). Few women remained breastfeeding at four or six months in 2000 (not unsurprising given statutory maternity leave was then only 18 weeks). Extended breastfeeding (in addition to supplementation with other food) is now also recommended after six months to over two years (World Health Organisation, 2003), but only unpaid maternity leave is available for up to a further six months and only to those women who have been with the same employer for at least a year prior to the birth. Those on low incomes or in families where the mother is the sole or major wage earner may not be able to afford the luxury of such a long period without pay, thus creating a potential social inequality for infant health. Plans to introduce extended paternity leave (increasing the current two weeks paid leave to six months unpaid leave) have met criticism from the Confederation of British Industries because of their potential negative impact especially on small businesses (Politics.co.uk, 2005). The value of men in the wage economy is still clearly viewed differently to that of women some 100 years after Newman described the labour force participation of women (with children) as a necessary evil. Given that most childbearing women will average two pregnancies, at least two years outside of the labour force is acceptable, but a total of 12 months is viewed as unacceptable for men.

Newman's recommendation to increase the regulation of childcare was adopted. Local authorise were responsible for this until 2001 when OFSTED, a national education monitoring body, took over (OFSTED, 2003). Childcare registration is required and the quality of childcare is monitored, but there has been no government policy to encourage the introduction of workplace childcare. Some employers do offer workplace crèche facilities and in certain cases they have discounted these, especially where it is hard to attract staff (e.g. London). As an alternative, some employers have offered schemes to discount childcare payments to registered childcare providers through salary sacrifice (Her Majesty's Revenue and Customs, 2004). In 2005 central government introduced tax relief towards registered childcare for all employees and in 2001 also increased benefits for those families who are working and paying for registered childcare, but are on low incomes (Inland Revenue, 2005). Government and often employers' policies now encourage women with children to participate in the labour market.

Newman's recommendation that a good, clean, affordable milk supply for women who do not breastfeed should be assured has also (largely) been accomplished. Despite occasional alarming reports on contaminated formula milk (e.g. with botulism), generally the supply is 'safe' and infant mortality from the use of formula in developed countries is rare (Scientific Panel on Biological Hazards, 2004; Department of Health, 2005). However, its nutritional content remains somewhat in question. Powdered formula milk requires sterile preparation with the addition of clean, cooled boiled water. Formula feeds which have been incorrectly made up put the infant at risk of either dehydration or under-nourishment (Renfrew *et al*, 2003). Also, formula milk does not vary in its composition in the way that breastmilk does, especially during the first six months of the infant's life. Recent work has recommended modifying the traditional National Center for Health Statistics/World Health Organisation curves of weight gain for infants (which were based on formula-fed infants). Breastfed infants have been shown to have slower weight gain after the first few months of feeding compared to formula-fed infants (de Onis and Onyango, 2003). Use of the existing curves may have encouraged recommending to mothers that they supplement or replace breastmilk with formula milk since babies would not appear to be developing in the correct manner. There is an increased risk of obesity in later life among formula-fed infants (Bandolier, 2005).

Until very recently, poorer mothers (mainly those on income support) in England and Wales were entitled to vouchers for free powdered formula milk, a scheme that had been in place since 1940. Breastfeeding mothers were offered vouchers to a much lower value than the formula ones for fresh milk for themselves (Department of Health, 2002). This scheme was replaced in 2005 with the Healthy Start Scheme to supply poor mothers with fresh milk, formula milk or vouchers for healthy food (see Chapter 11). This initiative was in response to the publication of the Acheson

Report on inequalities in health (Acheson, 1998). Given that this policy suggests that mothers on low incomes cannot decide on their own nutrition or what constitutes healthy food, they may feel disempowered by the move. It harks back to Newman's opinions on the ignorance of mothers. He had suggested feeding poor mothers and his recommendation has now come into force some 100 years later.

As was discussed above, the targeting of policies at particular social groups (e.g. lower social class women, or poor mothers on benefits) has been one way of focusing services both historically (see Chapter 9) and today without closely identifying individuals. Following the Acheson Report on inequalities in health (Acheson, 1998) the popularity of delivering services to geographically defined 'communities' was revitalised. Children were identified in the report as a group on which services should be focused. SureStart was developed as a government funded area level initiative which aims to achieve better health and well-being related outcomes for children, parents and communities by increasing the availability of childcare for all children; improving and emotional development of young children and supporting parents as both parents and in their aspirations towards employment. The aim was for this to be achieved, first, by helping service development in disadvantaged areas alongside financial help for parents to afford childcare and, second, by rolling out the principles driving the SureStart approach to all services for children and parents (SureStart, 2005). The SureStart Programme currently comprises early education for all with free part-time early education for three and four year olds; more and better childcare (at least 250,000 new childcare places by March 2006); the development of children's centres which offer childcare and other support services, where they are needed most (defined as being the most deprived wards) and other locally developed and implemented programmes. All children aged less than four and their parents who are resident in SureStart areas are eligible to receive the services provided regardless of parents' social status. Examples of the positive impact of local SureStart programmes include improvements in children's behaviour, language development, improved interactions between mothers and children and a general improvement in family support services.

Newman's legacy was that many of his recommendations have been developed into arguably successful policies. Unfortunately, casting blame on mothers and the need for their education remain themes of current policy. Knowledge of good motherhood is still perceived as partial and divided by social class. Infant health and welfare is now also frequently monitored at the national level, and though IMRs are very low, infant health remains a social concern. Statistics, including delivery type, breastfeeding initiation and immunisation uptake are routinely collected both at national level and also in sample surveys such as the Health Survey for England, the Scottish Health Survey and the Infant Feeding Survey (Blake, Herrick and Kelly, 2003; Shelton, 2005; IFS, 2000) and in the cohort studies discussed in Chapter 12. Risk factor data, such as income, single parenthood, ethnicity and mother's age, are also collected and the health of infants remains a social concern today. Regrettably, a small number of potentially avoidable infant deaths and large inequalities in infant health still occur in Britain due to social inequity 100 years after Newman identified infant mortality as a social problem.

References

Abel, E. and Hannigan, J. (1995) 'Maternal risk factors in fetal alcohol syndrome: provocative and permissive influences' *Neurotoxicology and Teratology* 17: 445-462.

Acheson, D. (1998) *Independent Inquiry into Inequalities in Health* (London: The Stationery Office).

Adair, R. (1996) *Courtship, Illegitimacy and Marriage in Early Modern England* (Manchester: Manchester University Press).

Albermann, E. (1991) 'Are our babies becoming bigger?' *Journal of the Royal Society of Medicine* 84: 257–260.

Albermann, E. (1994) 'Low birthweight and prematurity', in Pless, I. (ed) *The Epidemiology of Childhood Disorders* (Oxford: Oxford University Press): 49-65.

Alexander, M. (1904) *Some Kensington Problems* (London).

Anderson, J. (1894-95) 'An infant life table' *Public Health* 7: 396-398.

Anon. (1870) 'Report of the infant mortality committee' *Transactions of the Obstetrical Society of London* 9: 132-149.

Anon. (1894-95) 'The prevention of infantile mortality' *Public Health* 7: 297-298.

Ansell, C. (1874) *On the Rate of Mortality at Early Periods of Life, the Age at Marriage, the Number of Children to a Marriage, the Length of a Generation, and Other Statistics of Families in the Upper and Professional Classes* (London: National Life Assurance Society).

Apple, R. (1995) 'Constructing mothers: scientific motherhood in the nineteenth and twentieth centuries' *Social History of Medicine* 8: 161-178.

Armstrong, D. (1986) 'The invention of infant mortality' *Sociology of Health and Illness* 8: 211-232.

Armstrong, D. (2002) *A New History of Identity. A Sociology of Medical Knowledge* (Basingstoke: Palgrave Macmillan).

Armstrong, W. (1988) *The Farm Workers: a Social and Economic History, 1770-1850* (London: Batsford).

Ashby, H. (1884) 'Infant mortality' *Transactions of the National Association for the Promotion of Social Science*: 497-508.

Ashby, H. (1922) *Infant Mortality* (Cambridge: Cambridge University Press).

Baird, D. (1960) 'The evolution of modern obstetrics' *The Lancet* II: 557-564; 609-614.

Baker, D., Taylor, H., Henderson, J. and ALSPAC Study Team (1998) 'Inequality in infant morbidity: causes and consequences in England in the 1990s' *Journal of Epidemiology and Community Health* 52: 451-458.

Ballantyne, J. (1891) *An Introduction to the Diseases of Infancy* (Edinburgh: Oliver and Boyd).

Ballantyne, J. (1892) *The Diseases and Deformities of the Foetus: An Attempt Towards a System of Ante-Natal Pathology* (Two volumes) (Edinburgh: Oliver and Boyd).

Ballantyne, J. (1902, 1904a) *Manual of Antenatal Pathology and Hygiene. Volume I: The Foetus* (1902). *Volume II: The Embryo* (1904). (Edinburgh: William Green and Sons).

Ballantyne, J. (1904b) *Essentials of Obstetrics* (Edinburgh: William Green and Sons).

Ballantyne, J. (1905) *Essentials of Gynaecology* (Edinburgh: William Green and Sons).

Ballantyne, J. (1914) *Expectant Motherhood: Its Supervision and Hygiene* (Edinburgh: Cassell and Co.).

Bandolier (2005) *Can Breastfeeding Help Prevent Overweight Children and Adolescents?* http://www.jr2.ox.ac.uk.

Barker, D. (1994) *Mothers, Babies and Health in Later Life* (London: BMJ Publishing Group).

Barker, D. (1995) 'Fetal origins of coronary heart disease' *British Medical Journal* 311: 171-174.

Barker D., Forsén T., Uutela A., Osmond C., and Eriksson J. (2001) 'Size at birth and resilience to effects of poor living conditions in adult life: longitudinal study' *British Medical Journal* 323: 1-5.

Barker, T. and Drake, M. (eds) (1972) *Population and Society in Britain, 1850-1980* (London: Batsford Academic and Educational).

Bartley, M., Power, C., Blane, D., Davey Smith, D., and Shipely, M. (1994) 'Birth weight and later socioeconomic disadvantage: evidence from the 1958 British cohort study' *British Medical Journal* 309: 1475-1478.

Baxby, D. (1999) 'The end of smallpox' *History Today* 49: 14-16.

Berger, M. and Waldfogel, J. (2005) 'Maternity leave, early maternal employment and child health and development in the US' *The Economic Journal* 115: 29-47.

Berridge, V. (1979) 'Opium in the Fens in nineteenth-century England' *Journal of the History of Medicine and Allied Sciences* 34: 293-313.

Berridge, V. and Edwards, G. (1981) *Opium and the People: Opiate Use in Nineteenth-Century England* (London: St. Martin's Press).

Berry, B. and Schofield R. (1971) 'Age at baptism in pre-industrial England' *Population Studies* 25: 453-463.

Bevan, A. (1947) *In Place of Fear* (London: William Heinemann).

Beveridge, W. (1942) *Social Insurance and Allied Services* (The Beveridge Report) Part I Parliamentary Papers [Cd 6404] (London: HMSO).

Bevis, T. (1980) *Bring Out Your Dead. Disease and Unsanitary Conditions at March, 1849* (March: Westrydale Press).

Bideau, A., Desjardins, B. and Brignoli, H. (eds) (1997) *Infant and Child Mortality in the Past* (Oxford: Clarendon Press).

Blake, M. Herrick, K. and Kelly, Y. (2002) 'Maternal and infant health vol. 3' in Sproston, K. and Primatesta, P. (eds) *The Health Survey for England 2003* (London, The Stationery Office).

Blumin, S. (1968) 'The historical study of vertical mobility' *Historical Methods Newsletter* 1: 1-13.

Bongaarts, J. (1975) 'A method for the estimation of fecundability' *Demography* 12:

645-660.

Booth, C. (1893-1902) *Life and Labour of the People in London* (Several volumes) (London: Macmillan).

Bouquet, M. (1985) *Family, Servants and Visitors: The Farm Household in Nineteenth and Twentieth Century Devon* (Norwich: Geobook).

Bourgeois-Pichat, J. (1951) 'La mesure de la mortalité infantile' *Population* 6: 233-248, 450-80.

Branca, P. (1975) 'A new perspective on women's work: a comparative typology' *Journal of Social History* 9: 129-153.

British Paediatric Association (1994) 'Statement of the standing committee on nutrition. Is breastfeeding beneficial in the UK?' *Archives of Diseases in Childhood* 71: 376-380.

Bromley, C., Sproston, K. and Shelton, N. (2005) *The Scottish Health Survey 2003* (Edinburgh: Scottish Executive Edinburgh).

Brooks, B. (1986) 'Women and reproduction, 1860-1939', in Lewis, J. (ed) *Labour and Love, Women's Experience of Home and Family, 1850-1914* (Oxford: Blackwell): 149-171.

Brownlee, J. (1917) 'The relation of infantile mortality to mortality in subsequent life' *Journal of the Royal Statistical Society* 80: 222-248.

Brunton, L. (1905) 'The report of the Inter-Departmental Committee on Physical Degeneration' *Public Health* 19: 274-292.

Buchanan, I. (1983) 'Infant mortality in British coal mining communities 1880-1911' (unpublished University of London PhD thesis).

Budin, P. (1900) *Le Nourrisson: Alimentation et Hygiène, Enfants Débiles, Enfants Nés à Terme, Leçons Cliniques* (Paris: Doin) (Translated as *The Nursling* (London: Caxton, 1907)).

Bukowski, R., Ghan, D., Denning, J. and Saade, G. (2001) 'Impairment of growth in fetuses destined to deliver preterm' *American Journal of Obstetrics and Gynaecology* 185: 463-467.

Burnette, J. (1999) 'Labourers at the Oakes: changes in the demand for female day-labourers at a farm near Sheffield during the agricultural revolution' *Journal of Economic History* 59: 41-67.

Bynum, W. (1995) 'Sir George Newman and the American way' in Nutton, V. and Porter, R. (eds) *The History of Medical Education in Britain* (Amsterdam: Rodopi): 37-50.

Cassiday, R., Wright, C., Panter-Brick, C. and Parkinson, K. (2004) 'Do early feeding patterns relate to breast-feeding continuation and weight gain? Data from a longitudinal cohort study' *European Journal of Clinical Nutrition* 58: 1290-1296.

CBI [Confederation of British Industry] (2005) *Government's New Work & Families Bill Underestimates Impact On Smaller Firms* http://www/cbi.org.

Chadwick, E. (1842) *Report on the Sanitary Condition of the Labouring Population of Great Britain* (reprinted in Flinn, M. (ed) (1965) (Edinburgh: Edinburgh University Press).

Chadwick, E. (1860) 'Address' *Transactions of the National Association for the*

Promotion of Social Science: 574-606.

Chadwick, E. (1879) 'The crèche' *The Sanitary Record* 10: 29-31.

Chike-Obi, U., David, R., Coutinho, R. and Wu, S. (1996) 'Birthweight has increased over a generation' *American Journal of Edipemiology* 144: 563-569.

Clark, A. (2003) 'Family migration and infant mortality in rural Kent, 1876-1888' *Family and Community History* 6: 141-150.

Clarke, M. and Mason, E. (1985) 'Leatherwork: a possible hazard to reproduction' *British Medical Journal* 290: 1235-1237.

Cohen, G., Curet, L., Levine, R., Ewell, M., Morris, C., Catalano, P., Clokey, D. and Klebanoff, M. (2001) 'Ethnicity, nutrition and birth outcomes in nulliparous women' *American Journal of Obstetrics and Gynecology* 185: 660-667.

Collaborative Group on Hormonal Factors in Breast Cancer (2002) 'Breast cancer and breastfeeding: collaborative reanalysis of individual data from 47 epidemiological studies in 30 countries, including 50,302 women with breast cancer and 96,973 women without the disease' *Lancet* 260: 187-195.

Collins, J., Derrick, M., Hilder, L. and Kempley, S. (1997) 'Relation of maternal ethnicity to infant birthweight in East London, England' *Ethnicity and Disease* 7: 1-4.

Cone, T. (1980) 'Perspectives in neonatology' in Smith, G. and Vidyasagar, D. (eds) *Historical Review and Recent Advances in Neonatal and Perinatal Medicine* (Ebook, Mead Johnson Nutritional Division).

Cooke Taylor, R. (1874) 'What influence has the employment of mothers in manufactures on infant mortality?' *Transactions of the National Association for the Promotion of Social Science*: 569-584.

Coutts, F. (1911) 'Report of an inquiry as to condensed milks; with special reference to their use as infants' food' *Reports to the Local Government Board on Public Health and Medical Subjects,* N.S. no. 56, Food Reports no. 15 (London: HMSO).

Coutts, F. (1914) 'On the use of proprietary foods for infant feeding' *Reports to the Local Government Board on Public Health and Medical Subjects*, N.S. no. 80, Food Reports no. 20 (London: HMSO).

Coutts, F. (1918) 'Upon an inquiry as to dried milks; with special reference to their use in infant feeding' *Reports to the Local Government Board on Public Health and Medical Subjects,* N.S. no. 116, Food Reports no. 24 (London: HMSO).

Coxon, A. and Fisher, K. (1999) *Criterion Validation and Occupational Classification: The Seven Economic Relations and the NS SEC (*mimeograph, Institute for Economic and Social Research, University of Essex).

Cugalj, N. (1984) 'The prevention of ophthalmia neonatorum: historical background and current issues' in Jelliffe, D. and Jelliffe, E. (eds.) *Advances in International Maternal and Child Health* Vol. 4 (Oxford: Clarendon Press): 11-25.

CPHVA [Community Practitioners and Health Visitors Association] (2003) *The Health Visiting Service* http://www.amicus-cphva.org.

CPHVA [Community Practitioners and Health Visitors Association] (2005) *Response to the National Service Framework for Children, Young People and Maternity Services* http://www.amicus-cphva.org.

Cumming, J. (1874) 'On the neglect of infants in large towns, with some remarks on the crèche system' *Transactions of the National Association for the Promotion of*

Social Science: 723-729.

Curle, B. (1967) 'Some aspects of public health in Kensington in the nineteenth century' (unpublished typescript, Kensington Public Library).

Currie, M. (1998) 'Social policy and public health measures in Bedfordshire, within the national context 1904-1938' (unpublished University of Luton PhD thesis).

Darby, H. (1940) *The Draining of the Fens* (Cambridge: Cambridge University Press).

Darby, H. (1983) *The Changing Fenland* (Cambridge: Cambridge University Press).

Davey Smith, G., Dorling, D. and Shaw, M. (2001) *Poverty, Inequality and Health in Britain 1800-2000: a Reader* (Bristol: Policy Press).

David, R. (2001) 'Commentary: birthweights and bell curves' *International Journal of Epidemiology* 30: 1241-1243.

Davies, C. (1988) 'The health visitor as mother's friend: a woman's place in public health', *Social History of Medicine* 1: 39-59.

Davies, L. (2003) 'Faith Street, South Kirkby - "That troublesome place": infant mortality in a coal-mining community, 1894-1911' *Family and Community History* 6: 121-127.

Davies, W. (1913) 'Statistical comparison of the mortality of breast-fed and bottle-fed infants' *American Journal of Diseases of Children* 5: 234-247.

Davin, A. (1978) 'Imperialism and motherhood' *History Workshop Journal*: 9-65.

Davis, J. (1988) *Reforming London: the London Government Problem 1855-1900* (Oxford: Clarendon Press)

Dawson, I. and Jonas, E. (1982) 'Birthweight by gestational age and its effect on perinatal mortality in white and in Punjabi births: experience at a district general hospital in West London 1967-1975' *British Journal of Obstetrics and Gynaecology* 89: 896-899.

Department of Health (1999) *Reducing Health Inequalities: An Action Report. Our Healthier Nation* (London: Stationery Office).

Department of Health (2001) *£3 Million Boost to Help Pregnant Women Give Up Smoking* http://www.dh.gov.uk/PublicationsAndStatistics/PressReleases.

Department of Health (2002a) *Health and Social Care Reform Bill Reform of the Welfare Food Scheme* http://www.dh.gov.uk.

Department of Health (2002b) *Improvement, Expansion and Reform: the Next 3 Years. Priorities and Planning Framework 2003-2006* (London: Stationery Office).

Department of Health (2004a) *Healthy Start: Reform of the Welfare Food Scheme* Gateway Reference 2730 http://www.dh.gov.uk.

Department of Health (2004b) *Maternity Standard National Service Framework for Children, Young People and Maternity Services* http://www.dh.gov.uk.

Department of Health (2005a) *NHS Maternity Statistics, England: 2003-04* http://www.dh.gov.uk.

Department of Health (2005b) *Healthy Start: Consultation on Draft Regulations* (London: HMSO).

Department of Health (2006) *Alcohol and Health* http://www.dh.gov.uk.

Department of Health and Food Standard Agency (2006) *Revised Guidance on*

Preparation and Storage of Infant Formula Milk http://www.dh.gov.uk

Department of Trade and Industry (2005a) *Working Parents* http://www.dti.gov.uk.

Department of Trade and Industry (2005b) *More Money and Choice for Mums* http://www.gnn.gov.uk.

Department of Work and Pensions (2002) *Neligan's Digest - Chapter 6 Benefits for Children Supplement 41* http://www.dwp.gov.uk.

Dhawan, S. (1995) 'Birth weights of infants of first generation Asian women in Britain compared with second generation Asian women' *British Medical Journal* 311: 86-88.

Digby, A. (1996) 'Poverty, health and the politics of gender' in Digby, A. and Stewart, J. (eds) *Gender, Health and Welfare* (London: Routledge): 67-90.

Digby, A. and Stewart, J. (1996) 'Welfare in context' in Digby, A. and Stewart, J. (eds) *Gender, Health and Welfare* (London: Routledge): 1-31.

Dobson, M. (1992) 'Contours of death: disease and the environment in early modern England' *Health Transition Review* 2 Supplement: 77-95.

Dobson, M. (1994) 'Malaria in England a geographical and historical perspective' *Parassitologia* 36: 35-60.

Dorling, D. (1997) *Death in Britain: How Local Mortality Rates Have Changed, 1950s-1990s* (York: Joseph Rowntree Foundation).

Drake, M. (2003) 'Infant mortality: some family and community approaches' *Family and Community History* 6: 107-112.

Drake, M. (2005) 'The vaccination registers: what are they and what can we learn from them?' *Local Population Studies* 74: 36-54.

Drake, M. (2006) 'Illegitimacy and infant mortality' *Local Population Studies Society Newsletter* 76: 63-69.

Drake, M. and Finnegan, R. (eds) (1994) *Sources and Methods: A Handbook, vol.4, Studying Family and Community History, 19th and 20th Centuries* (Cambridge: Cambridge University Press).

Drake, M. and Razzell, P. (1997) *The Decline of Infant Mortality in England and Wales 1871-1948: A Medical Conundrum* (unpublished interim report).

Draper, E., Abrams, K. and Clarke, M. (1995) 'Fall in birth weight of third generation Asian infants' *British Medical Journal* 311: 876.

Dudfield, T. (1896) *Special Report on the First Quinquennial Census, 1 July 1896* (Kensington).

Dudfield, T. (various) *Kensington MOH Annual Report* (Kensington).

Dunn, P. (2005) 'Sir George Newman MD (1870-1948) and the prevention of perinatal disease' *Archives of Disease in Children Fetal and Neonatal Edition* 90: F278-F280.

Durbach, N. (2000) '"They might as well brand us": working class resistance to compulsory vaccination in Victorian England' *Social History of Medicine* 13: 45-62.

Dwork, D. (1987) *War is Good for Babies and Other Young Children: a History of the Infant and Child Welfare Movement in England 1898-1918* (London: Tavistock Publications).

Dyhouse, C. (1978) 'Working-class mothers and infant mortality in England, 1895-1914' *Journal of Social History* 12: 248-267.

Dyson, L., McCormick, F. and Renfrew, M. (2005) 'Interventions for promoting the initiation of breastfeeding' in *The Cochrane Database of Systematic Reviews 2006, 1 The Cochrane Collaboration* (London: John Wiley & Sons).

Eaton-Evans, J. and Dugdale, A. (1987) 'Effects of feeding and social factors on diarrhoea and vomiting in infants' *Archives of Disease in Childhood* 62: 445-448.

Elliston, G. (1874) *Medical Officer of Health Report for Ipswich* (Ipswich).

Elrington, H. (1922) *Kensington Past and Present* (London: Chelsea Publishing Co.).

Emmett, P., North, K. and Noble, S. (2000) 'Types of drinks consumed by infants at 4 and 8 months of age: a descriptive study' *Pediatrics* 3: 211-217.

Eyler, J. (1979) *Victorian Social Medicine* (Baltimore: Johns Hopkins University Press).

Eyler, J. (1997) *Sir Arthur Newsholme and State Medicine 1885-1935* (Cambridge: Cambridge University Press).

Fang, J., Madhavan, S. and Alderman, M. (1999) 'Low birthweight: race and maternal nativity - impact of community income' *Pediatrics* 103: E5.

Farr, W. (1837) 'Vital Statistics or the statistics of health, sickness, disease and death' (reprinted in Wall, R. (ed) (1974) *Mortality in Mid-Nineteenth Century Britain* (Farnborough: Gregg International)).

Farr, W. (1859) 'On the construction of life-tables, illustrated by a new life-table of the healthy districts of England' *Transactions of the Royal Society* 149: 837-878.

Farr, W. (1864) *English Life Table* (London: HMSO).

Farr, W. (1865) 'On infant mortality, and on alleged inaccuracies of the census' *Journal of the Statistical Society of London* 28: 125-149.

Farr, W. (1866) 'Mortality of children in the principal states of Europe' *Journal of the Statistical Society* 29: 1-35.

Farr, W. [Humphreys, N. (ed.)] (1885) *Vital Statistics* (London: Sanitary Institute of Great Britain).

Ferguson, T. (1958) *Scottish Social Welfare, 1864-1914* (Edinburgh: Livingstone Ltd).

Fildes, V. (1992) 'Breast-feeding in London, 1905-19' *Journal of Biosocial Science* 24: 53-70.

Fildes, V. (1998) 'Infant feeding practices and infant mortality in England, 1900-1919' *Continuity and Change* 13: 251-280.

Finlay, R. (1979) 'The accuracy of the London parish registers 1580-1653' *Population Studies* 32: 95-112

Finlay, R. (1982) 'Distance to church and registration experience' in Drake M. (ed) *Population Studies from Parish Registers* (Matlock: Local Population Studies): 71-84.

Flinn, M., Gillespie, J., Hill, N., Maxwell, A., Mitchison, R. and Smout, C. (1977) *Scottish Population History from the 17th century to the 1930s* (Cambridge: Cambridge University Press).

Fogel, R. (1989) 'The conquest of high mortality and hunger in Europe and America: timing and mechanisms' *Working Paper Series in Historical Factors in Long-run Growth* 16 (Cambridge, Mass.: National Bureau of Economic Research).

Friedman, D., Cohen, B., Mahan, C., Lederman, R., Vezina, R., and Dunn, V. (1993) 'Maternal ethnicity and birthweight among Blacks' *Ethnicity and Disease* 3: 255-

269.

Gairdner, W. (1860) 'On infantile death rates in their bearing on sanitary and social science' *Transactions of the National Association for the Promotion of Social Science*: 632-648.

Galley, C. (1998) *The Demography of Early Modern Towns. York in the Sixteenth and Seventeenth Centuries* (Liverpool: Liverpool University Press).

Galley, C. (2004) 'Social intervention and the decline of infant mortality: Birmingham and Sheffield, c. 1870-1910' *Local Population Studies* 73: 29-50.

Galley, C. and Shelton, N. (2001) 'Bridging the gap: determining long-term changes in infant mortality in pre-registration England and Wales' *Population Studies* 55: 65-77.

Galley, C. and Woods, R. (1999) 'On the distribution of deaths during the first year of life' *Population: an English Selection* 11: 36-59.

Garrett, E. (1998) 'Was women's work bad for babies? A view from the 1911 census of England and Wales' *Continuity and Change* 13: 281-316.

Garrett, E. and Davies, R. (2003) 'Birth spacing and infant mortality on the Isle of Skye, Scotland, in the 1880s; a comparison with the town of Ipswich, England' *Local Population Studies* 71: 53-74.

Garrett, E. and Reid, A. (1995) 'Thinking of England and taking care: family building strategies and infant mortality in England and Wales, 1891 - 1911' *International Journal of Population Geography* 1: 69-102.

Garrett, E., Reid, A., Schürer, K. and Szreter, S. (2001) *Changing Family Size in England and Wales* (Cambridge: Cambridge University Press).

Gergen, P., Fowler, J., Maurer, K., Davis, W. and Overpeck, M. (1998) 'The burden of environmental tobacco smoke exposure on the respiratory health of children 2 months through 5 years of age in the United States: third national health and nutrition examination survey, 1988 to 1994', *Pediatrics* 101: E8.

Gielgud, J. (1992) 'Nineteenth-century farm women in Northumberland and Cumbria: the neglected workforce' (unpublished University of Sussex DPhil thesis).

Gilbert, B. (1965) 'Health and politics: the British Physical Deterioration Report of 1904' *Bulletin of History of Medicine* 39: 143-153.

Gladstone, F. (1969) *Notting Hill in Bygone Days* (London: Bingley).

Glass R. and Bern C. (1994) 'Gastroenteritis' in Pless, I. (ed) *The Epidemiology of Childhood Disorders* (Oxford: Oxford University Press): 211-228.

Glass, D. (1973) *Numbering the People: the Eighteenth-century Population Controversy and the Development of the Census and Vital Statistics in Britain* (Farnborough: D.C. Heath).

Glass, D. and Eversley, D. (eds) (1965) *Population in History* (London: Aldine Pub.).

Glover, J. (1959) 'Newman, Sir George (1870-1948)' in Wickham Legg, L. and Williams, E. (eds) *The Dictionary of National Biography, 1941-1950* (Oxford: Oxford University Press): 624-625.

Glyde, J. (1850) *Moral, Social and Religious Condition of Ipswich in the Middle of the Nineteenth Century* (New Edition (1971) Wakefield: EP).

Godwin, H. (1978) *Fenland, its Ancient Past and Uncertain Future* (Cambridge:

Cambridge University Press).
Golden, J. (1996) *A Social History of Wet Nursing in America: from Breast to Bottle* (New York: Cambridge University Press).
Goldman, L. (1991) 'Statistics and the science of society in early Victorian Britain; an intellectual context for the General Register Office' *Social History of Medicine* 4: 415-433.
Gordon, M. (1902) 'The bacteriology of epidemic diarrhoea and its differential diagnosis from other similar diseases' *The Practitioner* 69: 180-194.
Gould J. and Leroy S. (1988) 'Socioeconomic status and low birthweight: a racial comparison' *Pediatrics* 82: 896-904.
Graham, D. (1994) 'Female employment and infant mortality: some evidence from British towns, 1911, 1931 and 1951' *Continuity and Change* 9: 313-346.
Graunt, J. (1662) *Natural and Political Observations Made upon the Bills of Mortality* (reprinted in Laslett, P. (ed) (1973) *The Earliest Classics: John Graunt and Gregory King* (Farnborough: Gregg International)).
Greenall, R. (2000) *A History of Northamptonshire* (Chichester: Philimore & Co Ltd).
Greenwood, M. (1901) T*he Law Relating to the Poor Law Medical Service and Vaccination* (London: Bailliere, Tindall and Cox).
Greenwood, M. (1948) *Some Pioneers of British Social Medicine* (London: Oxford University Press).
Gregory, I. (1998) 'Putting the past in its place: the Great British Historical GIS' in Carver, S. (ed) *Innovations in GIS 5* (London: Taylor & Francis): 210-221.
Gwinn, M., Lee, N., Rhodes, P., Layde, P. and Rubin, G. (1990) 'Pregnancy, breast feeding, and oral contraceptives and the risk of epithelial ovarian cancer' *Journal of Clinical Epidemiology* 43: 559-568.
Hall, D. and Harding, R. (1985) *Rushden; A Duchy of Lancaster Village* (Rushden: Buscott Publishing).
Hall, E. (2006) 'Aspects of infant mortality: Ipswich 1870-1910' (unpublished Open University PhD thesis).
Hamlyn, B., Brooker, S., Oleinikova, K. and Wands, S. (2002) *Infant Feeding 2000* (London: The Stationery Office).
Hamer, W. (1899) *Report on the Sanitary Condition of Kensington* Official Publication no 454, 10 November 1899 (London: London County Council).
Hammer, M. (1995) '"The building of a nation's health": the life and work of George Newman to 1921' (unpublished University of Cambridge Ph.D. thesis).
Hardie, A. (2005) 'Executive is accused of dangerous penny-pinching as baby visits cut' http://news.scotsman.com.
Harding, S. and Balarajan, R. (2001) 'Longitudinal study of socio-economic differences in mortality among South Asian and West Indian migrants' *Ethnicity and Health* 6: 121-128.
Harding, S. and Maxwell, R. (1997) 'Differences in the mortality of migrants' in Drever, F. and Whitehead, M. (eds) *Health Inequalities: Decennial Supplement* Series DS no.15 (London: The Stationery Office): 108-121.
Hardy, A. (1993) *The Epidemic Streets: Infectious Disease and the Rise of Preventive*

Medicine, 1856-1900s (Oxford: Oxford University Press).

Hardy, A. (1994) '"Death is the cure of all diseases": using the General Register Office cause of death statistics for 1837-1920' *Social History of Medicine* 7: 472-492.

Hardy, A. (2001) *Health and Medicine in Britain since 1860* (Basingstoke: Palgrave).

Harris, B. (2004) *The Origins of the British Welfare State* (Basingstoke: Palgrave).

Health and Safety Executive. (2003) *A Guide: New and Expectant Mothers who Work* (Suffolk: HSE Books).

Heath, H. (1907) *The Infant, the Parent and the State: a Social Study and Review* (London).

Heisey, D. and Nordheim, E. (1995) 'Modelling age-specific survival in nesting studies, using a general approach for doubly-censored and truncated data' *Biometrica* 51: 51-60.

Hendrick, H. (1994) *Child Welfare: England, 1872-1989* (London: Routledge).

Henry, L. (1967) *Manuel de Démographie Historique* (Paris: Droz-EPHE. Geneva).

Her Majesty's Revenue and Customs (2005) *Salary Sacrifice* http://www.hmrc.gov.uk.

Her Majesty's Revenue and Customs (2005) *Help with the Costs of Childcare Information for Parents and Childcare Providers* http://www.hmrc.gov.uk.

Hertz-Picciotto I. (2001) 'Commentary: when brilliant insights lead astray' *International Journal of Epidemiology* 30: 1243-1244.

Hewitt, M. (1958) *Wives and Mothers in Victorian Industry* (London: Rockliff).

Higgs, E. (1987) 'Women, occupations and work in the nineteenth century censuses' *Historical Workshop* 23: 59-80.

Higgs, E. (1995) 'Occupational censuses and the agricultural workforce in Victorian England and Wales' *Economic History Review* 48: 700-716.

Higgs, E. (2004) *Life, Death and Statistics. Civil Registration, Censuses and the Work of the General Register Office, 1836-1952* (Hatfield: Local Population Studies).

Hill, A. (1893) 'On the causes of infant mortality in Birmingham' *Practitioner* 51: 70-80.

Hill, T. (1905) 'Infant mortality' *Public Health* 19: 623-640.

Hills, R. (1967) *Machines, Mills and Uncountable Costly Necessities. A Short History of the Drainage of the Fens* (Norwich: Goose & Son).

HMSO [Her Majesty's Stationery Office] (various) *Annual Reports of the Registrar General* (London, HMSO).

Hoddinott, P. and Pill, R. (1999) 'Qualitative study of decisions about infant feeding among women in east end of London' *British Medical Journal* 318: 30-34.

Holdsworth, C. (1997) 'Women's work and family health: evidence from the Staffordshire potteries, 1890-1920' *Continuity and Change* 12: 103-128.

Hollingsworth, T. (1964) 'The Demography of the British Peerage' *Population Studies* 18 supplement.

Hollingsworth, T. (1977) 'Mortality in the British peerage families since 1600' *Population* 32 numéro special: 323-352.

Hope, E. (1899) 'Observations on autumnal diarrhoea in cities' *Public Health* 13:

660-665.

Hope, E. (1917) *Report on the Physical Welfare of Mothers and Children, Volume 1* (Liverpool: Carnegie United Kingdom Trust).

Horn, P. (1976) *Labouring Life in the Victorian Countryside* (Dublin: Gill and Macmillan).

Howarth, W. (1905) 'The influence of feeding on the mortality of infants' *The Lancet* 2: 210-213.

Howe, G. (1970) *A National Atlas of Disease Mortality in the United Kingdom* 2nd edition (London: Nelson).

Howie, P., Forsyth, J., Ogston, S., Clark, A. and Florey, C. (1990) 'Protective effect of breast feeding against infection' *British Medical Journal* 300: 11-16.

Howkins, A. (1991) *Reshaping Rural England 1850-1925* (London: Collins Harvill).

Huck, P. (1994) 'Infant mortality in nine industrial parishes in northern England, 1813-36' *Population Studies* 48: 513-526.

Hunter, H. (1864) 'Report by Dr. Henry Julian Hunter on the excessive mortality of infants in some rural districts of England' *Parliamentary Papers*, Appendix 14: 454-462.

Hunter, J. (2000) *The Making of the Crofting Community* 2nd edition (Edinburgh: John Donald).

Husbands, W. (1864) 'Infant Mortality' *Transactions of the National Association for the Promotion of Social Science*: 498-508.

Huxley R., Shiell A. and Law C. (2000) 'The role of size at birth and postnatal catch-up growth in determining systolic blood pressure: a systematic review of the literature' *Journal of Hypertension* 18: 815-381.

IFS [Infant Feeding Survey] (2002) *Infant Feeding Survey 2000 A Survey Conducted on Behalf of the Department of Health, the Scottish Executive, the National Assembly for Wales and the Department of Health, Social Services and Public Safety in Northern Ireland* (London: The Stationery Office) http://www.dh.gov.uk.

Inter-departmental Committee on Physical Deterioration (1904) *Report of the Inter-departmental Committee on Physical Deterioration* Vol. 1 Report and Appendix Parliamentary Papers [Cd 2175]; Vol. II List of Witnesses and Minutes of Evidence [Cd 2210]; Vol. III Appendix and General Index [Cd 2186] (London: HMSO).

International Center For Alcohol Policies (1999) *ICAP Reports Government Policies on Alcohol and Pregnancy* http://www.icap.org.

International Labour Organization (1919) *Maternity Protection Convention, 1919 Convention Concerning the Employment of Women before and after Childbirth* http://www.ilo.org.

James, T. (2003) 'Neonatal mortality in Northamptonshire: Higham Ferrers, 1880-1890' *Family and Community History* 6: 129-139.

Jarvelin, M., Elliott, P., Kleinschmidt, I., Martuzz, M., Grundy, C., Hartikainen, A. and Rantakallio, P. (1997) 'Ecological and individual predictors of birthweight in a northern Finland birth cohort 1986' *Paediatric and Perinatal Epidemiology* 11: 298-312.

Johansson, S. and Kasakoff, A. (2000) 'Mortality history and the misleading mean'

Historical Methods 33: 56-58.

Jones, H. (1894) 'The perils and protection of infant life' *Journal of the Royal Statistical Society* 57: 1-98.

Kammerman S. (2000) 'Parental leave policies: an essential ingredient in early childhood education and care policies' *Social Policy Report* 14: 5-15.

Kearns, G. (1988) 'The urban penalty and the population history of England', in Brändström, A. and Tedebrand, L.-G. (eds) *Society, Health and Population* (Stockholm: Almqvist and Wiksell International): 213-36.

Kelly, Y. and Watt, R. (2005) 'Breast feeding initiation and exclusive duration at 6 months by social class: results from the Millennium Cohort Study' *Public Health Nutrition*: 417-421.

Kelly, Y., Nazroo, J., McMunn, A., Boreham, R. and Marmot, M. (2001) 'Birthweight and behavioural problems in children: a modifiable effect?' *International Journal of Epidemiology* 30: 88-94.

Kelly, Y., Watt, R. and Nazroo, J. (2006) 'Racial/Ethnic differences in breastfeeding initiation and continuation in the United Kingdom and comparison with findings in the United States' *Pediatrics* 118: e1428-e1435

Kemmer, D. (1997) 'Investigating infant mortality in early twentieth century using civil registers: Aberdeen and Dundee compared' *Scottish Economic and Social History* 17: 1-19.

Kenwood, D. (1896-97) 'The sickness and mortality due to ignorance, and its remedy' *Public Health* 9: 192-196.

Kintner, H. (1986) 'Classifying causes of death during the late 19th early 20th century: the case of Germany's infant mortality' *Historical Methods*, 19: 45-54.

Kitson, P. (2004) 'Family formation, male occupation and the nature of parochial registration in England c. 1538-1837' (unpublished University of Cambridge PhD thesis).

Kitteringham, J. (1975) 'Country work girls in nineteenth century England' in Samuel, R. (ed.) *Village Life and Labour* (London: Routledge and Kegan Paul).

Kogan, M. (1995) 'Social causes of low birth weight' *Journal of the Royal Society of Medicine* 88: 611–615.

Konovalov, H., Kovetsky, N., Bobryshev, Y. and Ashwell, K. (1997) 'Disorders of brain development in the progeny of mothers who used alcohol during pregnancy' *Early Human Development* 48: 153-166.

Kramer, M. (1987) 'Determinants of low birth weight: methodological assessment and meta-analysis' *Bulletin of the World Health Organisation* 65: 663-737.

Kramer, M. and Kakuma, R. (2002) *The Optimal Duration of Breastfeeding; a Systematic Review* http://www.who.int/entity/nutrition/publicationsKramer, M., Seguin, L., Lydon, J. and Goulet, L. (2000) 'Socio-economic disparities in pregnancy outcome: why do the poor fare so poorly?' *Paediatric and Perinatal Epidemiology* 14: 194-210.

Kramer, M.S., Chalmers, B. *et al.* (2001) 'Promotion of breastfeeding intervention trial (PROBIT): a randomized trial in the Republic of Belarus' *Journal of the American Medical Association* 285: 413-420.

Kunitz, S. (1983) 'Speculations on the European mortality decline' *Economic*

History Review 36: 349-364.

Lambert, R. (1963) *Sir John Simon and English Social Administration* (London: MacGibbon and Kee).

Landers, J. (1993) *Death and the Metropolis. Studies in the Demographic History of London 1670-1830* (Cambridge: Cambridge University Press).

Laslett, P. (1977) 'Long-term trends in bastardy in England' in Laslett, P. (ed) *Family Life and Illicit Love in Earlier Generations* (Cambridge: Cambridge University Press): 102-155.

Lawton, R. (1968) 'Population changes in England and Wales in the later nineteenth century: an analysis of trends by registration districts' *Transactions of the Institute of British Geographers* 44: 55-74.

Lawton, R. (ed) (1978) *The Census and Social Structure* (London: Cass).

Laxton, P. and Williams, N. (1989) ' Urbanisation and infant mortality in England: a long-term perspective and review' in Nelson, M. and Rogers, J. (eds) *Urbanisation and the Epidemiologic Transition* Report from the Family History Group, Department of History Uppsala University 9: 109-131.

Lee, C. (1991) 'Regional inequalities in infant mortality in Britain, 1861-71: patterns and hypotheses' *Population Studies* 45: 55-65.

Levene, A. (2005) 'The estimation of mortality at the London Foundling Hospital, 1741-99' *Population Studies* 59: 87-98.

Lewes, F. (1991) 'The GRO and the provinces in the nineteenth century' *Social History of Medicine* 4: 479-496.

Lewis, J. (1980) *The Politics of Motherhood: Child and Maternal Welfare in England, 1900-1939* (London: Croom Helm).

Lewis, J. (1986) 'The Working Class Wife and Mother and State Intervention, 1870-1918' in Lewis, J. (ed.) *Labour and Love, Women's Experience of Home and Family, 1850-1914* (Oxford: Blackwell): 99-122.

Lewis, J. (1998) '"Tis a misfortune to be a great ladie": maternal mortality in the British aristocracy' *Journal of British Studies* 37: 26-53.

Li, R., Darling, N., Maurice, E., Barker, L., and Grummer-Strawn, L. (2005) 'Breastfeeding rates in the United States by characteristics of the child, mother, or family; the 2002 National Immunisation Survey' *Pediatrics* 115: e31-e37.

London County Council (1904) *Report of the Chief Officer of the Public Control Department as to Crèches or Day Nurseries* (London: LCC).

Loosmore, B. (1996) 'Smallpox' *Nursing Times* 92: 44-45.

Loudon, I. (1991) 'On maternal and infant mortality, 1900-1960' *Social History of Medicine* 4: 29-73.

Loudon, I. (1992) *Death in Childbirth: an International Study of Maternal Care and Maternal Mortality 1800-1950* (Oxford: Oxford University Press).

Loudon, I. (2000) *The Tragedy of Childbed Fever* (Oxford: Oxford University Press).

Lucas, A., Morley, R. and Cole, T. (1998) 'Randomised trial of early diet in preterm babies and later intelligence quotient' *British Medical Journal* 317: 1481-1487.

Lux, A., Henderson, A., Pocock, S. and ALSPAC Study Team (2000) 'Wheeze associated with prenatal tobacco smoke exposure: a prospective, longitudinal study' *Archives of Disease in Childhood* 83: 307-312.

Macfarlane, A., Mugford, M., Henderson, J., Furtado, A. and Dunn, A. (2000) *Birth Counts: Statistics of Childbirth and Pregnancy. Volume 2 – Tables* (London: The Stationery Office).

Mackay, J. (1992) *Kilmarnock* (Alloway: Darvel).

Maddison, A. (2005) *The World Economy: Historical Statistics.* http://www.eco.rug.nl/~Maddison.

Maclean, H. (1989) 'Implications of a health promotion framework for research on breast feeding' *Health Promotion International* 3: 355-360.

Malcolmson, P. (1970) 'The Potteries of Kensington: a study of slum development in Victorian London' (unpublished University of Leicester M. Phil thesis).

Malcolmson, P. (1975) 'Getting a living in the slums of Victorian Kensington' *London Journal* 1: 28-55.

Malcolmson, P. (1981) 'Laundresses and the laundry trade in Victorian England' *Victorian Studies* 24 439-462.

Malcolmson, P. (1986) *English Laundresses* (Urbana: University of Illinois Press).

Malthus, T. (1798) *An Essay on the Principle of Population* (reprinted (1970) Harmondsworth: Penguin).

Marks, L. (1996) *Metropolitan Maternity: Maternal and Infant Welfare Services in Early Twentieth Century London* (Amsterdam: Rodopi).

Marland, H. (1993) 'A pioneer in infant welfare: the Huddersfield scheme 1903-1920' *Social History of Medicine* 6: 25-50.

Marmot, M. (2001) 'Aetiology of coronary heart disease. Fetal and infant growth and socioeconomic factors in adult life may act together' *British Medical Journal* 323: 1261-1262.

Martin, J., Hamilton, B., Ventura, S., Menacker, P. and Park, M. (2002) 'Births: final data for 2000' *National Vital Statistics Reports* 50: 1-102.

Maternity Alliance (2005) *Talking about Food: How to Give Effective Healthy Eating Advice to Disadvantaged Pregnant Women A Publication for Health Professionals* http://www.maternityalliance.org.uk.

McCann, J. and Ames, B. (2005) 'Is docosahexaenoic acid, and n–3 long-chain polyunsaturated fatty acid, required for development of normal brain function? An overview of evidence from cognitive and behavioural tests in humans and animals' *American Journal of Clinical Nutrition* 82:281-295.

McCleary, G. (1905) *Infantile Mortality and Infants' Milk Depôts* (London: P. S. King & Son).

McCleary, G. (1933) *The Early History of the Infant Welfare Movement* (London: Lewis and Co.).

McCleary, G. (1935) *The Maternity and Child Welfare Movement* (London: Kay & Son Ltd.).

McCormick, A., Fleming, D. and Charlton, J. (1995) *Morbidity Statistics from General Practice. Fourth National Study 1991-1992.* Series MB5 no.3 (London: Stationary Office).

McDonald, A. and McDonald, J. (1986) 'Outcome of pregnancy in leatherworkers' *British Medical Journal* 292: 979-981.

McFeely, M. (1988) *Lady Inspectors: the Campaign for a Better Workplace* (Oxford:

Blackwell).

McRae, M. (1850) 'The case for state medical services for the poor: the Highland and Islands' www.rcpe.ac.uk/library/history/highlands.

Mein Smith, P. (1993) 'Mothers, babies, and the mothers and babies movement: Australia through depression and war' *Social History of Medicine* 6: 51-83.

Migone, A., Emanuel, I., Mueller, B., Daling, J. and Little, R. (1991) 'Gestational duration and birthweight in white, black and mixed-race babies' *Paediatric and Perinatal Epidemiology* 5: 378-391.

Miller, S. and Skertchley, S. (1878) *The Fenland Past and Present* (Wisbech: Leach & Son).

Mills, D. (1978) 'The quality of life in Melbourn, Cambridgeshire, in the period 1800-50' *International Review of Social History* 23: 382-404.

Mills, D. (1984) 'The nineteenth century peasantry of Melbourn, Cambridgeshire' in Smith, R. (ed) *Land, Kinship and Lifecycle* (Cambridge University Press, Cambridge): 481-518.

Mills. D. and Mills, J. (1989) 'Occupation and social stratification revisited: the census enumerators' books of Victorian Britain' *Urban History Newsletter*: 63-77.

Millward, R. and Bell, F. (2001) 'Infant mortality in Victorian Britain: the mother as medium' *Economic History Review* 54: 699-733.

Mitchell, B. (1988) *British Historical Statistics* (Cambridge: Cambridge University Press).

Modood, T., Berthoud, R., Lakey, J., Nazroo, J., Smith, P., Virdee, S. and Beishon, S. (1997) *Ethnic Minorities in Britain: Diversity and Disadvantage* (London: Policy Studies Institute).

Mooney, G. (1994a) 'Did London pass the "sanitary test"? Seasonal infant mortality in London, 1870-1914' *Journal of Historical Geography* 20: 158-174.

Mooney, G. (1994b) 'Stillbirths and the measurement of infant mortality rates c. 1890-1930' *Local Population Studies* 53: 42-52.

Mooney, G. (1997) 'Professionalization in public health and the measurement of sanitary progress in nineteenth-century England and Wales' *Social History of Medicine* 10: 53-78.

Moore, S. (1904) *County Borough of Huddersfield. Report of the Medical Officer of Health on Infantile Mortality* (Huddersfield: Huddersfield Health Committee).

Moore, W., Bannister, R., Ward, B., Hillier, V. and Bamford, F. (1995) 'Fetal and postnatal growth to age 2 years by mother's country of birth' *Early Human Development* 42: 111-121.

Morgan, N. (2002) 'Infant mortality, flies, and horses in later-nineteenth-century towns, a case study of Preston' *Continuity and Change* 17: 97-132.

Nash, J. (1905) 'The waste of human life' *Journal of the Royal Sanitary Society* 26: 494-498.

National Collaborating Centre for Women's and Children's Health (2003) *Antenatal Care Routine. Care for the Healthy Pregnant Woman Clinical Guideline* (London: RCOG Press).

National Collaborating Centre for Primary Care (2005) *Postnatal Care: Routine Postnatal Care of Women and their Babies Full Guideline First Draft for*

Consultation http://www.nice.org.

National Institute for Health and Clinical Excellence (NICE). (2004) *CG13 Caesarean Section: Information for the Public* http://www.nice.org.uk.

Nazroo, J. (1998) 'Genetic, cultural or socio-economic vulnerability? Explaining ethnic inequalities in health' *Sociology of Health and Illness* 20: 710-730.

Nazroo, J. (2001) *Ethnicity, Class and Health* (London: Policy Studies Institute).

Nazroo, J. (2003) 'The structuring of ethnic inequalities in health: economic position, racial discrimination and racism' *American Journal of Public Health* 93: 277-284.

Nazroo, J. (2004) 'Ethnic disparities in aging health: What can we learn from the United Kingdom?' in Anderson, N. Bulatao, R. and Cohen, B. (eds) *Critical Perspectives on Racial and Ethnic Differentials in Health in Late Life* (Washington, D.C.: National Academies Press): 677-702.

Nazroo, J., Becher, H., Kelly, Y. and McMunn, A. (2001) 'Children's health' in Erens. B., Primatesta, P. and Prior, G. (eds) *Health Survey for England - The Health of Minority Ethnic Groups '99 Volume 1: Findings.* Series HS no.9 (London: The Stationary Office).

Newmach, W. (1861) 'The progress of economic science during the last thirty years' *Journal of the Statistical Society of London* 24: 451-471.

Newman, G. (1895) *On the History of the Decline and Final Extinction of Leprosy as an Endemic Disease in the British Islands* (London: New Sydenham Society).

Newman, G. (1899) *Bacteria* (London: John Murray).

Newman, G. (1903) *Report on the Milk Supply of Finsbury* (London: Thomas Bean).

Newman, G. (1904) 'Editorial notes - The report of the physical deterioration committee' *Friends' Quarterly Examiner* 38: 441-452.

Newman, G. (1905) *A Special Report on an Infants' Milk Depot* (London: Thomas Bean).

Newman, G. (1906) *Infant Mortality: a Social Problem* (London: Methuen).

Newman, G. (1907a) *Special Report on Infant Mortality in Finsbury* (London: Thomas Bean).

Newman, G. (1907b) *The Health of the State* (London: Headley Brothers).

Newman, G. (1918) *Some Notes on Medical Education in England* (London: HMSO).

Newman, G. (1920) *Public Opinion in Preventive Medicine* (London: Ministry of Health).

Newman, G. (1923) *Recent Advances in Medical Education in England* (London: HMSO).

Newman, G. (1924) *Public Education in Health* (London: HMSO).

Newman, G. (1925) *On the State of Public Health. Annual Report of the Chief Medical Officer of Health for the Year 1924* (London: HMSO).

Newman, G. (1927) *Interpreters of Nature* (London: Faber & Gywer).

Newman, G. (1928) *Citizenship and the Survival of Civilization* (New Haven: Yale University Press).

Newman, G. (1930) 'The application of Quaker principles in medical practice'

Friends Quarterly Examiner 64: 57-70.

Newman, G. (1931) *Health and Social Evolution* (London: Allen & Unwin).

Newman, G. (1932) *The Rise of Preventive Medicine* (London: Oxford University Press).

Newman, G. (1939) *The Building of a Nation's Health* (London: Macmillan).

Newman, G. (1941) *English Social Services* (London: Collins).

Newman, G. (1946) *Quaker Profiles* (London: Bannisdale Press).

Newman, G. (various) *Annual Report of the Medical Officer of Health for the London Borough of Finsbury* (London: Thomas Bean).

Newsholme, A. (1889) *The Elements of Vital Statistics in their Bearing on Social and Public Health Problems* (London: George Allen and Unwin).

Newsholme, A. (1891) 'The vital statistics of Peabody buildings and other artisans' and labourers' block dwellings' *Journal of the Royal Statistical Society* 54: 70-99.

Newsholme, A. (1899) 'A contribution to the study of epidemic diarrhoea' *Public Health* 12: 139-213.

Newsholme, A. (1901) 'The epidemiology of scarlet fever in relation to the utility of isolation hospitals' *Transactions of the Epidemiological Society* 20: 48-69.

Newsholme, A. (1902) 'The public health aspects of summer diarrhoea' *The Practitioner* 69: 161-180.

Newsholme, A. (1902-03) 'Remarks on the causation of epidemic diarrhoea' *Transactions of the Epidemiological Society* 22: 34-43.

Newsholme, A. (1905) 'Infantile mortality. A statistical study from the public health point of view' *The Practitioner* 75: 489-500.

Newsholme, A. (1906a) 'Review of: *Infant Mortality; a Social Problem*. By George Newman, M.D., &c. 356 pp. Methuen and Co., 1906. Price 7s. 6d. net' *Journal of the Royal Statistical Society* 69: 610-11 (Written as AN.).

Newsholme, A. (1906b) 'Domestic infection in relation to epidemic diarrhoea' *Journal of Hygiene* 6: 139-148.

Newsholme, A. (1906c) 'An inquiry into the principal causes of the reduction in death-rate from phthisis during the last forty years, with special reference to the segregation of phthisical patients in general institutions' *Journal of Hygiene* 6: 304-384.

Newsholme, A. (1910) *Supplement to the 39th Annual Report of the Medical Officer of the Local Government Board: Infant and Child Mortality* [Cd 5263] (London: HMSO).

Newsholme, A. (1913) *Supplement to 42nd Annual Report of the Medical Officer of the Local Government Board: Second Report on Infant and Child Mortality* [Cd 6909] (London: HMSO).

Newsholme, A. (1914) *Supplement to 43rd Annual Report of the Medical Officer of the Local Government Board: Third Report on Infant Mortality Dealing with Infant Mortality in Lancashire* [Cd 7511] (London: HMSO).

Newsholme, A. (1917-18) *Supplement to 45th Annual Report Annual Report of the Medical Officer of the Local Government Board: Report on Child Mortality at Ages 0-5 in England and Wales* [Cd 8496] (London HMSO).

Newsholme, A. (1923a) 'The measurement of progress in public health: with special reference to the life and work of William Farr' *Economica* 3: 186-202.

Newsholme, A. (1923b) *Elements of Vital Statistics in their Bearing on Social and Public Health Problems* 3rd edition (London: G. Allen & Unwin).

Nicholls, A. (2000) 'Fenland ague in the nineteenth century' *Medical History* 44: 513-530.

Nicolson, A. (1994, first published 1930) *History of Skye* (Portree: Maclean Press).

Niemi, M. (2000) 'Public Health discourses in Birmingham and Gothenburg, 1890-1920' in Sheard, S. and Power, H. (eds) *Body and City: Histories of Urban Public Health* (Aldershot: Ashgate): 123-142.

Nissel, M. (1987) *People Count: a History of the General Register Office* (London: HMSO).

North, A. and MacDonald, H. (1977) 'Why are neonatal mortality rates lower in small black infants than in white infants in similar birth weight' *Journal of Pediatrics* 90: 809-810.

O'Campo, P., Xue, X., Wang, M. and O'Brien, C. (1997) 'Neighborhood risk factors for low birthweight in Baltimore: A multilevel analysis' *American Journal of Public Health* 87: 1113-1118.

O'Connor, T., Davies, L., Dunn, J., Golding, J. and ALSPAC Study Team (2000) 'Distribution of accidents, injuries, and illnesses by family type' *Pediatrics* 106: E68.

Oakley, A. (1984) *The Captured Womb: A History of the Medical Care of Pregnant Women* (Oxford: Blackwell).

OFSTED [Office for Standards in Education] (2006) *How we Regulate Childcare* http://www.ofsted.gov.uk

Oliver, T. (ed) (1902) *Dangerous Trades: the Historical, Social, and Legal Aspects of Industrial Occupations as Affecting Health* (London: John Murray).

de Onis, M. and Onyango, A. (2003) 'The Centers for Disease Control and Prevention 2000 growth charts and the growth of breastfed infants' *Acta Paediatrica* 92: 413-419.

ONS [Office for National Statistics] (2003) *Mortality Statistics: Childhood, Infant and Perinatal. Review of the Registrar General on Deaths in England and Wales, 2001* Series DH3 no.34 (London: ONS).

ONS [Office for National Statistics] (2005) *Deaths 2002: Childhood, Infant and Perinatal Mortality: Live Births, Stillbirths and Infant Deaths by Area of Residence (numbers and rates)* http://www.ons.gov.uk

Parker, J., Schoendorf, K. and Kiely, J. (1994) 'Associations between measures of socioeconomic status and low birth weight, small for gestational age and premature birth in the United States' *Annals of Epidemiology* 4: 271-278

Pearl, M., Braveman, P. and Abrams, B. (2001) 'The relationship of neighborhood socioeconomic characteristics to birthweight among 5 ethnic groups in California' *American Journal of Public Health* 91: 1808-1814.

Pennybacker, S. (1995) *A Vision for London, 1889-1914: Labour, Everyday Life and the LCC Experiment* (London: Routledge).

Peretz, E. (1992) 'Maternal and child welfare in England and Wales between the wars: a comparative regional study' (unpublished University of Middlesex PhD thesis).

Perkins, J. (1976) 'Harvest technology and labour supply in Lincolnshire and the East Riding of Yorkshire 1750-1850, Part 1' *Tools and Tillage* 3: 46-58.

Pinchbeck, I. (1930) *Women Workers and the Industrial Revolution, 1750-1850* (London: Routledge).

Pittenger, D. (1973) 'An exponential model of female sterility' *Demography* 10: 113-121.

Playfair, W. S. (1886) *A Treatise on the Science and Practice of Midwifery* 6th edition (two volumes) (London: Smith, Elder & Co.).

Politics.co.uk (2005) *CBI: Paternity Leave Plans will have Negative Impact* http://www.politics.co.uk

Pope, S., Whiteside, L., Brooks-Gunn, J., Kelleher, K., Rickert, V., Bradley, R. and Casey, P. (1993) 'Low birthweight infants born to adolescent mothers. Effects of co-residency with grandmother on child development' *Journal of the American Medical Association* 269: 1396-1400.

Porter, C. (1894-95) 'Preventable infant mortality with special reference to the influence of the condition of life in factory towns' *Public Health* 7: 162-167.

Power, C. (1994) 'National trends in birth weight: implications for future adult disease' *British Medical Journal* 308: 1270-1271.

Preston, S. and Haines, M. (1991) *Fatal Years: Child Mortality in Late Nineteenth-Century America* (Princeton, NJ: Princeton University Press).

Primatesta, P. and Prior, G. (eds) (2001) *Health Survey for England - The Health of Minority Ethnic Groups '99. Volume 1: Findings.* Series HS no.9. (London: The Stationery Office).

Pringle, A. (various) *Medical Officer of Health Report for Ipswich* (Ipswich).

Prochaska, F. (1989) 'A mother's country: Mothers' meetings and family welfare in Britain 1850-1950' *History* 74: 379-399.

Prochaska, F. (1995) *Royal Bounty: the Making of a Welfare Monarchy* (New Haven: Yale University Press).

Protheroe, L., Dyson, L., Renfrew, M., Bull, J. and Mulvihill, C. (2003) *The Effectiveness of Public Health Interventions to Promote the Initiation of Breastfeeding* (London: Health Development Agency).

Raleigh, S. and Balarajan, R. (1995) 'The health of infants and children among ethnic minorities' in Botting, B. (ed) *The Health of our Children, Decennial Supplement* (London: Stationary Office): 82-94.

Ranade, W. (1997) *A Future for the NHS? Health Care for the Millennium* (London: Addison Wesley Longman)

Ranger, W. (1856) *Report to the President of the General Board of Health on Sanitary Conditions in Ipswich* (Ipswich).

Ravensdale, J. (1974) *Liable to Floods. Village Landscape on the Edge of the Fens, AD 450-1850* (London: Cambridge University Press).

Razzell, P. (1977) *The Conquest of Smallpox* (Firle: Caliban Books).

Reeder, D. (1968) 'A theatre of suburbs: some patterns of development in West London, 1801-1911', in Dyos, H. (ed) *The Study of Urban History* (London: Arnold): 253-258.

Rehydrate (2006) 'Focus on diarrhoea, dehydration and rehydration' http://rhydrate.

org.

Reid, A. (1999) 'Infant and child health and mortality in Derbyshire from the Great War to the mid-1920s' (unpublished University of Cambridge PhD thesis).

Reid, A. (2001a) 'Health visitors and child health: did health visitors have an impact?' *Annales de Demographie Historique*: 117-137.

Reid, A. (2001b) 'Neonatal mortality and stillbirths in early twentieth century Derbyshire, England' *Population Studies* 55: 213-232.

Reid, A. (2002) 'Infant feeding and post-neonatal mortality in Derbyshire, England, in the early twentieth century' *Population Studies* 56: 151-166.

Reid, A. (2004) 'Child care and maternal health: intermediaries between socio-economic and environmental factors and infant and child mortality?' in Breschi, M. and Pozzi, L. (eds) *The Determinants of Infant and Child Mortality in Past European Populations* (Udine: Forum): 139-152.

Reid, A. (2005) 'The effects of the 1918-1919 influenza pandemic on infant and child health in Derbyshire' *Medical History* 49: 29-54.

Renfrew, M., Ansell, P. and Macleod, K. (2003) 'Formula feed preparation: helping reduce the risks; a systematic review' *Archives of Diseases in Childhood* 88: 855-858.

Reynolds, G. (1987) *A Demographic and Socio-Economic Study of March, 1550-1750* (unpublished Open University MSc thesis).

Richards, H. (1903) 'The factors which determine the local incidence of fatal infantile diarrhoea' *Journal of Hygiene* 3: 325-346.

Roberts E. (1997) 'Neighborhood social environments and the distribution of low birthweight in Chicago' *American Journal of Public Health* 87: 597-603.

Roberts, M. (1979) 'Sickles and scythes: women's work and men's work at harvest time' *History Workshop* 7: 3-28.

Robertson, J. (1904) *Special Report of the Medical Officer of Health on Infant Mortality in the City of Birmingham* (Birmingham).

Robinson, M. (ed.) (1985) *The Concise Scots Dictionary* (Aberdeen: Aberdeen University Press).

Robson, W. (1948, 2nd edition) *The Government and Misgovernment of London* (London: Allen & Unwin).

Roitt, I. (1988) *Essential Immunology* (Oxford: Blackwell Scientific Publications).

Rooth, G. (1980) 'Low birthweight revised' *The Lancet* 1: 639-641.

Rose, L. (1986) *The Massacre of the Innocents: Infanticide in Britain 1800-1939* (London: Routledge).

Rose, L. (1988) *Rogues and Vagabonds* (London: Routledge).

Ross, E. (1993) *Love and Toil: Motherhood in Outcast London, 1870-1918* (Oxford: Oxford University Press).

Ross, E. (1986) 'Labour and love: rediscovering London's working class mothers 1870-1918', in Lewis, J. (ed) *Labour and Love, Women's Experience of Home and Family, 1850-1914* (Oxford: Blackwell): 73-96.

Rowntree, S. (1901) *Poverty: A Study of Town Life* (London: Longmans, Green & Co.).

Ruhm, C. (2000) 'Parental leave and child health' *Journal of Health Economics* 19: 931-960.

Rush, D. and Cassano, P. (1983) 'Relationship of cigarette smoking and social class to birthweight and perinatal mortality among all births in Britain, 5-11 April 1970' *Journal of Epidemiology and Community Health* 37 (1983): 249-255.

Ryan, A., Wenjun, Z. and Acosta, A. (2002) 'Breastfeeding continues to increase into the new Millennium' *Pediatrics* 110: 249-255.

Sacker, A., Quigley, M. and Kelly, Y. (2006) 'Breastfeeding and developmental delay: findings from the Millennium Cohort Study' *American Journal of Public Health* (in press).

Sandilands, J. (various) *Kensington MOH Annual Report 1909* (Kensington).

Saxena, S., Mejeed, A. and Jones, M. (1999) 'Socioeconomic differences in childhood consultation rates in general practice in England and Wales: prospective cohort study' *British Medical Journal* 318: 642-646.

Sayer, K. (1995) *Women of the Fields: Representations of Rural Women in the Nineteenth Century* (Manchester: Manchester University Press).

Schofield, R. (1986) 'Did the mothers really die? Three centuries of maternal mortality in "The World We Have Lost"' in Bonfield, L. Smith, R. and Wrightson, K. (eds) *The World We Have Gained: Histories of Population and Social Structure* (Oxford: Blackwell): 231-260.

Schofield, R. and Wrigley, E. (1983) 'English population history from family reconstitution' *Population Studies* 37: 157-184.

Scripter, M. (1970) 'Nested-means map classes for statistical maps' *Annals of the Association of American Geographers* 60: 385-393.

Searle, G. (1971) *The Quest for National Efficiency* (Oxford: Blackwell).

Shadwell, A. (1902) *Drink, Temperance and Legislation* (London: Longmans, Green & Co.).

Shadwell, A. (1906) *Industrial Efficiency: A Comparative Study of Industrial Life in England, Germany and America* (Two volumes) (London: Longmans, Green & Co.).

Sharpe, P. (1996) *Adapting to Capitalism: Working Women in the English Economy, 1700-1850* (New York: St. Martin's Press).

Shelton, N. (2005) 'Infant measurements and breastfeeding' in Bromley, C., Sproston, K. and Shelton, N. *The Scottish Health Survey 2003* (Edinburgh: Scottish Executive) http://www.scotland.gov.uk.

Shiono, P., Klebanoff, M., Graubard, B., Berendes, H. and Rhoads, G. (1986) 'Birth weight among women of different ethnic groups' *Journal of the American Medical Association* 255: 48-52.

Shryock, H. and Siegel, J. (1976) *The Methods and Materials of Demography* (New York: Academic Press).

Simon, J. (1872) 'Instructions to Vaccination Officers issued by the Local Government Board, 21st December 1871' *First Report of the Local Government Board* (Parliamentary Papers 1872 XXV111: 77-81).

Simon, J. (1897) *English Sanitary Institutions* 2nd edition (London: Smith Elder).

Sinclair, C. (2000) *Jock Tamson's Bairns: a History of the Records of the General Register Office for Scotland* (Edinburgh: GROS).

Slattery M. and Morrison J. (2002) 'Preterm delivery' *Lancet* 360: 1489-1497.

Smith, F. (1979) *The People's Health, 1830-1910* (London: Croom Helm).

Smith, R. and Newton G. (forthcoming) 'Infant mortality and environment in intra-

and extra-mural London c. 1650-1730' *The London Journal*.

Smith, S. (2003) '"Who you are or where you are?": determinants of infant mortality in Fulham, 1876-1888' *Family and Community History* 6: 113-120.

Sneddon, S. (2002) 'Infant and early childhood mortality in the Fens, 1850-1900' (unpublished University of London PhD thesis).

Snell, K. (1985) *Annals of the Labouring Poor: Social Change and Agrarian England, 1660-1900* (Cambridge: Cambridge University Press).

Southall, H. (2006) 'A vision of Britain through time; making sense of 200 years of census reports' *Local Population Studies* 76: 76:89.

Speechley, H. (1999) 'Female and child agricultural day labourers in Somerset, c.1685-1870' (unpublished University of Exeter PhD thesis).

Sproston, K. and Primatesta, P. (eds) (2002) *The Health Survey for England 2003* (London: The Stationery Office).

Squire, R. (1924) *Thirty Years in the Public Service: An Industrial Retrospect* (London, Nisbet).

Steel, T. (1994) *The Life and Death of St. Kilda: the Moving Story of a Vanished Island Community* (London: Harper Collins).

Strange, J-M. (2005) *Death, Grief and Poverty in Britain, 1870-1914* (Cambridge: Cambridge University Press).

Sturdy, S. (2004) 'Newman, Sir George (1870-1948)' in Matthew, C. and Harrison, B. (eds) *The Dictionary of National Biography Vol. 40* (Oxford: Oxford University Press): 636-640.

Sun, J. (1995) 'Empirical estimation of a distribution function with truncated and doubly interval-censored data and its application to AIDS studies' *Biometrica* 51: 1096-1104.

SureStart. (2005) *Welcome to SureStart* http://www.surestart.gov.uk.

Swithinbank, H. and Newman, G. (1903) *Bacteriology of Milk* (London: John Murray).

Sykes, J. (1910) 'Mothercraft' *Journal of the Royal Sanitary Institute* 31: 573-588.

Szreter, S. (1991a) 'Introduction: The GRO and the historians' *Social History of Medicine* 4: 401-414.

Szreter, S. (1991b) 'The GRO and the public health movement in Britain, 1837-1914' *Social History of Medicine* 4: 435-463.

Szreter, S. (1996) *Fertility, Class and Gender in Britain, 1860-1940* (Cambridge: Cambridge University Press).

Szreter, S. and Mooney, G. (1998) 'Urbanization, mortality, and the standard of living debate: new estimates of the expectation of life at birth in nineteenth-century British cities' *Economic History Review* 51: 84-112.

Tanaka S. (2005) 'Parental leave and child health across OECD countries' *Economic Journal* 115: 7-28.

Tanner, A. (1998) 'Thomas Orme Dudfield: the model medical officer of health' *Journal of Medical Biography* 6: 79-85.

Tarnier, C., Chantreuil, G. and Budin, P. (1888-1901) *Trait de L'art des Accouchements* (Four volumes) (Paris: Steinheil).

Taylor, R. (1874) 'What influence has the employment of mothers in manufactures on infant mortality' *Transactions of the National Association for the Promotion of*

Social Science: 569-584.

Thirsk, J. (1957) *English Peasant Farming. The Agrarian History of Lincolnshire from Tudor to Recent Times,* (London: Routledge & Kegan Paul).

Thomas, D. (1898-99) 'On infantile mortality' *Public Health* 11: 810-816.

Thomas, J. (1883) *First Annual Report made of the Wellingborough District for the Wellingborough Rural Sanitary Authority* PRO MH12 19098/83.

Toogood, F. (1910) 'The role of the crèche or day nursery', in Kelynack, T. (ed) *Infancy* (London: Robert Culley): 77-87.

Tressell, R. (1914, 1955) *The Ragged Trousered Philanthropists* (London: Lawrence and Wishart).

Tripe, J. (1873) 'The sanitary statistics of the Metropolis for the ten years, 1861-1870' *British Medical Journal* (27 September): 371-373.

Turnbull, B. (1976) 'The empirical distribution function with arbitrarily grouped censored and truncated data' *Journal of the Royal Statistical Society* B 38: 290-295.

UNICEF [United Nations Children's Fund] (2001) *Global Database Breastfeeding Indicators* http://www.childinfo.org.

UNICEF [United Nations Children's Fund] (2004) *Low Birthweight: Country, Regional and Global Estimates* (New York: UNICEF).

U.S. Surgeon General. (2005*) United States Department of Health and Human Services Releases Advisory on Alcohol Use in Pregnancy* http://www.hhs.gov/surgeongeneral

Valentine, V. (2005) 'Shaping Public Health in Victorian London: the Battle for Notting Dale' (unpublished University College London MA thesis).

Vangen, S., Stoltenberg, C., Skjaerven, R., Magnus, P., Harris, J. and Stray-Pedersen, B. (2002) 'The heavier the better? Birthweight and perinatal mortality in different ethnic groups' *International Journal of Epidemiology* 31: 654-660.

Verdon, N. (2001) 'The employment of women and children in agriculture: a reassessment of agricultural gangs in nineteenth century Norfolk' *Agricultural History Review* 49: 41-55.

Verdon, N. (2002a) 'The rural labour market in the early nineteenth century: women's and children's employment, family income, and the 1834 Poor Law Report' *Economic History Review* 55: 299-323.

Verdon, N. (2002b) *Rural Women Workers in Nineteenth Century England* (Woodbridge: Boydell Press).

Vögele, J. (1998) *Urban Mortality Change in England and Germany, 1870-1913* (Liverpool: Liverpool University Press).

Wadsworth, M., Butterworth, S., Montgomery, S., Ehlin, A. and Bartley, M. (2003) 'Health', in Ferrie E., Bynner J. and Wadsworth M. (eds) *Changing Britain, Changing Lives. Three Generations at the End of the Century* (London: Institute of Education): 207-236.

Wainwright, E. (2003) '"Constant medical supervision": locating reproductive bodies in Victorian and Edwardian Dundee' *Health and Place* 9: 163-174.

Walker-Smith, J. (1998a) 'Sir George Newman, infant diarrhoeal mortality and the paradox of urbanism' *Medical History* 42: 347-361.

Walker-Smith, J. (1998b) 'Diarrhoea in two worlds: the messages of Ballard and

Newman' *Journal of the Royal Society of Medicine* 91: 314-316.

Waller, M. (1994) *The Fenland Project, Number 9: Flandrian Environmental Change in Fenland* (East Anglian Archaeology, Report no.70) (Cambridge: Fenland Project Committee, Cambridgeshire Archaeological Committee).

Waller, P. (1983) *Town, City and Nation* (Oxford: Clarendon Press).

Walsh, M. (2002) 'Smallpox: the disease and strategies for its control' *Nursing Times* 98: 26-27.

West, F. (1974) 'Infant mortality in the East Fen parishes of Leake and Wrangle' *Local Popualtion Studies* 13: 41-44.

Wheeler, W. (1898) *The History of the Fens of South Lincolnshire* 2nd edition (London: Newcomb, Simpkin, Marshall & Co.).

Whitbread, N. (1972) *The Evolution of the Nursery-infant School* (London: Routledge & Kegan Paul).

Whitelegge, A. and Newman, G. (1905) *Hygiene and Public Health* (London: Cassell and Co.).

Whitley, G. (1864) 'Report by Dr. George Whitley as to the quantity of ague and other malarious diseases now prevailing in the principal marsh districts of England', *Parliamentary Paper, Sixth Report of the Medical Officer of the Privy Council* Appendix 13: 446-453.

WHO [World Health Organization] (1992) *International Classification of Diseases and Related Health problems. Tenth revision.* vol. 1. (Geneva: World Health Organization).

WHO [World Health Organization] (2002) *Infant and Young Child Nutrition; Global Strategy on Infant and Young Child Feeding. Report by the Secretariat A55/15* (Geneva: World Health Organization).

WHO [World Health Organization] (2003) *Global Strategy for Infant and Young Child Feeding* (Geneva: World Health Organization).

Wilcox, A. (2001) 'On the importance – and the unimportance – of birthweight', *International Journal of Epidemiology* 30: 1233-1241.

Wilcox, A. and Russell, I. (1983a) 'Birthweight and perinatal mortality: I. On the frequency distribution of birthweight' *International Journal of Epidemiology* 12: 314-318.

Wilcox, A. and Russell, I. (1983b) 'Birthweight and perinatal mortality: II. On weight-specific mortality' *International Journal of Epidemiology* 12: 319-325.

Wilcox, A. and Russell, I. (1986) 'Birthweight and perinatal mortality: III. Towards a new method of analysis' *International Journal of Epidemiology* 15: 188-196.

Wilcox, M., Gardosi, J., Mongelli, M., Ray, C. and Johnson, I. (1993) 'Birth weight from pregnancies dated by ultrasonography in a multicultural British population' *British Medical Journal* 307: 588-591.

Wilcox, M., Smith, S., Johnson, I., Maynard, P. and Chilvers, C. (1995) 'The effect of social deprivation on birthweight, excluding physiological and pathological effects' *British Journal of Obstetrics and Gynaecology* 102: 918-924.

Williams, N. (1992) 'Death in its season: class, environment and the mortality of infants in nineteenth century Sheffield' *Social History of Medicine* 5: 71-94.

Williams, N. (1996) 'The reporting and classification of causes of death in mid-nineteenth century England: the example of Sheffield' *Historical Methods* 29:

58-71.

Williams, N. and Galley, C. (1995) 'Urban-rural differentials in infant mortality in Victorian England', *Population Studies* 49: 401-420.

Williams, N. and Mooney, G. (1994) 'Infant mortality in an 'Age of Great Cities': London and the English provincial cities compared, c. 1840-1910' *Continuity and Change* 9: 185-212.

Wills, N. (1970) *Woad in the Fens* (Spalding: The Author).

Wilson, A. (1995) *The Making of Man-midwifery: Childbirth in England 1660-1770* (London: UCL Press).

Wilson, A., Forsyth, J., Greene, S., Irvine, L., Hau, C. and Howie, P. (1998) 'Relation of infant diet to childhood health: seven year follow up of cohort of children in Dundee infant feeding study' *British Medical Journal* 316: 21-25.

Wilson, C. (1982) 'Marital fertility in pre-industrial England, 1550-1849' (unpublished University of Cambridge PhD thesis).

Wingate, P. with Wingate, R. (1988) *The Penguin Medical Encyclopedia* 3rd edition (London: Penguin).

Winter, J. (1977) 'The impact of the First World War on civilian health in Britain' *Economic History Review* 30: 487-507.

Withers, C. (1988) *Gaelic Scotland: the Transformation of a Cultural Region* (London: Routledge).

Wolfe, R. and Sharp, L. (2002) 'Anti-vaccinationists past and present' *British Medical Journal* 325: 430-432.

Woodbury, R. (1922) 'The relation between breast and artificial feeding and infant mortality' *American Journal of Hygiene* 2: 668-687.

Woodhead, G. (1907) 'Introduction', in Llewellyn Heath, H. (ed.) *The Infant, the Parent and the State: A Social Study and Review* (London, King): xi-xv.

Woods, R. (1978) 'Mortality and sanitary conditions in the "best governed city in the world" – Birmingham, 1870-1910' *Journal of Historical Geography* 4: 35-56.

Woods, R. (1982) 'The structure of mortality in mid-nineteenth century England and Wales' *Journal of Historical Geography* 8: 373-394.

Woods, R. (1991) 'Public health and public hygiene: The urban environment in the late nineteenth and early twentieth centuries', in Schofield, R., Reher, D. and Bideau, A. (eds) *The Decline of Mortality in Europe* (Oxford: Clarendon Press): 233-247.

Woods, R. (1995) *The Population of Britain in the Nineteenth Century* (Cambridge: Cambridge University Press).

Woods, R. (1997) 'Infant mortality in Britain: A survey of current knowledge on historical trends and variations' in Bideau, A., Desjardins, B. and Brignoli, H.-P. (eds) *Infant and Child Mortality in the Past* (Oxford: Clarendon Press): 74-88.

Woods, R. (2000) *The Demography of Victorian England and Wales* (Cambridge: Cambridge University Press).

Woods, R. (2003) 'Urban-rural mortality differentials: an unresolved debate' *Population and Development Review* 29: 29-46.

Woods, R. (2005) 'The measurement of historical trends in fetal mortality in England and Wales' *Population Studies* 59: 147-162.

Woods, R. (2006a) *Children Remembered: Responses to Untimely Death in the Past*

(Liverpool: Liverpool University Press).

Woods, R. (2006b) 'Mortality in eighteenth-century London: a new look at the bills' *Local Population Studies*, 77: 12-23.

Woods, R. and Hinde, A. (1987) 'Mortality in Victorian England: models and patterns' *Journal of Interdisciplinary History* 18: 27-54.

Woods, R. and Shelton, N. (1997) *An Atlas of Victorian Mortality* (Liverpool: Liverpool University Press).

Woods, R. and Shelton, N. (2000) 'Disease environments in Victorian England and Wales' *Historical Methods* 33: 73-81.

Woods, R. and Woodward, J. (eds) (1984) *Urban Disease and Mortality in Nineteenth-century England* (London: Batsford).

Woods, R., Watterson, P. and Woodward, J. (1988) 'The causes of rapid infant mortality decline in England and Wales, 1861-1921, Part I' *Population Studies* 42: 343-366.

Woods, R., Watterson, P. and Woodward, J. (1989) 'The causes of rapid infant mortality decline in England and Wales. Part II' *Population Studies* 43: 113-132.

Woods, R., Williams, N. and Galley, C. (1997) 'Differential mortality patterns among infants and other young children: The experience of England and Wales in the nineteenth century' in Corsini, C. and Viazzo, P. (eds) *The Decline of Infant and Child Mortality. The European Experience, 1750-1990* (The Hague: Martin Nijhoff Publishers): 48-62.

Wright, C., Parkinson, K. and Drewett, R. (2004) 'Why are babies weaned early? Data from a prospective population based cohort study' *Archives of Disease in Childhood* 89: 813-816.

Wright, P. (1988) 'Babyhood: The social construction of infant care as a medical problem in England in the years around 1900' in Lock, M. and Gordon, D. (eds) *Biomedicine Examined* (Dordrecht: Kluwer): 299-329.

Wrigley, E. (1977) 'Births and Baptisms: The use of Anglican baptism registers as a source of information about the numbers of births in England before the beginning of civil registration' *Population Studies* 31: 281-312.

Wrigley, E. (1985) 'Urban growth and agricultural change: England and the continent in the early modern period' *Journal of Interdisciplinary History* 15: 683-728.

Wrigley, E. (1997) 'How reliable is our knowledge of the demographic characteristics of he English population in the early modern period' *The Historical Journal* 40: 571-595.

Wrigley, E. (1998) 'Explaining the rise in fertility in the 'long' eighteenth century' *Economic History Review* 51: 435-464.

Wrigley, E. (2005) *Poverty, Progress and Population* (Cambridge: Cambridge University Press).

Wrigley, E., Davies, R., Oeppen, J. and Schofield, R. (1997) *English Population History from Family Reconstitution, 1580-1837* (Cambridge: Cambridge University Press).

Index

Aberdeen 121
Acheson, D. 257, 260
Adair, R. 55
Addison, C. 19
Africans 234, 236, 240
AIDS 70
Albermann, E. 232, 236
Anderson, J. 26
Ansell, C. 39, 45
Apple, J. 47
Armstrong, D. 25, 169, 188
Armstrong, W. 95
Ashby, H. 23, 27, 81
Australia 35
Austria 34, 41, 251

Baird, D. 49
Baker, D. 242, 246
Balarajan, R. 242, 246
Ballantyne, J. 37, 45, 48
Bandolier 259
Bangladesh 231, 234, 236, 240, 244
Barker, D. 246
Barnsley 44
Bartley *et al.* (1994) 246
Battersea, London 30, 214, 218
Baxby, D. 103
Bedfordshire 3, 35, 44
Bell, F. 47, 116
Bentley, B. 28
Bern, C. 246
Berger, M. 258
Berridge, V. 94, 96
Berry, M. 54
Bertillon, J. 45
Bevan, A. 221
Beveridge, W. 221
Bevis, T. 81
Bideau, A. 6, 45, 253
Bideford, Devon 62
Bills of Mortality, London 67
Birmingham 26, 28, 39, 44, 45, 217
birthweight 230-
Blackburn 44, 120
Blake, M. 260
Blumin, S. 156, 166
Board of Education 3, 18-19, 21, 23, 29
Boer War 4, 28
Bongaarts, J. 54
Booth, C. 39, 45, 170, 183
Boston, Lincs. 91
Botting, B. 256
Bouquet, M. 95
Bourgeois-Pichat, J. 59
Branca, P. 115
breastfeeding 9, 27, 37, 40-41, 95, 142, 149-, 178, 182, 185, 191-, 198, 205, 209, 229, 252, 259
Bridford, Devon 8, 64
Brighton 37, 40, 41, 44, 251
Bristol 216
British Paediatric Association 232
Broadbent, B. 30
Brooks, B. 178
Brownlee, J. 26
Brunton, L. 28
Buchanan, I. 198, 201
Buckinghamshire 217
Budin, P. 45
Bukowski, R. 236
Burnette, J. 95
Burns, J. 28
Bury 44
Bynum, W. 19, 22

Cambridge 193
Cambridgeshire 8, 80, 82, 83, 217
Caribbean 231, 234, 236, 240
Cassano, P. 234
Cassiday, R. 238
causes of death (see also individual causes) 8, 126, 160, 161, 162
Chadwick, E. 24, 82, 187
Chalmers, Dr (MOH Glasgow) 44

Chesterfield 201
Chike-Obi *et al.* (1996) 246
Chile 35
Clark, A. 121
Clarke, M. 116
Cohen, G. 244
Collet, C. 250
Collins, J. 236
Cone (1980) 230
Confederation of British Industry (CBI) 258
Cooke Taylor (1874) 185
Cornwall 8
Corsini, C. 6, 8
Coutts, F. 205
Coxon, A. 234
Croydon 44
Cugalj (1984) 203
Cumming, J. 187
Curle, B. 171
Currie, M. 150

Darby, H. 81, 86, 94
Davey Smith, G. 218
David, R. 231
Davies, C. 188
Davies, L. 121
Davies, R. 143
Davies, W. 205
Davin, A. 28, 47
Davis, J. 172, 175
Dawson, I. 236
Department of Health 213-14, 226, 229, 238, 256, 258-59
Department of Trade and Industry (DTI) 258
Department of Work and Pensions 254
Derby 40-41, 44, 194, 203, 251
Derbyshire 13, 194-
Dhawan, S. 246
diarrhoea 4, 10, 12, 37, 41, 97, 121, 129, 132, 149-, 182, 251
Dobson, M. 81, 94
Dorling, D. 13, 213-, 217
Dorset 41, 120, 251
Drake, M. 9, 12, 83, 99, 101, 103-04, 121, 140, 149-, 156
Draper, E. 246
Dudfield, T. 171, 173-74, 176, 178-79, 181, 183
Dugdale, A. 246
Dundee 44, 121
Durbach, N. 100-02
Durham 120
Dwork, D. 4, 192-93, 213-24, 218, 220-21
Dyhouse, C. 47, 183, 192
Dyson, L. 257

East Anglia 24, 150
East Riding, Yorkshire 82
Eaton-Evans, J. 246
Edinburgh University 18
Edwards, G. 94, 96
Elliston, Dr (MOH Ipswich) 151, 159, 166
Ely, Cambs. 62
Emmett *et al.* (2000) 238
endogenous mortality 8, 11, 59-, 66, 75, 77
Erlington, H. 171
exogenous mortality 8, 11, 59-, 66
Eyler, J. 18, 23, 28, 45

Fang, J. 234
Farr, W. 23-26, 40-41, 45, 62, 71, 82
Fens, The 8, 11, 64, 79-
Ferguson, T. 141
Fildes, V. 178, 232
Finland 230
Finlay, R. 67
Finnegan, R. 83
Finsbury, London 3-4, 10, 17-18, 23, 26-27, 33, 39-41, 44, 150, 193, 251
First World War 3, 158, 170, 188, 253
Fisher, K. 234
Flinn, M. 136
foetal mortality 9, 11, 36, 48-49, 69-, 75-77, 116, 143, 195, 200, 253, 255-56
Fogel, R. 69
Foundling Hospital, London 68
France 6, 33, 45, 74, 251
Friedman, D. 236

Gainsborough, Lincs. 8, 64, 91
Gairdner, W. 24
Galley, C. 3-, 8, 10, 17-, 26, 28, 81-82, 85, 94, 97, 121, 142-43
Garrett, E. 3-, 9, 12, 79, 89, 105, 119-, 121, 143, 163
Gergen, P. 246
Germany 39

Gielgud, J. 95
Gladstone, F. 185
Glasgow 40, 44, 217
Glass, D. 56, 83
Glass, R. 246
Glover, J. 17, 19
Glyde, J. 165
Godwin, H. 85-86, 95
Golden (1996) 232
Gordon, M. 26
Gould, J. 234
Graham, D. 116
Grant, Dr (MOH Inverness) 141
Grantham, Lincs. 91
Graunt, J. 82
Greenall, R. 115
Greenhow, E. 24
Greenwood, M. 24, 100-02
Gregory, I. 84
Grimsby 86-87, 91
Gwinn, M. 232

Haimes, M. 47
Hall, D. 113
Hall, E. 9, 12, 140, 149-
Hamer, W. 175
Hamlyn, B. 233, 238, 240
Hammer, M. 17-18, 22, 29, 31
Hardie, A. 257
Harding, S. 113
Harding, S. 246
Hardy, A. 82, 126
Harris, B. 28
Health and Safety Executive 256
health visitors 9, 12, 43, 121, 191-, 220
Heath (1907) 185, 187
Heisey, D. 70
Henniker, B. 26
Henry, L. 54, 61
Herrick, K. 260
Hertfordshire 120
Hertz-Picciotto, I. 231
Hewitt, M. 185-86
Higgs, E. 7, 24, 26, 86, 115
Higham Ferrers, Northants. 100-, 104-17
Hill, A. 26, 28
Hills, R. 86, 94
Hinde, P. 82
HM Revenue and Customs 259
Hoddinott, P. 241
Holdsworth, C. 116, 181
Hollingsworth, T. 70, 73, 76
Hope, E. 26, 41, 44, 193
Horn, P. 86
Howarth, W. 41, 206
Howe, G. 84
Howie, P. 232
Howkins, A. 95
Huck, P. 82
Huddersfield 30, 43-44, 192-93
Hull 44
Humber Estuary 80
Humphreys, N. 25
Hunter, H. 24, 80-81, 86, 95
Hunter, J. 123
Huxley, R. 246

IFS 260
India 224, 231, 234, 236, 240
influenza 75
Inland Revenue 259
International Labour Organization (ILO) 257
Ipswich 150-
Ireland 33-34, 40

James, T. 9, 11, 99-, 121
Jarvelin, M. 234
Jenner, W. 100
Johnsson, S. 79
Jones, H. 23, 25

Kammerman, S. 257
Kasakoff, A. 79
Keats, J. 21
Kelly, Y. 13, 229-, 238, 239, 240, 246, 260
Kemmer, D. 121, 137
Kensington, London 12, 169-, 198
Kenwood, D. 26
Kilmarnock 12, 119-, 121, 128
King's College London 18
King's Lynn 86
Kinross, 120
Kitson, P. 56
Kitteringham, J. (1975) 95
Kogan, M. 236
Konovalov, H. 256
Kramer, M. 232, 236

Kunitz, S. 77

Laird, Dr (MOH Cambridge) 193
Lambert, R. 24
Lanarkshire 217
Lancashire 120
Landers, J. 8, 65-67, 76-77
Laslett, P. 55
Latin America 237
laundrywork 179, 185
Lawton, R. 82, 84
Laxton, P. 67-68
Lee, C. 79
Leeds 11
Leicester 120
Leicestershire 116, 120
Leominster, Hereford. 17-18
Leroy, S. 234
Levene, A. 68
Lewes, F. 85
Lewis, Isle of 142
Lewis, J. 28, 47, 74, 76, 186, 192-93
Li *et al*. (2005) 240
Lincoln 86, 91
Lincolnshire 79-, 80, 82-83, 85, 87, 89-90, 92
Lister, J. 21
Liverpool 11, 41, 44, 62, 216
Local Government Board (LGB) 23, 28, 33, 44, 99, 119, 198-99, 205
London 11, 18, 24, 29, 30, 35, 39-41, 44, 65-68, 74, 76-77, 80, 169-, 224, 226, 249
London County Council (LCC) 174, 186-89
Long Eaton, Derbys. 203, 208-10
Loosmore, B. 99-100
Loudon, I. 47, 74, 253
Lucas, A. 232
Lux, A. 246

MacDonald, H. 231
Macfarlane, A. 230, 232, 242
Mackay, J. 123, 137
Maclean, H. 241
Macleod, D. 144
Malcolmson, P. 172, 179
Malthus, T. R. 82
Manchester 11, 44, 192, 218
March, Cambs. 8, 64, 81
Marks, L. 171, 176, 181, 183, 188-89, 198
Marland, H. 30, 43, 188, 192, 194
Marmot, M. 246
Martin, J. 236
Mason, E. 116
maternal mortality 11, 68-74
Maternity Alliance 257
McCann and Ames (2005) 232
McCleary, G. 28, 30, 45, 192, 213-14, 216
McCormick, A. 242
McDonald, A. 116
McDonald, J. 116
McFeely, M. 171, 185
McRae, M. 130
Medical Officers of Health 3-4, 10, 18, 20, 24, 26, 28, 33, 37, 40-41, 44, 100, 106, 150, 170, 192
Medical Research Council (MRC) 19
Medway Estuary 80
Mein Smith, P. 194
Melbourn, Cambs. 81
midwives 41, 75, 147, 200, 203
Migone, A. 236
Miller, N. 230
Miller, S. 96, 94-95
Mills, D. 81, 156, 166
Mills, J. 156, 166
Millward, R. 47, 116
Ministry of Health 3, 19, 21-23, 193
Montreal 116
Mooney, G. 9, 12, 24, 49, 80, 82, 85, 96, 169-, 179, 216
Moore, D. 193
Moore, S. 28, 30
Moore, W. 236
Morant, R. 19, 23
Morgan, N. 112, 140
Morrison, J. 232
mortality, see endogenous, exogenous, foetal, maternal, neonatal, causes of death
Moscow 230
motherhood 4, 10, 27, 33, 38, 42, 45, 186, 191-
Mugford, M. 230, 232, 142
Murphy, S. 44

Nash, J. 28

National Health Service (NHS) 13, 102, 226, 230
Nazroo, J. 242, 246
neonatal mortality 110-15, 143-47
Newman, George (1870-1948) background and career 17-31; *Infant Mortality* 3-5, 9-11, 14, 33-49, 53, 79, 97, 99, 117, 119-21, 142-43, 147-48, 149-50, 170, 191-92, 214, 216, 218, 226, 228, 231, 249-52, 254, 256-57, 260
New York 41
New Zealand 34-36
Newsholme, A. 18, 22-23, 26, 28, 31, 37, 41, 44-45, 49, 81-82, 179, 199-200, 205, 209
Newton, G. 68
Nicholls, A. 94, 96
Nicolson, A. 123
Niemi, M. 170, 176, 183, 188
Nissel, M. 83
Niven, Dr (MOH Manchester) 44, 218
Nordheim, E. 70
Norfolk 80, 82-83, 95, 217
North, A. 231
Northampton 115-16
Northamptonshire 11, 99-
Norway 34, 236, 251
Norwich 80, 86
Notting Dale Special Area, London 12, 169-
Nottingham 44

O'Campo, P. 234
O'Connor, T. 246
Oakley, A. 47
Obstetrical Society 25
Oeppen, J. 11, 53-
Office of National Statistics (ONS) 217
OFSTED 259
Ogle, W. 26, 45, 121
Oliver, T. 45
Open University 121
Ordnance Survey 146
Orkney 120
Oxfordshire 217

Padua School of Medicine 21
Pakistan 231, 234, 236, 240
Paris 44-45
parish registers 7, 11, 53-, 63
Parker *et al.* (1994) 234
Parliamentary Papers 4, 24, 28, 100, 149
Pasteur, L. 21
Pearl, M. 244
peerage 53, 70-73, 76
Pennybacker, S. 186-87
Peretz, E. 193, 198
Perkins, J. 95
Physical Deterioration, Inter-Departmental Committee on 4, 10, 28-29, 33, 45, 167, 186-87
Pill, R. 241
Pinchbeck, I. 95
Pittenger, D. 54
Playfair, W. 45
Poor Law 100-01, 130, 172, 185
Pope, S. 246
Porter, C. 26
Power, C. 232
Preston 44, 120
Preston, S. 47
Primatesta, P. 244
Pringle, A. 152, 166-67
Prior, G. 244
Prochaska, F. 176, 188
Protheroe, L. 242

Raleigh, S. 242
Ranade (1997) 230
Ranger, W. 165
Ravensdale, J. 81
Razzell, P. 77
Registrars General 3, 7, 9, 12, 23, 25-26, 31, 45, 61-64, 68, 80-81, 83, 120-21, 132-33, 147, 151
Reid, A. 9, 12, 115, 121, 142, 191-, 198, 204, 207
Reynolds, G. 81
Richards, H. 26
Roberts, C. 173
Roberts, E. 234
Roberts, M. 95
Robertson, J. 39, 44-45
Robinson, M. 142
Robson, W. 175
Roitt, I. 101
Rooth, G. 231
Rose, L. 174, 185

Ross, E. 47, 179, 183
Rowntree, B. S. 39, 45, 216, 218
Ruhm, C. 258
Rush, D. 234
Rushden, Northants. 107-09, 111-14, 117
Russell, I. 231
Russia 34, 251
Rutland 223
Ryan *et al.* (2002) 240

Sacker *et al.* (2006) 232
Salford 44-45, 192
Sandilands, J. 180-83, 188
Saxena, S. 242
Sayer, K. 96
Schofield, R. 6, 54, 68, 72, 75, 253
Scotland 6, 9, 12, 33, 35, 41, 49, 119-, 146, 148, 232-33, 251, 260
Scripter, M. 84
Searle, G. 23
seasonality 139-40, 154-55
Second World War 216, 225
Shadwell, A. 39, 45
Sharp, L. 100
Sharpe, P. 95
Sheffield 28, 44, 80, 121, 203, 251
Shelton, N. 3-, 8, 11, 13, 35, 79, 80-85, 103, 121, 126, 133, 135, 249-, 260
Shetland 120
Shiono, P. 234
shoemakers 110-17
Shryock, H. 83
SIDS 9, 228
Siegel, S. 83
Simon, J. 24, 38, 100-01, 104
Sinclair, C. 124, 129
Skertchley, S. 86, 94-95
Skye, Isle of 12, 119-, 121
Slattery, M. 232
smallpox (see also vaccination registers) 75, 77, 99, 102-03
Smith, F. 79, 194
Smith, R. 11, 53-, 68
Smith, S. 121
Sneddon, S. 11, 79-, 82-83, 94-96
Snell, K. 95
Social Workers' Association 27
Somalia 103
South Asia 233, 237
Southall, H. 83, 85
Spalding, Lincs. 91
Speechley, H. 95
Squire, R. 179, 181
St Bartholomew's Hospital, London 3, 18
St Helens 44
St Kilda 144
St Pancras, London 44
Staffordshire 44, 120
Stamford, Lincs. 91
Stark, Dr (of Scotland) 41, 142
Steel, T. 144
stillbirths, see foetal mortality
Stockport 44
Strange, J.-M. 47
Sturdy, S. 17
Suffolk 217
Sun, J. 70
SureStart 260
Surgeon General (USA) 256
Sweden 41, 74, 237
Swithinbank, H. 20
Sydenham, T. 21
Sykes, J. 44, 183, 186-87
Szreter, S. 24, 26, 83, 216

Tanala, S. 258
Tanner, A. 9, 12, 169-, 171
Tarnier, C. 48
Tatham, J. 45
Taylor, C. 86
Thetford, Norf. 80
Thirsk, J. 86, 95
Thomas, A. 193
Thomas, D. 26
Thorp, Adelaide (Lady Newman) 18
Toogood, F. 170
Tressell, R. 215
Tripe, J. 24
tuberculosis 21, 48, 75
Turnbull, B. 70
typhoid 96

USA 39, 231, 236, 240, 244, 251, 256

vaccination registers 9, 11, 99-, 150, 156
Valentine, V. 172
Vangen, S. 236
venereal disease 21, 40, 48, 129

Verdon, N. 95-96
Viazzo, P. 6, 8
Vögele, J. 125-26

Wadsworth, M. 232, 233
Wainwright, E. 169, 179, 187, 189
Walker-Smith, J. 30, 37
Waller, M. 85-86, 88
Waller, P. 26
Walsh, M. 103
Wash, The 80, 86
Waterbeach, Cambs. 81
Watt, R. 238, 239, 246
Watterson, P. 80-81, 165
Webb, S. and B. 23
Wellingborough, Northants. 106
West Riding, Yorkshire 120, 217
West, F. 8, 82
Wheeler, W. 86
Whitbread, N. 186
Whitelegge, A. 20
Whitely, G. 86, 94
whooping cough 129
Wilcox, A. 231-32, 234
Williams, N. 67-68, 80-83, 85, 94, 96-97, 125-26, 133
Wills, N. 95
Wilson, C. 54, 72
Wiltshire 120
Wingate, P. 144
Wingate, R. 144
Winter, J. 193
Wolfe, R. 100
Woodhead, G. 183
Woods, R. 3-, 8, 10-11, 26, 30-31, 33-, 34-35, 45, 47, 67, 71, 75-76, 79-85, 103, 121, 126, 133, 135, 142-43, 165
Woodward, J. 80-81, 165
World Health Organization (WHO) 103, 229, 231, 241, 258, 259
Wright, C. 238
Wright, P. 23
Wrigley, E. A. 7-9, 53-54, 69, 75, 80, 143

Yarmouth 80, 86
Ylppo, A. 230
York 8, 39, 45, 216, 218

For Product Safety Concerns and Information please contact our EU representative GPSR@taylorandfrancis.com
Taylor & Francis Verlag GmbH, Kaufingerstraße 24, 80331 München, Germany

www.ingramcontent.com/pod-product-compliance
Lightning Source LLC
Chambersburg PA
CBHW071345290426
44108CB00014B/1447

* 9 7 8 1 1 3 8 3 5 8 1 7 1 *